WEB HISTORY

Steve Jones
General Editor

Vol. 56

PETER LANG
New York • Washington, D.C./Baltimore • Bern
Frankfurt • Berlin • Brussels • Vienna • Oxford

WEB HISTORY

EDITED BY Niels Brügger

PETER LANG
New York • Washington, D.C./Baltimore • Bern
Frankfurt • Berlin • Brussels • Vienna • Oxford

Library of Congress Cataloging-in-Publication Data

Web history / edited by Niels Brügger.
p. cm. — (Digital formations; v. 56)
Includes bibliographical references and index.
1. World Wide Web—History. I. Brügger, Niels.
TK5105.888.W37325 004.67'8—dc22 2009038455
ISBN 978-1-4331-0469-5 (hardcover)
ISBN 978-1-4331-0468-8 (paperback)
ISSN 1526-3169

Bibliographic information published by **Die Deutsche Nationalbibliothek.**
Die Deutsche Nationalbibliothek lists this publication in the "Deutsche Nationalbibliografie"; detailed bibliographic data is available on the Internet at http://dnb.d-nb.de/.

This book was published with support from
The Danish Council for Independent Research—Humanities.

The paper in this book meets the guidelines for permanence and durability of the Committee on Production Guidelines for Book Longevity of the Council of Library Resources.

29 Broadway, 18th floor, New York, NY 10006
www.peterlang.com

Printed in the United States of America

CONTENTS

FOREWORD

History—With a Future

Charles Ess

In 2010—i.e., towards the end of the second decade of the bewildering phenomenon called "the World Wide Web"—it is perhaps difficult to recall a time when the web was *not* an increasingly important component of our lives (at least within the developed countries). In my view and experience, this difficulty is in part due to the web itself, along with its related technologies. To begin with, these technologies are strongly biased towards the present and the all-too-imminent future—i.e., a future that crashes in on us at every moment (especially as we in the developed countries are "always on," i.e., more or less always connected via the internet one way or another), with ever-new and ever-expanding possibilities, facilities, information, etc., etc.—all coupled with the obligation of cognitively and affectively responding to all of this in some way. Moreover, as any number of theorists and commentators have noted, the web seems to collapse our earlier notions of time and space—and given the dizzying rate of technological change and diffusion of the internet and the web, a calendar year in "internet time" is a very long time: some eighteen "internet years" seems long enough to contain the equivalent of geological epochs.

In this context, it is not surprising that the *history* of the web—better, the *idea,* and thereby, the concomitant methodologies, disciplines, and techniques of web archiving and history—has emerged more at the margins rather than at the center of the rapidly expanding literatures of research and reflection on the web. Again, in my view, this is in some measure thanks to the breathtaking speed with which innovations and significant changes flood our awareness and potential uses of these technologies. Like Pooh being carried downstairs by Christopher Robin, his head bumping with every step: we might think (bump) that there must be (bump) a better way (bump)[1]—but the constant bumping of rapid change and novelty so overwhelms our attention and abilities that we are all but forced to neglect even the most basic and essential questions and issues. Clearly, the history of the web—and how that history is best preserved for research and scholarship—belongs among those essential questions and issues. If nothing else: without a strong and somewhat stable record of that history, 'present' and 'future' lose one of their primary reference points and thus much of their meaning.

Most happily, Niels Brügger and the contributors to this volume have managed to avoid the bumping stairs of constant change long enough to develop both theory and emerging best practices of what can now confidently stand as a field in its own right—namely, the discipline of web history and archiving. This volume stands as the essential handbook of this field—especially as it accomplishes four crucial functions.

First, it provides the orientation, overview, and history needed to introduce the rest of us to the field. As we would expect, Brügger's own introduction and opening essay in Part One on web historiography and archiving methodologies accomplish this orientation and introit first of all: these are complemented in essential ways by Kirsten Foot and Steven Schneider's chapter on web historiography, rounding out Part One of the volume.

Second, both Part One and the subsequent parts of the book serve to document the current shape and state of the field—i.e., "Web Cultures" (Part Two), "Web Industries and Media Institutions" (Part Three), and, finally, "Preserving and Presenting" (Part Four). Both individually and collectively, the contributors provide important exemplars and case studies of Web archiving and history in practice, coupled with practitioners' reflections on methodological and theoretical issues, central challenges, attempted resolutions, observations of what works and what does not, and so forth.

Third, one of the most important characteristics of a good theory is its fertility—it takes us to new and interesting places. This is a particularly crucial function for this volume as the handbook of a new field. And in fact, the theoretical and methodological reflections offered here, along with case study examples, exhibit a manifold fertility. To begin with, the chapters as collected into their respective parts thereby work, as we would expect, to document and articulate diverse examples and perspectives on the central foci of each part. At the same time, each chapter further links up in important ways with the others across the volume, suggesting further syntheses of the work collected here that will fruitfully contribute to the ongoing development of the field. Of equal importance, many of the chapters point beyond the boundaries of the field of web history and archiving in equally fruitful ways, suggesting diverse ways in which other disciplines and methodologies constituting internet studies may synthesize and synergistically interact with those of web history and archiving. For example, Charles van den Heuvel's chapter introduces us to Paul Otlet, whose visionary work in the first half of the twentieth century (i.e., well before anything like the web had been conceived) on ways of documenting and disseminating knowledge "beyond the book" thereby illustrates how earlier efforts to overcome the problems and limitations of traditional archiving and library practices prefigure their electronic and computer-based correlates in the contemporary world. Similarly, Albrecht Hofheinz's history of Allah.com connects in rich ways to current work on religious uses of the internet, including interest in how the internet may thereby be contributing to something of an Islamic "Reformation" in Europe. In metaphors: for those interested in the field of web archiving, the book will be an essential first stop and gold mine for the further development of the field as it both lays the field's foundations and provides resources and inspiration for additional development and work, both within and beyond its own boundaries.

In fact (and this is the fourth function), the volume—both within specific chapters and in Brügger's epilogue—explicitly points the way to further work in and development of the field. Brügger's closing "call to action" (as I think of it)—i.e., his mapping out the central challenges and next steps in the development of the field—includes attention to the need for web history beyond the current tendency to focus within specific national boundaries. This call is exactly right, in my view. My only critical observation of the volume is that, although it is reasonably international in its scope and representation—whole continents and cultural traditions are missing here. In fairness, this is not surprising for a young field whose focus depends on a technology most fully diffused and available only in developed countries. And it seems clear that further development of the field along the lines Brügger outlines in the epilogue will naturally extend to histories of the web in all of its global manifestations.

In these multiple ways, then, the volume works as a watershed publication that introduces and documents the emergence and foundations of a new field, provides much of the resources required for further development, and begins to catalyze this development by way of the wide range of intersections and complementaries offered up in the chapters collected here. In these many ways, Niels Brügger and his contributors have created a pioneering volume—one focusing on a clearly foundational set of methodological and theoretical issues surrounding the complex and often difficult matters captured under the rubric 'website history,' and providing a clear vision of what is needed in order to further the development of web history and archiving as a field in its own right. Those of us keenly interested in these remarkable phenomena—internet and web—owe a great debt of gratitude to Brügger and his contributors for helping us turn from the constant bumping of incessant technological change in order to reflect on the essential matters of website history and archiving and for providing the foundations of what promises to be a flourishing and certainly essential field.

In doing so, this volume makes the broader point that close attention to history is essential *especially* for phenomena otherwise so strongly marked by rapid change and an inclination to dismiss the past. Directly contrary to that inclination, Brügger and his colleagues make clear that such history has a future.

NOTE

1. Milne, A.A. (1926). *Winnie-the-Pooh*. New York: E.P. Dutton, p. 3.

INTRODUCTION

Web History, an Emerging Field of Study

Niels Brügger

The World Wide Web—or the web—is now at an age at which it may be time to address its past. During the past two decades we have witnessed an extraordinary dissemination of the web to still more parts of the globe and to still more areas of our individual lives and our societies at large. In addition, on the Internet the web has gradually become the centre of gravity in many respects, either by absorbing or by supplementing formerly separate uses and applications such as bulletin boards, newsgroups, listservs, chat, and email.

But if we stop for a moment, turn around, and face the past, is it then possible to provide an adequate overview of what actually happened, why, and with what consequences? Can we identify some patterns in the variety of events on and in relation to the web? And can we uncover historical developments, transformations, and driving forces? Answers to these questions are not only important for historical reasons but also because the web of the past must be taken into consideration in order to fully understand the web of the present, as well as the new web forms of the future.

However, web history is still an emerging and unclearly defined field of study. The uncertainty of the field is mirrored in the present volume. Instead of promoting a clear-cut conception of what web history should be and should be about, the volume broadly and tentatively explores some of the possible ways of approaching the web of the past. Therefore, the web histories as well as the theoretical and methodological discussions which follow will prove successful if they can place web history on the research agenda of internet research and if they can serve as a stepping stone to future discussions of the past of the web.

WEB

It is not obvious what a history of the web should be about. We have to make a number of choices when selecting and delimiting our analytical object. The following considerations suggest some of the possible analytical choices to be made.

Web?

One of the first tasks of the web historian is to reflect on what the 'web' denotes. Because the point of departure for this book is that web history is an emerging and as yet undefined field, 'web' is understood in a very broad sense so as not to exclude any relevant perspectives and/or research areas in advance. Therefore, the web is simply understood as a sub-domain of the internet—the variety of internet activities based on the use of the http protocol, the html mark-up language, and the URL resource locator (and their historical transformations) as well as the various internet phenomena which can be 'nested' in the web (for instance, other protocols and languages, specific applications and plug-ins, and the like).

On the one hand, this technological definition is sufficiently open and flexible enough to constitute the lowest common denominator shared by the variety of cultural, political, and societal forms which most people would consider part of the web, but, on the other, it is also sufficiently precise and operational to draw a clear demarcation line between the internet and the web, thus opening web history as a field of study in its own right within internet history. As a side-effect it also points out a relatively clear starting date for the history: August 1991, the date when the protocol was released on the internet. This broad definition can then be narrowed down and tailored according to specific analytical needs, and on a general level it should, of course, be questioned and discussed in future works.

However, the definition also gives rise to the following two comments. First, although the 'internet' and the 'web' are not the same, the two are intertwined. In many cases, the history of the internet must be part of historical analyses of the web because the internet is a precondition for the web; conversely, the history of the web can also shed light on the history of the internet.

Second, although the beginning of the web can be determined with a fair amount of precision, the period prior to this date is by no means irrelevant to the history of the web. For instance, many of the use patterns, communicative forms, and functionalities of the web are anticipated in media types and technologies which were invented years before the advent of the web, thus laying the foundations for the web (see, e.g., Berners-Lee, 1999; Boczkowski, 2001, 2004; Banks, 2008; Carey & Elton, 2009).

Web Strata

Taking this broad but precise definition of 'the web' as our point of departure, we must then have to reflect on whether it is fruitful to identify some general analytical levels on the web. I would suggest that within the variety of web material that constitutes 'the web', one can distinguish the following five analytical strata: the web as a whole, the web sphere, the individual website, the individual webpage, and the individual textual web element on a webpage.

A history of the web as a whole could be a history of specific web standards, applications, institutions, or forms of use which transcend the web, such as the communicative infrastructure of networked web servers, the establishment of the World Wide Web Consortium (W3C) (see, e.g., Gillies & Cailliau, 2000), web hacktivism, or the use of cookies in web browsers (Elmer, 2002). A history of a web sphere, on the other hand, would involve identifying and selecting a number of websites (or other web material) considered of relevance to a particular event, concept, or theme—for instance, a variety of web activities in relation to a political event such as the U.S. elections of 2000, 2002, and 2004 (see Schneider and Foot, 2006b).[1]

We could also choose to regard the individual website as the unifying entity of a historical analysis of the web by writing the history of a broadcasting company's website, such as the website of the national Australian broadcaster ABC (see, e.g., Burns, 2000, 2002, 2003; Martin 1999, 2005, 2008).[2] Another approach could be to place the individual webpage at the centre of our historical study, in which case we might account for the web design on individual web pages (Engholm, 2002, 2003). And, finally, we could set out to write the history of specific web elements, such as the history of advertisements/banner ads on the web (e.g., Li & Zhunag, 2007), or of the use of images on newspapers' websites (e.g., Knox, 2009).

Following this five-level stratification, we can focus our web history on any of the five different analytical clusters of web material, just as we can examine their interrelations—for instance, websphere/website, website/webpage—because the individual levels serve as contexts for one another.

Focal Points and Societal Spheres of Use

No matter which of the five web strata outlined above we set out to analyze we also have to choose one or more focal points. For instance, we could decide to focus on the technology of the web (software, hardware), its producers, the content and the users, or on institutions, regulation, legal or ethical issues, or policy or economic issues, just to mention some of the most obvious focal points.

Furthermore, it may be fruitful to reflect on which societal spheres of web use we want our history to be about. This could, for instance, be the web used by actors who are related to the state, civil society, or market. Or it could be the history of the blurred boundaries between these spheres which are caused by the development of the web.

HISTORY

Writing web history implies not only a number of choices as to the nature of the field of study and our approach to it; we also face a number of historiographical choices. This short introduction does not allow for a detailed discussion of historiography as a scholarly discipline and of historical research in general. However, a few short remarks must be made about the difference between the past and history as well as about web archiving.

Past and History

It may be fruitful to distinguish between 'the past' and 'history'. The past refers to the traces which human as well as non-human activities and events have left behind; history, on the other hand, is the scholarly reconstruction of these activities and events on the basis of a careful and critical evaluation of the available traces. The past is just there, and to us—in the present—it often appears as nothing but a meaningless, accidental, and confused mass of traces which the historian must piece together in a coherent narrative about the past. In this sense, history and story are closely related, because history is a scholarly based story about—and told with—mute traces from the past. Historians tell (hi)stories; the past does not.

However, although history deals with the past the writing of history is always already entangled in and marked by the present of its writing: the historian asks, "Why are these traces from the past important or relevant at exactly this point in time in the present?" Thus, the historian's (hi)stories attest to this tension between the past and the present (cf. also the discussion in Peters, 2009, pp. 13–15).

Questions of History

On a general level the historian's (hi)story ventures to answer three clusters of questions. First, the story can set out to tell us *what* happened, what existed, what changed, and *when*. Second, the historian can set out to tell a story about *why* things happened, existed, or changed—and who/what made this happen. Answering the first set of questions will involve making a chronology of the facts—at that date this and that happened there—whereas answering the second cluster of questions will involve pointing out the forces which have caused what happened to happen at a specific point in time and place. And the two questions are mutually interdependent: A chronology can be interesting in itself, but in order to be enlightening it must be coupled with answers to 'why?' And answers to the question of why something happened may be interesting in themselves, but we have to know what actually happened and when in order to fully understand why.

Third, and finally, the historian can use his history to relate the past to the present and the future, answering questions such as, "What can we use historical study for today?" And, "To what extent can we use history to understand the present and anticipate the future?"

Web History

The above-mentioned difference between past and history also makes it possible to determine when a study of the web is, in fact, a historical study. I would suggest that it is historical when it deliberately sets out to turn the past into history. Two comments can serve to elaborate this point of view.

First, a study of the web of the present is not a historical study of the Web. Although this may seem somewhat self-evident it may be important to remind ourselves that today, in 2009, when we read a 1999 study about the web of 1999, this is not a historical study of the web, but rather a study of the web which belongs to the past. However, this does not mean that the study is irrelevant to us today since this trace of the web in the form of a scholarly study can be used as a source for our historical study of the web of 1999.

Second, a study of the web of the past is not necessarily a historical study of the web; it must deliberately be conceived as a historical study by its author(s) by setting out to answer clusters of questions similar to the ones mentioned above. For instance, if a book about the web in 1999 is published in 2002, this temporal distance does not necessarily make it a historical study; it may rather be the result of a well-known practical problem, namely, the fact that the lengthy production period of scholarly texts tends to make them 'historical' when they are published. In this sense the present which is studied can differ from the present of publication without thereby making the study historical. Therefore, in order to be a genuine historical study, it must be conceived as historical by explicitly conceiving its object as 'the past' rather than as part of a present of the past which 'stretches' into the present of publication. Again, these kinds of studies are also relevant as sources of the past.

WEB ARCHIVING

As maintained above, the aim of historical research is to produce scholarly knowledge on the basis of the available sources. Although archived web material is by no means the only source used in the writing of web history, it is an important source type, and therefore we have to reflect on what web archiving is and what characterizes the archived web document.[3]

In the main, up till the advent of the web, historians have concentrated on reconstructing the past by interpreting the sources, while in most cases the sources themselves were just found and preserved and were therefore not subject to any kind of 'construction.' However, the problems involved in finding, collecting, and preserving web material differ in many ways from those characterizing the archiving of other types of traces, including other media types.

No matter how an archived web document has been created, and no matter the archive in which it is found, web historians cannot expect it to be an identical copy on a 1:1 scale of what was actually on the web at a given time. There are two reasons for this: the archived web document is an actively created and subjective re-construction and it is almost always deficient.

The simple fact that a choice has to be made between different archiving forms and strategies implies that the archived web document is based on a subjective decision by either an individual or an institution. For instance, it has to be decided which form of archiving is to be used, where the archiving should start, how far from the start URL the archiving is to continue, whether specific file types are to be included/excluded, whether material is to be collected from other servers, and so on. In addition, the archived web document is a re-construction in the sense that it is re-created on the basis of a variety of archived web elements that are re-assembled and re-combined in the archive. Thus, the archived web document is the result of an active process and it does not exist prior to the act of archiving.

In addition, an archived web document is almost always deficient when compared to what was once on the web. There are two reasons for this. First, the archived web document is very likely to be deficient due to technical reasons (soft- or hardware). For example, words, images/graphics, sounds, moving images can be missing, or some of the possibilities for interaction may be non-functional in the archived web document. Second, one of the major reasons for deficiencies is related to time and to what could be called the dynamics of updating: the fact that the web content might have changed during the process of archiving, and we do not know if, where, and when this happens (cf. Foot & Schneider, 2004, p. 115; Brügger, 2005, pp. 21–27; Masanès, 2006, pp. 12–16).

In summary, web historians who intend to use archived web material as a source must therefore be prepared for the archived web documents to have the characteristics outlined above, and, as a consequence, they must not expect an archived website archived at the same date to necessarily be identical in different archives (cf. Brügger, 2010). In this respect, the process of archiving creates a unique version rather than a copy, and it is a version of an original which we can never expect to find in the form it actually took on the web; we can neither find an original among the different versions nor reconstruct an original based on the different versions.[4] Therefore, in general, web scholars must be proactive if they want to provide archived web material enabling them to answer their research questions, either by doing the archiving themselves or by contacting an archiving institution with their specific needs. But for the web historian it may often be impossible to act proactively, as the Web material he wants to study is gone, which is why in most cases he is forced to make do with what can be found in existing web archives. In addition, due to the characteristics mentioned above, the web historian must treat the archived web document differently than well-known media types and material on the live web, especially as regards source reliability.[5]

EXISTING WEB HISTORIES: AN OVERVIEW

As maintained above, web history can be considered an emerging field of study within internet studies. Historical studies of the web have been made, but the existing literature is characterized by the following three general tendencies. First, the number of historical studies of the web is very limited compared to web studies in general. Second, the little which has been written is not inscribed in a common research tradition dealing with web history based on a set of shared theoretical and methodological assumptions or discussions and on a self-reflexive approach to web history as a field of study in its own right; rather the contributions to the history of the web are often isolated historical supplements to a number of heterogeneous and already established fields of study such as political communication or media studies. Evidence of this tendency is that in most cases references to other existing historical web studies are very much the exception to the rule. Third, the number of historical web studies remains limited: a few studies emerged towards the end of the 1990s, then the number grew slowly in the first years after 2000 and increased a bit from 2005/06 and onwards.

Despite the limited number of studies and the heterogeneous nature of existing historical web studies, it is possible to identify some thematic patterns. The following overview is divided into two main groups—on the one hand, studies of the history of the web as an object of study in its own right, on the other, studies of other topics, including a short history of the web. Focus will mainly be on the first group, where nine clusters of texts are identified, but the last group will also be briefly touched on.[6]

The History of the Web as an Object of Study in Its Own Right

Apparently, scholars working within the field of political studies were the first to show an interest in the history of the web. As early as in 1998 the two Dutch scholars Gerrit Voerman and J.D. de Graaf published an article about the history of the web sites of political parties in the Netherlands from 1994 to 1998 (Voerman & de Graaf, 1998). In the following years, a number of articles were published in the field of web history: an article about the adoption of the web by American political candidates from 1996 to 1998 (D'Alessio, 2000), one about the evolution of the use of the web as a political campaign medium in Japan (Tkach-Kawasaki, 2003), another about the websites of members of the U.S. House of Representatives from 1996 and 2001 (Jarvis & Wilkerson, 2005), and another about web campaigning by U.S. presidential primary candidates in 2000 and 2004 (Foot & Schneider, 2006a), a longitudinal content and structural analysis of German party websites in the 2002 and 2005 national elections (Schweitzer, 2008), and a study of the web as a democratizing agent from 1994 to 2003 in 152 countries (Groshek, 2009). The first book-length historical study of how the web has been used for political communication was Kirsten Foot and Steven Schneider's *Web Campaigning* (Foot & Schneider, 2006b), about the changing campaign web practice during the U.S. elections of 2000, 2002, and 2004.

A second cluster of historical web-related texts includes histories about the web in general. In 2000 the first book-length history of the origins and development of the web was published: James Gillies and Robert Cailliau's *How the Web Was Born* (Gillies & Cailliau, 2000). This book offers valuable information about the technical infrastructure and institutionalization of the web, covering the period up till 1995. It is followed in 2005 by Kevin Hillstrom's *Defining Moments: The Internet Revolution*, which, despite the name, is almost exclusively about the history of the web (Hillstrom, 2005).

The third cluster of historical studies of the web revolves around the web as related either to existing media types such as electronic or print media or to themes which cut across the media types such as news and journalism.

The first historical studies of a broadcaster's web activities are Fiona Martin's and Maureen Burns's work on the Australian Broadcasting Corporation's (ABC) first years online, starting with Martin's article about the first four years of the development of ABC Online (Martin, 1999), and Burns's article about the prehistory of ABC Online (Burns, 2000) and culminating with the dissertations *ABC Online* (Burns, 2003; published as a book in 2008 [Burns, 2008b]) and *Digital Dilemmas: The Australian Broadcasting Corporation and Interactive Multimedia Publishing 1992–2002* (Martin, 2008; see also Burns, 2002, 2008a, 2008c, 2008d, and Martin, 2005). In a 2003 dissertation Robert Stepno studied the development of the news presentation and web page design of the television station website WRAL-TV.com from 1995 to 1999 (Stepno, 2003), and recently three dissertations have been published: Hallvard Moe has studied how the Norwegian NRK, the British BBC, and the German ARD and ZDF have approached the web (Moe, 2008a, 2008b), Tomi Lindblom has examined the development of the web and new media in three Finnish broadcasting companies (Lindblom, 2009), and Einar Thorsen has evaluated the BBC's news and the website dedicated to the 2005 UK parliamentary general election, entitled "Election 2005" (Thorsen, 2009).

The first historical study of print media as related to the web is Pablo Boczkowski's dissertation *Affording Flexibility: Transforming Information Practices in Online Newspapers* (Boczkowski, 2001), about the online development of American newspapers from pre-web electronic publishing to the web in the 1990s (later it developed into the book *Digitizing the News* [Boczkowski, 2004]). Three years later, in 2004, two dissertations were published: Carina Ihlström's *The Evolution of a New(s) Genre* (Ihlström, 2004), which analyzes how the online newspaper genre has evolved in Sweden from the mid-1990s and onward by focusing on the producer, the online newspaper, and the users (see also Ihlström & Henfridsson, 2005); and Jonas Lundberg's *Shaping Electronic News* (Lundberg, 2004), about the implications of going from hypertext news to hypermedia by focusing on design, producers, and users in a Swedish context. And in 2009 John Knox published his historical study of the use of images on newspaper websites (Knox, 2009).

Historical studies have also been published about media-oriented themes which cut across the 'old' media types. In 2005 Annika Bergström published her dissertation *nyhetsvanor.nu* (newshabits.nu) (Bergström, 2005), about the use of news and current affairs reporting on the web in Sweden from 1998 to 2003; in 2006 Stuart Allan published *Online News* (Allan, 2006), about online journalism and news across media types in a U.S. and British context; in 2007 Aaron Barlow published *The Rise of the Blogosphere*, which provides a historical account of the blog as an alternative to professional news production; and in 2008 An Nguyen published a study of the traditional media's online migration since the 1990s (Nguyen, 2008).

The fourth cluster of historical studies of the web relates to the commercial use of the web, and it includes the following publications: Jeffrey Blevins's study of a commercial web portal, the Walt Disney Company's Go Network internet portal from its debut in December 1998 to its closure in January 2001 (Blevins, 2001); Morrison and Firmstone's longitudinal study of the role of trust in relation to e-commerce in 1997–1999 (Morrison & Firmstone, 2000); Maria Frostling-Henningsson's dissertation *Internet Grocery Shopping* on internet grocery shopping in Sweden 1998–2003 (Frostling-Henningsson, 2003), Lennon et al.'s longitudinal study of rural consumers' adoption of online shopping in 2000 and 2003 (Lennon et al., 2007); Li & Zhunag's analysis of the dominant cultural values in the internet advertising of the top U.S. web sites in 2000, 2003 and 2007 (Li & Zhunag, 2007); and, finally, several of the chapters in *The Internet and American Business* (Aspray & Ceruzzi, 2008), an edited volume about the history of the internet and commerce after 1992.

A fifth cluster of texts tells the story of the 'big men' in the history of the web, including Melissa Stewart's *Tim Berners-Lee* (Stewart, 2001), Ann Gaines's *Tim Berners-Lee and the Development of the World Wide Web* (Gaines, 2002), and Harry Henderson's *Pioneers of the Internet* (Henderson, 2002). Within this topic, a number of books have been published on, among others, Tim Berners-Lee (the web), Marc Andreessen and Jim Clark (Netscape), Jeff Bezos (Amazon), Sergey Brin and Larry Page (Google), and Jerry Yang and David Filo (Yahoo).

A sixth cluster revolves around the history of privacy matters: Greg Elmer's study of web browser cookies and Netscape from 1995 and onwards (Elmer, 2002), Milne et al.'s longitudinal study of the readability of online privacy notices posted by leading websites 2001–2003 (Milne et al., 2006), and Anthony Miyazaki's longitudinal study of how the use of cookies influences online users' behavior (Miyazaki, 2008).

The seventh cluster is constituted by historical studies of web design and it includes John Nerone and Kevin G. Barnhurst's study of the design of newspaper websites (Nerone & Barnhurst, 2001); Ida Engholm's study of the development of web design from an art historical perspective, first in a short article (Engholm, 2002), later in the dissertation *WWW's designhistorie* (WWW's Design History) (Engholm, 2003); Lynne Cooke's analysis of changes in the design of print news presentations and news presentations on television and the web (Cooke, 2005); Simon Heilesen's examination of the development of web design (Heilesen, 2007); and Megan Sapnar Ankerson's discussion of the difficulties of writing a history of web design (Ankerson, 2009).

In the eighth cluster are a number of articles and books about one of the methodological and practical issues related to the writing of web history, namely, the archiving and preserving of web material. The most important books treating this topic are *Archiving Websites* (Brügger, 2005), *Web Archiving* (Masanès, 2006), and *Archiving Websites* (Brown, 2006).

And, finally, a ninth cluster of web historical texts can be identified, although it can barely be labeled a cluster as it is constituted by the remaining miscellaneous and isolated studies: a dissertation about the technological development of the web's first search engine, WebCrawler, from 1994 to 1997 (Pinkerton, 2000); a study of the history of a big web company, Google (Vise & Malseed, 2005; Stross, 2008); a study of the history of the web in a specific geographic area, Australia (Goggin, 2005); a study which may be termed an indirect history of the web: Janna Q. Anderson's *Imagining the Internet,* about the history of how the future of the web was imagined between 1990 and 1995 (Anderson, 2005); and a historical content analysis of blogs (Herring et al., 2007).

Studies of Other Topics Including a Short History of the Web

In the literature mentioned above, web history has been the primary object of study. However, the history of the web may also be found in publications about other topics which include a short history of the web. In the texts in this second main group, the history of the web is subordinate to another subject, and it is considered either a phase in the history of something else or a stepping stone aimed at putting the topic in a historical perspective. Although these types of publications may include valuable information on the history of the web, they very often do not provide in-depth historical analyses. Three different types of text can be singled out.

The first type includes histories about the internet in general where the history of the web is only considered the last—short—chapter in the long history of the internet. This approach can be seen in J.-C. Guédon's *Internet* (Guédon, 1996, pp. 52–59), Art Wolinsky's *The History of the Internet and the World Wide Web* (Wolinsky, 1999, pp. 38–52), John Naughton's *A Brief History of the Future* (Naughton, 2002, pp. 231–254), Janet Abbate's *Inventing the Internet* (Abbate, 2000, pp. 212–218), and in Hilary W. Poole's *The Internet. A Historical Encyclopedia* (Poole, 2005).

The second type includes a number of general media histories where the history of the web is seen as the continuation of the history of other types of media. Examples of this are Brian Winston's *Media Technology and Society* (Winston, 2000, pp. 333–334), Briggs & Burke's *A Social History of the Media* (Briggs & Burke, 2002, pp. 307–312), and Helmut Schanze's *Handbuch der Mediengeschichte* (Schanze, 2001, pp. 546–552).

The third type includes a great number of very heterogeneous texts about web phenomena that include a short history of the web with a view to shedding historical light on the subject treated. A couple examples of this type are Bruce Bimber and Richard Davis's short account of web campaigning before 2000, which constitutes a stepping stone for their analysis of the use of the web in the U.S. elections in 2000 (Bimber & Davis, 2003, pp. 19–28), and danah m. boyd & Nicole B. Ellison's study of social network sites, which includes a short history of social network sites on the web (boyd & Ellison, 2007).

Non-scholarly Histories about the Web

In addition to the two main groups of literature identified above—studies of the history of the web as an object of study in its own right, and studies of other topics, including a short history of the web—numerous non-scholarly histories about the web can be found. Some important examples are general histories of the web such as David Hudson's *Rewired. A Brief (and Opinionated) Net History* (Hudson, 1997); autobiographies by individuals somehow involved in the development of the web, from Tim Berners-Lee's *Weaving the Web* (Berners-Lee, 1999) to Mike Daisey's account of being employed in amazon.com in the mid-1990s (Daisey, 2002); journalistic approaches like the business journalist Roger Lowenstein's *Origins of the Crash*, about the dot-com era (Lowenstein, 2004) and the journalist Paul Andrews's chronicle *How the Web Was Won* about Microsoft and the web (Andrews, 1999); and, finally, a comicwriters history of webcomics (Campbell, 2006). The fact that these kinds of texts are not scholarly does not mean that they are of no interest for the writing of web history; on the contrary, they may constitute important sources for the web historian.

ORGANIZATION OF THE BOOK

Two overall approaches to the emerging field of web history are possible. Either we set out to discuss how web history can be done or we produce a number of concrete examples of web histories. In the first case, the focus is on the fundamental theoretical and methodological challenges involved in doing web history, whereas in the latter the focus is on the actual historical study of the web. Because web history is a relatively new field of study as well as a new discipline, both approaches are equally important if we want to get an impression of what web history could be. Thus, this duality is mirrored in the overall organization of the book.

The following chapters are divided into four parts. The first part essentially treats one of the fundamental theoretical questions related to doing web history: what is our object of study? The next two parts present a number of historical case studies of the web, focusing on web cultures on the one hand and web industries and media institutions on the other. Finally, part four addresses a number of methodological questions related to web preservation as well as to how web history may be presented on the web in collections and in web museums. However, despite this overall structure of the book, the two approaches overlap in a number of the individual chapters: several of the theoretical and methodological chapters also feature historical analyses, just as the case studies also offer valuable theoretical and methodological insights.

The first part, "Web History and the Object of Study," addresses the theoretical foundation of web history as a discipline within internet studies by questioning the object of study. In chapter 1, "Website History: An Analytical Grid," Niels Brügger sets out to develop a systematic conceptual framework which can be used as an analytical grid to guide web scholars in approaching historical studies of websites. It is argued that the conceptual framework used to identify the forces driving the history of the web should be open and that it should be based on a broad understanding of website history. With a view to evaluating the possible use of the concepts, they are applied in a concrete analysis of the early history of the Danish public service broadcaster DR's website dr.dk, and the range of their uses in other possible cases is briefly discussed and, among other things, illustrated by a informal history of facebook.com.

In chapter 2, "Object-Oriented Web Historiography," Kirsten Foot and Steven Schneider maintain that developmental analyses of any aspect of the web, whether engaged in contemporaneously or retrospectively, entail dynamics within and between the (co)producers of web artifacts, the production practices and techniques, and the web artifacts themselves, and that these dynamics make it difficult but very important for scholars to identify and situate their object(s) of analysis historically and theoretically. Taking as their starting point the term 'object orientation,' which is employed in a range of fields from computer programming to museology, Foot and Schneider present three distinct notions of object that are relevant to researching and writing web history, and they propose an object-oriented approach to web historiography that encompasses all three.

Under the heading of "Web Cultures," Part Two presents four case studies of how the use of the web has developed in the realm of web culture at large.

In chapter 3, "Evolution of U.S. White Nationalism on the Web," Alexander Halavais traces the development of 'white nationalist' and other racist websites from the mid-1990s. Contrary to the stereotypes of white racism in the United States, racists were among the first to embrace computer networks as a medium for spreading their messages and recruiting new members. Halavais shows how white nationalists were not only sophisticated in their thinking about how to deploy these new technologies, but they made effective use of "social media" well before many other organizations, and he suggests that the aim of white racists to become part of the mainstream has been at least partially realized via the web during the period analyzed.

In chapter 4, "A History of Allah.com," Albrecht Hofheinz analyzes the development of the website Allah.com and related websites which presented themselves as the 'official site' of an Islamic religious brotherhood in the mystical tradition of the Sudanese Sammaniyya. The website Allah.com turns out to be the work of an Egyptian Muslim immigrant in the United States who uses it to promote his own understanding of Islam. Hofheinz argues that Allah.com is an example of how the web furthers an ever-growing assertiveness among modern Muslims to construct their 'own' Islam, to assume authority to speak out and represent this Islam to the whole world, and to assume personal responsibility for spreading this understanding to a generalized audience—"everyone."

In chapter 5, "Historicizing Webcam Culture: The Telefetish as Virtual Object," Ken Hillis focuses on the early history of webcam culture to assess the complicated meanings of the term 'object' in virtual settings. Specifically, he examines the use of web cameras among some English-speaking, First World gay/queer men from the late 1990s to the early 2000s. These men constitute a historical vanguard that turned to the web as a new form of media centrality in order to perform personal identity claims. Hillis's analysis of the practices and techniques developed by these men indicates the historical emergence of a new form of fetish, the online moving image of the webcam operator as digital fetish—termed the telefetish—still manifest in contemporary web applications.

Webcams are also on the agenda in chapter 6, "Self-portrayal on the Web." In this chapter, Dominika Szope addresses the phenomenon of self-portrayal on the web by juxtaposing a relatively new presentation of the self on YouTube—GreenTeaGirlie—with an example from 1996 which can be regarded as a predecessor of today's self-portrayals on YouTube, namely, the webcam of Jennifer Ringley, the so-called JenniCam. It is argued, on the one hand, that the early webcam recordings such as the JenniCam can be regarded as predecessors of today's self-portrayals since they were produced by means of social software, and, on the other hand, that the new self-portrayals such as GreenTeaGirlie can shed new light on the old forms like the JenniCam.

Part Three, "Web Industries and Media Institutions," presents four case studies focusing on the organization of web industries and old media institutions on the web.

In chapter 7, "Web Industries, Economies, Aesthetics: Mapping the Look of the Web in the Dot-com Era," Megan Sapnar Ankerson investigates the economic and industrial organization of new media industries during the dot-com era (1994–2001), relating shifts in the business of web production to changes in web style and design practices. Using Hyman Minsky's classical model of a speculative bubble as a periodization framework, the strategies of web industries and the changing notions of 'quality' web design are situated within the historical context of the dot-com boom. Tracking the business and design of the web through the stages of displacement, boom, euphoria, and bust, Ankerson argues that we can better understand popular post-crash movements like 'web 2.0' by investigating websites as cultural forms that respond to particular social, economic, and industrial contexts.

In chapter 8, "Analysing and Comparing the Histories of Web Strategies of Major Media Companies—Case Finland," Tomi Lindblom examines the development of new media and web strategies of three major media companies in Finland—Alma Media, Sanoma and the Finnish Broadcasting Company Yleisradio—from 1994 to 2004. Based on a discussion of how to classify media companies' new media strategies, Lindblom analyzes and compares the success of the different strategies used by the three Finnish media companies. It is concluded that none of the companies has been absolutely more successful than any of the others in their new media strategy in the period 1994–2004. However, the biggest success seems to have been enjoyed by Yleisradio.

In chapter 9, "BBC News Online: A Brief History of Past and Present," Einar Thorsen examines four key historical periods of the BBC News website: the Corporation's online activities leading up to the official launch in 1997, the early years of the website until it became firmly established as one of the world's most popular news sites, the period from 2004 to 2005 when 'user-generated content' was propelled from being a culinary add-on to taking center stage, and finally the present-day experimentation with technology, the iPlayer and the first stages in the realization of integrated multimedia storytelling. The chapter concludes by highlighting areas of concern related to preserving the BBC News website as a historical web artefact and the human processes associated with its evolution.

In chapter 10, "(R)evolution under Construction: The Dual History of Online Newspapers and Newspapers Online," Vidar Falkenberg discusses the development of online newspapers. The main argument is that it is necessary to look at two intermingled processes in parallel—the history of the online presence of printed newspapers and the making of an online newspaper fulfilling functions similar to those of traditional newspapers. The chapter revolves around the evolution of the newspaper within the revolution of what is termed the media matrix—all existing media at a given point in time. The case of online newspapers in Denmark is used to illustrate how the online presence of the newspapers and the emergence of the online newspaper are co-evolving as part of newspaper development in addition to a more radical change in the media matrix.

The chapters in the last part, Part Four, "Preserving and Presenting," revolve around some of the methodological issues connected with the preservation of web material as well as with the presentation of the history of the web in collections and in web museums on the web.

In chapter 11, "The Aesthetics of Web Advertising: Methodological Implications for the Study of Genre Development," Iben Bredahl Jessen discusses the methodological implications of studying the actual appearances of advertising on the web. Inspired by an empirical study of web advertising, Jessen proposes a method of collecting and documenting unstable content on the web in order to study the development of web advertising as a genre. Furthermore, she proposes mapping web advertising forms based on their aesthetic characteristics. The mapping provides a basis for a typology which may be useful for studying the historical development of web advertising, and her methodological approach can serve as inspiration for studies of similar text forms on the web.

In chapter 12, "Web Archiving in Research and Historical Global Collaboratories," Charles van den Heuvel discusses the need for an infrastructure to enhance access to and enrichment of web archives in the form of annotations, with the aim of making them more valuable for both institutions and researchers. This contextual knowledge production not only adds value to web archives, it also forms part of our collective memory and needs to be preserved together with the original content. In the nineteenth and twentieth centuries, documentalists such as Paul Otlet (1868–1944) began exploring methods for ordering and annotating ephemeral material for research. Van den Heuvel maintains that these pre-web annotation initiatives are of interest for future strategies for selecting and preserving dynamic, ephemeral forms of digital cultural heritage, such as web archiving.

In chapter 13, "Collecting and Preserving Memories from the Virginia Tech Tragedy: Realizing a Web Archive," Brent Jesiek and Jeremy Hunsinger address some of the methodological issues related to the building and maintaining of the April 16 Archive created following the Virginia Tech tragedy. Hunsinger and Jesiek argue that the profoundly social character of this archive also demands that the process of valuation and provenance be opened up to accommodate a diverse population of users and contributors. Nonetheless, the goal of creating an open, social, and pragmatic memory bank requires that a range of difficult issues be taken into consideration, including questions related to accession and deaccession, balancing archival and memorial missions, organizational politics, and the use of new archiving platforms that enable us to creatively re/negotiate our individual and social memories.

In chapter 14, "Research-based Online Presentation of Web Design History: The Case of webmuseum.dk," Ida Engholm examines how a museum for the web—a web museum—can be established and how it differs from a traditional museum. These questions are exemplified through a specific case, webmuseum.dk, a Danish online museum for website design history. In conclusion, Engholm discusses what sort of web history the form of this museum is capable of conveying and how it may assist historians wishing to study web history.

Finally, the volume is rounded off by a short epilogue, "The Future of Web History" in which Niels Brügger suggests three issues to discuss in the future: broadening the conception of web archiving, doing web history beyond national boundaries, and establishing a research infrastructure.

It is hoped that the volume as a whole will prove to be a fruitful mixture of theoretical reflections, case studies, and methodological considerations. The overall aim is to give the reader an idea of what web history could be about if it is considered a field of study in its own right and in this way to broadly and tentatively explore some of the possible ways of approaching the past of the web.

ACKNOWLEDGMENTS

This book project would never have been initiated if it were not for the web history conference "Web_Site Histories: Theories, Methods, Analysis" held in Aarhus, Denmark, in October 2008. The majority of the chapters are revised versions of papers presented at this conference, and the rest are written by scholars who either attended the conference or submitted a paper but were prevented from attending.

I would like to thank a number of individuals and organizations who played a key role in the success of the conference and, accordingly, in laying the ground for the present volume (in order of appearance): Vidar Falkenberg, Ph.D., who did an excellent job co-organizing the conference; the sponsors at Aarhus University whose support made the conference possible—the Centre for Internet Research, The Department of Information and Media Studies, and 'The Knowledge Society' research priority area; the Association of Internet Researchers for generously allowing us to label the conference as 'associated' with the AoIR 9.0 conference in Copenhagen; Kirsten Foot and Steve Schneider for accepting without hesitation the invitation to give the keynote lecture; the international board of reviewers for their meticulous reviewing of the abstracts; all the participants for giving memorable presentations and engaging in lively discussions, and the chairs and the student helpers for making the conference run smoothly.

This book is also deeply indebted to two editors without whom the book would not have taken shape: Acquisitions Editor Mary Savigar at Peter Lang Publishing and Series Editor Steve Jones, who both immediately believed in the project and have supported my work with the manuscript in a number of ways. I owe them many thanks.

In addition, all the contributors to the volume deserve to be thanked for writing thought-provoking chapters and for having been so patient with the editor's endless stream of instructions.

And last, but not least, I would like to express my gratitude to Charles Ess, who, with the courtesy that is his trademark, immediately agreed to write the foreword.

Permission to reprint the illustrations that appear throughout the book is gratefully acknowledged:

Chapter 14: The websites (Figures 1–3) are designed by Oncotype and Danish Centre for Design Research.

NOTES

1. The concept of 'web sphere' was coined by Schneider and Foot and is defined as "not simply a collection of websites, but as a set of dynamically defined digital resources spanning multiple websites deemed relevant or related to a central event, concept, or theme" (Foot & Schneider, 2006b, p. 20).
2. In Brügger 2009a, it is maintained "that from a textual point of view, the website is a coherent textual unit that unfolds in one or more interrelated browser windows, the coherence of which is based on semantic, formal and physically performative interrelations" (Brügger, 2009a, p. 122).

3. These short remarks on web archiving are elaborated in more detail in Brügger, 2009a, 2009b (see also Brügger, 2005; Masanès, 2006; Brown, 2006).
4. These characteristics of the archived web document are also important when we study today's web, as in practice most web studies preserve the web in order to have a stable object to study and refer to when the analysis is to be documented (except for studies of the live web).
5. I have elsewhere outlined some of the principles for a web site philology (Brügger, 2008).
6. The following overview is compiled based on four criteria. To be included the studies must be: (a) scholarly (rather than personal evidence or the like), (b) about the web (rather than about the internet), (c) about the past (rather than a present of the past which 'stretches' into the present of publication), and (d) made publicly available to a larger public (thus excluding conference papers, but including dissertations). In addition, contributions to the present volume are not included. The publications mentioned are probably close to being a comprehensive list of existing historical studies of the web based on the above-mentioned criteria (in the cases where a publication is only an example of a category, this is specifically indicated); however, historical studies of the web may exist that have escaped my attention.

REFERENCES

Abbate, J. (2000). *Inventing the internet*. Cambridge, MA.: MIT Press.

Allan, S. (2006). *Online news: Journalism and the internet*. Maidenhead: Open University Press.

Anderson, J.Q. (2005). *Imagining the internet: Personalities, predictions, perspectives*. Lanham, MA: Rowman & Littlefield.

Andrews, P. (1999). *How the web was won: How Bill Gates and his internet idealists transformed the Microsoft empire*. New York: Broadway Books.

Ankerson, M.S. (2009). Historicizing web design: Software, style, and the look of the web. In J. Staiger & S. Hake, *Convergence media history* (pp. 192-203). New York/London: Routledge.

Aspray, W. & Ceruzzi, P.E. (Eds.) (2008). *The internet and American business*. Cambridge, MA: MIT Press.

Banks, M.A. (2008). *On the way to the web: The secret history of the internet and its founders*. Berkeley: Apress.

Barlow, A. (2007). *The rise of the blogosphere*. Westport, CT/London: Praeger.

Bergström, A. (2005). *nyhetsvanor.nu: Nyhetsanvändning på internet 1998 till 2003* (newshabits.nu: The use of news on the internet 1998–2003). Göteborg: JMG Institutionen för journalistik och masskommunikation, Göteborgs universitet.

Berners-Lee, T. (1999). *Weaving the web: The past, present and future of the World Wide Web by its inventor*. London: Orion.

Bimber, B. & Davis, R. (2003). *Campaigning online: The internet in U.S. elections*. New York: Oxford University Press.

Blevins, J.L. (2001). *The political economy of an internet portal: A case study of Disney's Go Network*. Columbus, OH: Ohio University Press.

Boczkowski, P.J. (2001). *Affording flexibility: Transforming information practices in online newspapers*. Ithaca: Cornell University Press.

Boczkowski, P.J. (2004). *Digitizing the news: Innovation in online newspapers*. Cambridge, MA: MIT Press.

boyd, d.m. & Ellison, N.B. (2007). Social network sites: Definition, history, and scholarship. *Journal of Computer-mediated Communication*, 13(1), retrieved January 23, 2009, from http://jcmc.indiana.edu/vol13/issue1/boyd.ellison.html.

Briggs, A. & Burke, P. (2002). *A social history of the media: From Gutenberg to the internet*. Cambridge: Polity.

Brown, A. (2006). *Archiving websites: A practical guide for information management professionals*. London: Facet.

Brügger, N. (2005). *Archiving websites: General considerations and strategies*. Århus: the Centre for Internet Research.

Brügger, N. (2008). The archived website and website philology: A new type of historical document? *Nordicom review*, 29(2), 155–175.

Brügger, N. (2009a). Website history and the website as an object of study. *New media & society*, 11(1–2), 115–132.

Brügger, N. (2009b). Web archiving—between past, present and future. In M. Consalvo & C. Ess (Eds.), *The handbook of internet studies*. London: Blackwell.

Brügger, N. (2010, forthcoming). *Archived websites between copies and versions: Test of versions in existing web archives*. Århus: the Centre for Internet Research.

Burns, M. (2000). ABC Online: A prehistory. *Media International Australia*, 97, 91–103.

Burns, M. (2002). Nostalgia for the future: Nation, memory and technology at ABC Online. *Southern Review*, 35, 63–73.

Burns, M. (2003). *ABC Online: Becoming the ABC*. School of Arts Media and Cultural Studies, Griffith University, retrieved January 2008 from http://www4.gu.edu.au:8080/adt-root/uploads/approved/adt-QGU20040520.111544/public/02Whole.pdf

Burns, M. (2008a). Remembering Public Service Broadcasting: Liberty and security in early ABC Online interactive sites. *Television & New Media*, 9(5), 392–406.

Burns, M. (2008b). *ABC Online: Becoming the ABC. The first five years of the Australian Broadcasting Corporation Online*. Saarbrücken: VDM Verlag Dr. Müller.

Burns, M. (2008c). Public Service Broadcasting meets the internet at the Australian Broadcasting Corporation (1995–2000). *Continuum*, 22(6), 867–881.

Burns, M. (2008d). A short wave to globalism: Radio Australia Online. *Journal of Australian Studies*, 32(3), 335–347.

Campbell, T. (2006). *A history of webcomics v1.0: 'The golden age': 1993-2005*. San Antonio: Antarctic Press.

Carey, J. & Elton, M.C.J. (2009). The other path to the web: The forgotten role of Videotex and other early online services. *New media & society*, 11(1–2), 241–260.

Cooke, L. (2005). A visual convergence of print, television, and the internet: Charting 40 years of design change in news presentation. *New Media & Society*, 7(1), 22–46.

Daisey, M. (2002). *21 Dog years: Doing time @ Amazon.Com*. New York: The Free Press.

D'Alessio, D. (2000). Adoption of the World Wide Web by American political candidates, 1996–1998. *Journal of Broadcasting & Electronic Media*, 44(4), 556–568.

Elmer, G. (2002). The case of web browser cookies: Enabling/disabling convenience and relevance on the web. In G. Elmer (Ed.), *Critical perspectives on the Internet* (pp. 49–62). Lanham, MD: Rowman & Littlefield.

Engholm, I. (2002). Digital style history. The development of graphic design on the internet. *Digital Creativity*, 13(4), 193–211.

Engholm, I. (2003). *WWW's designhistorie: Website udviklingen i et genre- og stilperspektiv* (WWW's design history: The website development from the perspective of genre and style). Copenhagen: The IT University Press.

Foot, K.A. & Schneider, S.M. (2004). The web as an object of study. *New Media & Society*, 6(1), 114-122.

Foot, K.A. & Schneider, S.M. (2006a). Web campaigning by U.S. presidential primary candidates in 2000 and 2004. In A.P. Williams & J.C. Tedesco (Eds.), *The internet election: Perspectives on the web in campaign 2004* (pp. 21-36). Lanham, MD: Rowman & Littlefield.

Foot, K.A. & Schneider, S.M. (2006b). *Web campaigning*. Cambridge, MA: MIT Press.

Frostling-Henningsson, M. (2003). *Internet grocery shopping: A necessity, a pleasurable adventure, or an act of love? A longitudinal study 1998–2003 of 23 Swedish households*. Stockholm: Stockholms universitet, Samhällsvetenskapliga fakulteten, Företagsekonomiska institutionen.

Gaines, A. (2002). *Tim Berners-Lee and the development of the World Wide Web*. Hockessin, DA: Mitchell Lane.

Gillies, J. & Cailliau, R. (2000). *How the web was born: The story of the World Wide Web*. Oxford: Oxford University Press.

Goggin, G. (2005). Net acceleration: The advent of everyday internet. In G. Goggin (Ed.) *Virtual nation: The internet in Australia* (pp. 55–70). Sydney: UNSW Press.

Groshek, J. (2009). The democratic effects of the internet, 1994–2003: A cross-national inquiry of 152 countries. *The International Communication Gazette*, 71(3), 115–136.

Guédon, J.-C. (1996). *Internet: Le monde en réseau*. Paris: Découvertes.

Heilesen, S. (2007). A short history of designing for communication on the web. In S. Heilesen & S.S. Jensen, *Designing for networked communications: Strategies and development* (pp. 118-136). London: Idea Group.

Henderson, H. (2002). *Pioneers of the internet*. San Diego, CA: Lucent.

Herring, S.C., Scheidt, L.A., Kouper, I. & Wright, E. (2007). Longitudinal content analysis of blogs: 2003–2004. In M. Tremayne (Ed.), *Blogging, citizenship, and the future of media* (pp. 3–20). London: Routledge.

Hillstrom, K. (2005). *Defining moments: The internet revolution*. Detroit: Omnigraphics.

Hudson, D. (1997). *Rewired: A brief (and opinionated) net history*. Indianapolis, IN: Macmillan Technical.

Ihlström, C. (2004). *The evolution of a new(s) genre.* Göteborg: Gothenburg Studies in Informatics, Report 29, September.

Ihlström, C. & Henfridsson, O. (2005). Online newspapers in Scandinavia: A longitudinal study of genre change and interdependency. *Information technology & people,* 18(2), 172–192.

Jarvis, S.E. & Wilkerson, K. (2005). Congress on the internet: Messages on the homepages of the U.S. House of Representatives, 1996 and 2001. *Journal of Computer-mediated Communication,* 10(2), retrieved 24 February 2009, from http://jcmc.indiana.edu/vol10/issue2/jarvis.html.

Knox, J.S. (2009). Punctuating the home page: Image as language in an online newspaper. *Discourse & Communication,* 3(2), 145–172.

Lennon, S.J., Kim, M., Johnson, K.K.P., Jolly, L.D., Damhorst, M.L. & Jasper, C.R. (2007). A longitudinal look at rural consumer adoption of online shopping. *Psychology & Marketing,* 24(4), 375–401.

Li, X. & Zhunag, L. (2007). Cultural values in internet advertising: A longitudinal study of the banner ads of the top U.S. web sites. *Southwestern Mass Communication Journal,* 23(1), 57–72.

Lindblom, T. (2009). *Uuden median murros Alma Mediassa, Sanoma Osakeyhtiössoä ja Yleisradiossa 1994–2004.* Helsinki: Helsinki University Press.

Lowenstein, R. (2004). *Origins of the crash: The great bubble and its undoing.* New York: Penguin.

Lundberg, J. (2004). *Shaping electronic news: A case study of genre perspectives on interaction design.* Linköping: Linköpings Universitet.

Martin, F. (1999). Pulling together the ABC: The role of ABC Online. *Media International Australia,* 93, 103–118.

Martin, F. (2005). Net worth: The unlikely rise of ABC Online. In G. Goggin (Ed.), *Virtual nation: The internet in Australia* (pp. 193–208). Sydney: UNSW Press.

Martin, F. (2008). *Digital dilemmas: The Australian Broadcasting Corporation and interactive multimedia publishing 1992–2002.* Lismore: Southern Cross University Press.

Masanès, J. (Ed.) (2006). *Web archiving.* Berlin: Springer.

Milne, G.R., Culnan, M.J. & Greene, H. (2006). A longitudinal assessment of online privacy notice readability. *Journal of Public Policy & Marketing,* 25(2), 238–249.

Miyazaki, A.D. (2008). Online privacy and the disclosure of cookie use: Effects on consumer trust and anticipated patronage. *Journal of Public Policy & Marketing,* 27(1), 19–33.

Moe, H. (2008a). Public service media online? Regulating public broadcasters' internet services. A comparative analysis. *Television & New Media,* 9(3), 220–238.

Moe, H. (2008b). *Public broadcasters, the internet, and democracy: Comparing policy and exploring public service media online.* Bergen: University of Bergen Press.

Morrison, D.E. & Firmstone, J. (2000). The social function of trust and implications for e-commerce. *International Journal of Advertising,* 19(5), 599–623.

Naughton, J. (2002). *A brief history of the future: The origins of the internet.* London: Phoenix.

Nerone, J. & Barnhurst, K.G. (2001). Beyond modernism: Digital design, Americanization and the future of newspaper form. *New Media & Society,* 3(4), 467–482.

Nguyen, A. (2008). Facing "The fabulous monster": The traditional media's fear-driven innovation culture in the development of online news. *Journalism Studies*, 9(1), 91–104.

Peters, B. (2009). And lead us not into thinking the new is new: A bibliographic case for new media history. *New Media & Society*, 11(1–2), 13–30.

Pinkerton, B. (2000). *WebCrawler: Finding what people want*. Seattle: University of Washington Press.

Poole, H.W. (Ed.) (2005). *The internet: A historical encyclopedia*. Santa Barbara, CA: ABC-Clio.

Schanze, H. (Ed.) (2001). *Handbuch der Mediengeschichte*. Stuttgart: Kröner.

Schweitzer, E.J. (2008). Innovation or normalization in e-campaigning?: A longitudinal analysis of German party websites in the 2002 and 2005 national elections. *European Journal of Communication*, 23(4), 449–470.

Stepno, R.B. (2003). *Presenting Happy Valley: A case study of online newswork roles, design and decision-making at WRAL-TV.com, 1996–1999*. Chapel Hill: The University of North Carolina Press.

Stewart, M. (2001). *Tim Berners-Lee: Inventor of the World Wide Web*. New York: Ferguson.

Stross, R. (2008). *Planet Google: How one company is transforming our lives*. London: Atlantic Books.

Thorsen, E. (2009). *News, citizenship and the internet: BBC News Online's reporting of the 2005 general election*. Bournemouth: Bournemouth University Press.

Tkach-Kawasaki, L.M. (2003). Clicking for votes: Assessing Japanese political campaigns on the web. In K.C. Ho, R. Kluver, K.C.C. Yang (Eds.), *Asia.com: Asia encounters the internet* (pp. 159–174). London: RoutledgeCurzon.

Vise, D.A. & Malseed, M. (2005). *The Google story*. New York: Delacorte.

Voerman, G. & de Graaf, J.D. (1998). De websites van de nederlandse politieke partijen, 1994–1998. In *Documentatiecentrum Nederlandse Politieke Partijen (DNPP) Jaarboek 1997* (pp. 238–69). Groningen: DNPP.

Winston, B. (2000). *Media technology and society. A history: From the telegraph to the internet*. London: Routledge.

Wolinsky, A. (1999). *The history of the internet and the World Wide Web*. Berkeley Heights, NJ: Enslow

PART ONE

Web History and the Object of Study

CHAPTER 1

Website History: An Analytical Grid

Niels Brügger

The aim of this chapter is to develop a systematic conceptual framework which can be used as an analytical grid to guide web scholars in approaching historical studies of websites.

The need for the development of analytical concepts emerges from the fact that scholarly studies cannot rely on the terminology and categories in the source material being used in a consistent and precise manner. The producers of a website may call it 'home page,' 'web pages,' 'web documents,' 'site,' etc., just as the users may use other terms for the same phenomenon. And all of this may very well change over time: What was once called a "home page" may now be called a "website." In everyday language this is normally not a problem, but within a scholarly context we have to use concepts which are consistent, explicit, and independent of the analysed object.

The following attempt to develop a conceptual framework is general in scope, with a view to making the framework usable in as many cases as possible. This demand means, first of all, that the framework may seem somewhat abstract at first sight, and, second, that it cannot be expected to fit a specific analytical case without being modified. The proposed analytical grid can therefore be considered part of a theoretical-methodological toolbox, and it may serve as the starting kit for an analytical bricolage suitable for analyzing a variety of individual cases.

However, the use of the conceptual framework can only be justified if it is analytically productive, that is, if it makes it possible to produce a more comprehensive and systematic analysis than if it was not used. So, in order to evaluate the usefulness of the concepts put forward in this discussion, the proposed framework will be applied to an analysis of an ongoing project, "The history of dr.dk, 1996–2006." There will also be a brief discussion of how the concepts may be applied more widely, with a particular focus on the history of facebook.com.[1]

THE HISTORY OF DR.DK, 1996–2006

Let us start by taking a look at the research project "The history of dr.dk, 1996–2006." DR (formerly Danmarks Radio [Denmark's Radio]) is the Danish Broadcasting Corporation. It was founded in 1925 as a public service organization, and it is an independent, license-financed public institution. DR is headed by a board of 11 members elected for a four-year period, the Minister for Culture appoints three members, the Danish parliament six, and the permanently employed staff appoint two. Among other things, the board establishes the guidelines for DR's activities and appoints the members of the executive board, which undertakes the daily operations.

In January 1996 DR launched the website DR Online—www.dr.dk—and in the following years dr.dk played an ever more important role in Denmark's web presence, both as regards size and number of users. It also became an important actor in the development of new ways of using the web as well as in cross-media integration of the web with older mass media.[2]

The research project "The history of dr.dk, 1996–2006" began in 2007 with the overall aim of writing the history of the first ten years of dr.dk. It was decided to approach the project by trying to find answers to two interrelated questions: What were the *driving forces* behind the creation and development of dr.dk from 1996 to 2006, and what were the consequences of these for the website? In addition the project set out to identify and create the theoretical and methodological tools needed to analyse dr.dk.

The question of what driving forces were behind the creation and development of dr.dk will guide the remainder of this paper, while the chapter as a whole can be viewed as an answer to some of the issues implied in the question of what theoretical and methodological tools are necessary to analyse dr.dk.

GENERAL APPROACH

Before introducing the conceptual framework in greater detail, I shall make a few preliminary theoretical remarks on the two terms used in 'website history.'

A Pluri-deterministic Historical Approach

As the first research question above indicates, the main purpose of the research project is to localize, map, and explain the *driving forces* behind the creation and development of a website, dr.dk. Although this question is related to a specific research project, the illumination of the forces which drive history will probably be part of other historical studies of the web to some extent. Historical studies are often the result of a fundamental wonderment: Why was there something and not just nothing? Why was there not something else? And why did it change? In other words, a genuine part of historical studies is often answering the question: What forces made things happen?

In general, throughout the intellectual history of historical studies this question has been answered either by focusing on individual actors (or groups of actors) or on impersonal structures (economic, technological, cultural, etc.) (cf., for instance, Tosh, 1984, pp. 28–56). And in many cases the determining factors which the historian should identify were decided on beforehand (e.g., individuals, structures), just as the number of possible determining factors has often been limited.

In this chapter a more open position is taken on identifying the forces which drive history. This is encompassed in the following three points. First, we cannot know what kind of forces drive history, but we can assume that individuals, groups, and structures—and probably several other kinds of entities—play a role. Second, we cannot assume a mono-determinism but rather a pluri-determinism, where the focus is on multiple driving forces as well as on the network of their interrelations and balance of power in the past. Third, we should not begin with the aim of determining factors which were decided on beforehand and which we then set out to prove through appropriate recovery of sources. Instead, we should take the available sources as the starting point for identifying the driving forces, which actually emerge out of them, following careful study. However, this last point does not mean that we can choose and analyse the sources without any idea of what we are looking for (for instance, a hypothesis will always determine our way of approaching the source). The point I would like to make is that if we set out to look for economic and cultural determinants and the source we have before our eyes reveals only institutional determinants, we must accept this and thereby approach sources with an open mind.

These fundamental arguments regarding the kind and the number of the driving forces as well as where to start our analysis will guide the following attempt to develop a conceptual framework which tries to balance, on the one hand, the need for as precise, varied and systematic an explanation of the driving forces as possible, and, on the other hand, the desire that it should not be decided in advance that some driving forces are the only determining factors.

Media Theoretical Approach

The fundamental theoretical point in the present discussion of website history is that a website is a *mediated* phenomenon. Therefore, it is best approached as a medium, on the basis of insights from media studies. However, media studies is a very broad discipline (or field) based on a variety of theories and methods and analytical foci (economic, policy, social interaction, culture, texts, reception, etc.).

When faced with a given media artefact, media studies can choose to frame the analysis by focusing on five major constituents: sender, medium, text, receiver and context, and on the mutual relations among these. For instance, with regard to 'television,' the media scholar can decide to study the media organisation (sender), the technical devices used for production, distribution, and reception (medium), the reception of the programs (receiver), specific programs or the flow of programs (text), and the societal context (macro-context) or the immediate context of production or use (micro-context). And finally, the relations between the above-mentioned aspects can be studied—for instance, how the technical aspect of the medium affects the producers, the text, and its reception (cf. figure 1).[3]

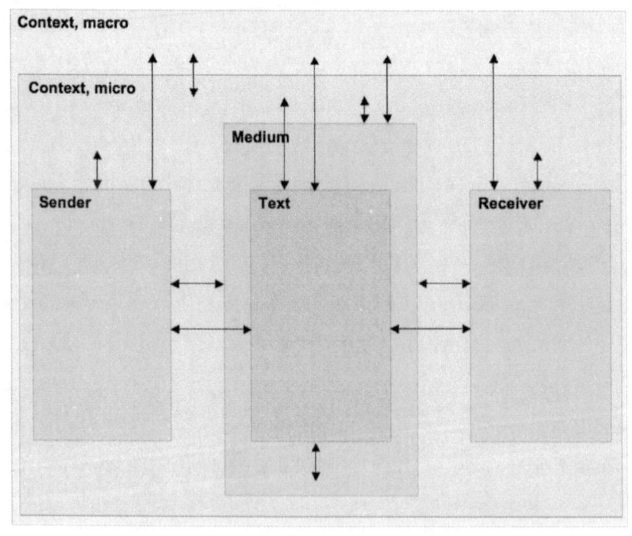

Figure 1. Ways of focusing media studies

When this approach is applied to website history the first task is to clarify what is understood by 'website' since 'the website' is what provides cohesion throughout the analysis, both what we understand by 'website' in general and how we delimit the specific website whose history we are writing. I would argue that the analytical object 'website' must first and foremost be delimited as a *mediated textual artefact*, that is, a specific nexus of signifying textual units within the medium internet (the term 'text' is understood in a broad sense, referring to all forms of expression such as written words, still images, moving images, and sound, cf. Brügger, 2009a).[4] Thus as a minimum 'website history' would be the writing of the history of this mediated textual entity as it appears on the web or in a web archive.

However, when trying to identify the driving forces behind the creation and development of this artefact, we cannot limit our focus to the artefact itself. On the one hand, the artefact is not created by itself, it is created by someone/something, and, on the other hand, it can be argued that the artefact creates its sender, receiver, and context (for instance, the sender in the form of an organizational structure, buildings, etc.). In other words, the artefact is part of a recurrent process of (re)-creation with its surroundings, where the artefact is created by the surroundings just as the surroundings are created by the existence of the artefact. Therefore, if we want our analysis to be as exhaustive as possible, focusing on the artefact is only a first step, namely, the one which provides cohesion to our analysis, but we must also incorporate sender, receiver, and context as well as the relations between these constituents and the artefact.

Thus, 'website history' becomes a broad discipline, the aim of which is to write not just the history of the mediated textual phenomenon which was once on the web or is now in an archive but also the history of the relevant entities which surrounded it and with which it was interrelated in a recurrent process of (re-)creation. On a general level, website history is therefore a discipline which aims at writing the history of the complex *strategic situation* in which the artefact is entangled.

The following is an outline of a systematic conceptual framework which tries to meet the demands of a pluri-deterministic historical approach as well as of a broad understanding of website history.

AN OUTLINE OF A CONCEPTUAL FRAMEWORK

Following the line of thought that a website is above all a mediated phenomenon and that it could be approached as part of media studies, a relevant question is how the constituents—sender, medium, text, receiver, and context and their interrelations—are in fact articulated. I would argue: (a) that they are articulated in the source material as a variety of *elements*, (b) between which different *relations* are possible, (c) and that elements and relations thereby form a *configuration* which can be understood as a *strategic situation*. In the following, each of these three points will be elaborated and then put in context in a short analysis of the two years before the launch of dr.dk in 1996.

Elements

An element is an entity which can articulate one of the constituents mentioned above (sender, medium, etc.). Because elements will always be closely related to a specific and concrete website history, it is impossible to set up an exhaustive list of elements. However, what can be done is to illustrate some of the elements which can be identified in relation to the history of dr.dk.

Elements of the 'sender' constituent can, for instance, include the management of dr.dk, sub-divisions (web production, news, etc.), individual employees, physical localities, available resources (economic, technical, etc.), written strategies, and work routines. The contextual elements can be said to cluster in a macro and a micro context, respectively; the macro context can, for instance, be the media legislation, the (media) market, foreign sources of inspiration, whereas the micro context can be DR's board and executive board, DR's other broadcast activities (radio, television, mobile media), DR's program policies, the concept of cross-mediality, etc. Elements of the 'medium' constituent can include software types and the technical and organizational structure of the internet/web. The 'text' constituent can, for instance, be articulated through the structure, the layout or the possibilities of interaction of the website, and finally the elements of the 'receiver' constituent can include feedback from users (user surveys, formal/informal feedback fora) and the users' computer equipment.

Two comments should be made on how the elements can be viewed. First, as the examples above show, elements can be human (individuals or groups) as well as non-human entities such as material objects (technological artefacts, physical environment, resources, etc.) or ideas (of, for instance, public service, cross-media production, digital strategies, etc.). Second, the same entity can appear and be understood as an element of more than one constituent—for instance, as both sender and receiver. An example of this is a debate forum on a website where the same individual alternates between sender and receiver.[5] Thus, one of the tasks of the website analysis is to map the elements.

Relations

In many cases—but not in all—the elements will be related to each other in various ways. As with the elements, the character of the relations will always be based on the nature of the concrete study as well as on the available sources. However, a few examples will help illustrate what is meant by relation. The following should therefore not be considered an exhaustive typology of relations, but just examples which should be supplemented and specified in more detail. Furthermore, the relations mentioned may not be clearly delimited and they may overlap, just as all types of relations cannot be applied to all kinds of elements (for instance, some may only be applied to human or to non-human entities).

The relation between two (or more) elements can, for instance, be a relation of creation—a web designer creates the textual elements of the website, the management creates strategies, the internet creates the media environment for the textual elements, etc.—or, for instance, the relation between two elements can be conducive, enhancing, stabilizing, restraining, destructive, subversive, or there can exist various relations of opposition (disagreement, dispute, conflict, fight, negotiation) or of motivation (invitation, request, demand, order, etc.).

The subject of the relation—the entity which is created, enhanced, restrained etc.—can be a variety of things: power, identity (individual, group), culture, organizational structures, resources, routines, content, technology, and the like. And, finally, the relations can manifest themselves in different ways: formally/informally, manifest/hidden, explicitly/implicitly, consciously/unconsciously, etc.

Considered as a whole, the network of elements and relations constitutes a specific *configuration* (cf. figure 2).

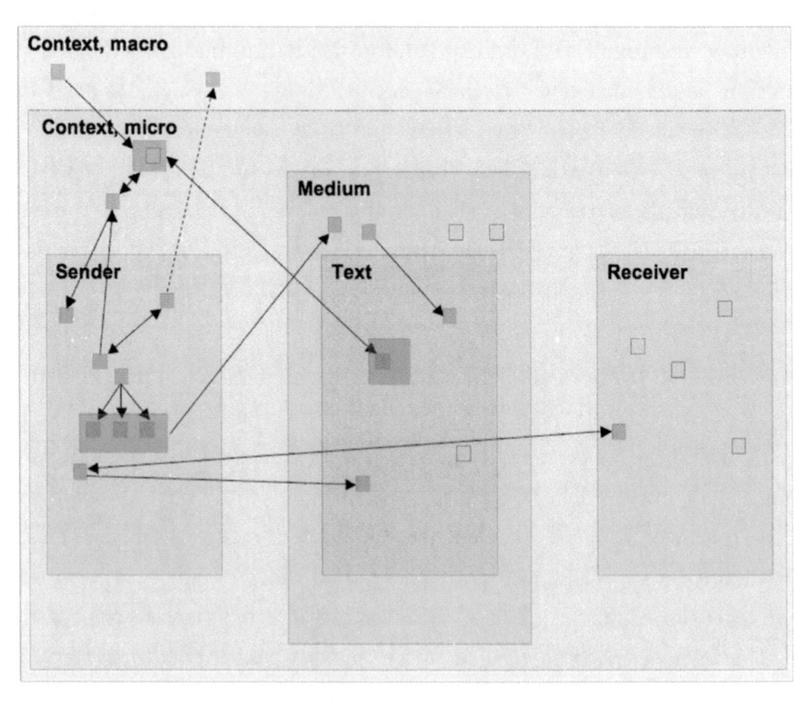

Figure 2: A configuration of elements and relations

The following three supplementary comments can help to clarify what is meant by elements and relations.

First, an element can be part of several relations (both from and to the element), and in this sense an element can be a node in a network of relations.

Second, the elements can cluster if their relations are based on some kind of similarity. For instance, the elements on the constituent 'medium' will often cluster in three groups, dependent on whether they relate to the production, distribution, or consumption of the media text. Elements of the 'sender' constituent may cluster as management, department, sub-division, or individual employee. These clusters are not defined in advance, and they can be nested in one another, for instance, a relation of conflict may exist between individuals in a department, but as a whole the department can act as one element which can then be part of other relations. Dependent on the characteristics of the elements and the clusters, detailed theoretical approaches may be needed to explain their interrelations. For instance, theories of media organisations could be the focus of an analysis of how elements of the constituents 'sender' and 'micro context' interact as various forms of organisations (cf. Küng-Shankleman, 2000; Küng, 2008, pp. 180–195), just like other theories may be used to explain how elements on the constituents 'medium' and 'text' interact as a website (cf. Brügger, 2009a).

Third, individual elements as well as clusters of elements can be related to other elements or clusters as both background and foreground—a cluster can constitute the background on which other elements and relations appear—for instance, the web production department can be considered the background for a conflict between two employees in the department. In this connection, two things should be noted. On the one hand, the background/foreground can be dynamic entities, since one and the same element/relation can be backgrounded in one perspective but foregrounded in another, depending on the relation brought into focus. For instance, the conflict between two employees in the web production department might be the background for actions taken by the web production department; thus, elements and relations can oscillate between figure and ground. On the other hand, both figure and ground are articulated through elements and relations and are not just there; the background upon which a given element/relation is foregrounded is in itself constituted by elements and relations.

One of the tasks of the website analysis is therefore not only to map the elements but also their relations.

Elements, Actors, and Driving Forces

An element is not necessarily related to other elements, but some elements influence other elements by their relations. I call these elements *actors*. An actor is an element (human as well as non-human) which influences one or more element(s)/actor(s), and the forms of the influence are a function of the relation to the other element/actor. I shall make three comments to clarify the distinction between element and actor.

First, the word 'actor' is understood in a weaker sense than is often the case within the social sciences. In the present context an actor does not necessarily become an actor by performing an act in the strict sense of the word but just by 'influencing,' that is, by having an effect on another element's specific nature. But this influence can, of course, also be an act, for instance, based on some kind of will, thus indicating a continuum with, at the one end, a very vague but detectable influence to, at the other end, an intentional will, the first being applicable to both human and non-human entities, whereas the latter is normally only applied to human entities.

Second, an entity can be understood as both an element and an actor, depending on whether it is related to all other elements in the configuration or not.

Third, the transformation from element to actor does not necessarily happen all of a sudden, it can also be an ongoing process; I call these kinds of relations *incipient relations* (cf. the dotted arrows in figure 3). And whether the transformation from element to actor turns out to be successful or not can, for instance, depend on its enhancing or restraining relations to other elements.

What, then, should be understood by driving force? I would argue that a driving force is an actor which has a *decisive* influence on other elements, actors, or even on the whole configuration (the bold arrows in figure 3).

Three comments can help specify what is meant by driving forces. First, as opposed to the difference between element and actor, the distinction actor/driving force is a difference of quantity (more/less influence) and not a difference of quality (does/does not have influence). Second, several driving forces can be at stake in a configuration, they may point in different directions, they may conflict, etc. Third, the transformation from actor to driving force can also be an ongoing process of transformation, that is, an incipient relation.

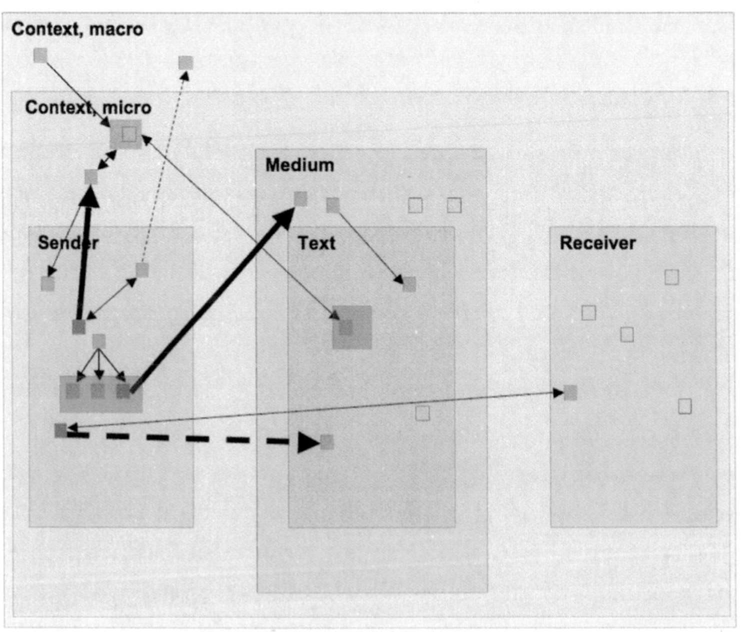

Figure 3: The strategic situation of elements, actors, and driving forces

So far, the main point of the conceptual framework can be summarized as follows. The configuration of elements, actors, and driving forces as well as their relations constitute and are constituted by a network which can be considered a *strategic situation*. And the aim of the website analysis is to map and explain this strategic situation to determine the forces that drive the creation and development of the website in question.[6]

Historical Analysis of the Strategic Situation

The task of identifying and explaining the driving forces of the development of a website has until now been approached in a somewhat ahistorical way. How can a historic perspective be integrated in the analysis?—In other words: How can the outlined conceptual framework be used in writing website *history*?

I would propose introducing a well-known distinction between two different approaches to studying the past, the *synchronic* and *diachronic* approaches. On the one hand, in the synchronic approach the website historian sets out to identify the driving forces which make things look as they do at a certain point in time, while in the diachronic approach, on the other hand, he sets out to explain the driving forces which make things change from one point in time to another.

Both perspectives are historical in the sense that they address the past, but the synchronic analysis seeks an answer to the question of why the strategic situation is as it is at one (or more) point(s) in time, and the diachronic analysis deals with the question of the differences between the two situations as well as the agent(s) of change between the two—in other words: what makes history 'progress'? The distinction between synchronic and diachronic driving forces can be illustrated in figure 4 (the synchronic forces are the bold arrows in each of the two frames, whereas the diachronic forces are illustrated by the two arrows on the left).

The following three comments should help to clarify the historic perspective. First, the transformation of elements or relations or the introduction of new elements in an existing configuration can influence other parts of the configuration or maybe the configuration as a whole, thus the elements can constitute a diachronic driving force. This is one of the reasons why the mapping of elements which do not play a role in a configuration at a certain point in time can be important since they may be transformed into actors in a later configuration. Second, there may be more diachronic driving forces at stake as agents of change. And, third, the diachronic driving forces may also exist in an incipient state.

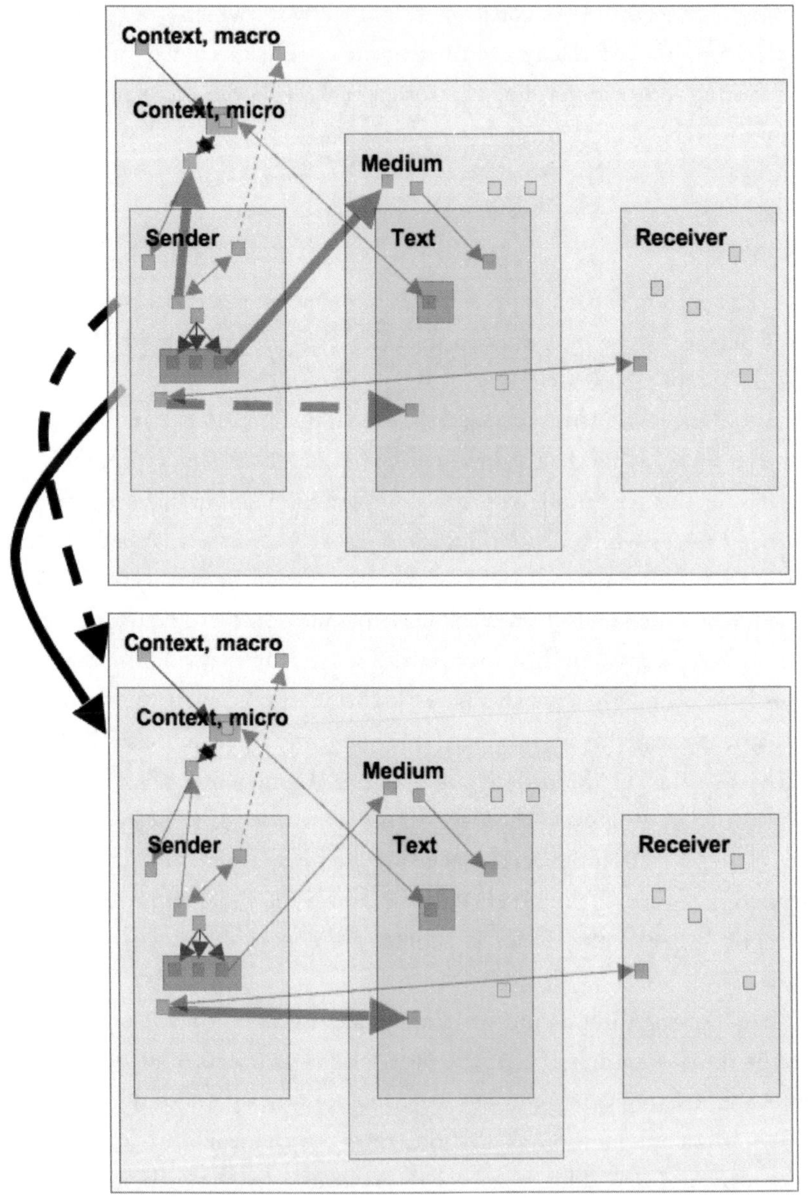

Figure 4: The synchronic and diachronic forces in the past

Therefore, website history is not just the history of the driving forces which manifested themselves as actual driving forces—synchronic as well as diachronic—but also, as indicated above, the history of the whole strategic situation—that is, the history of all elements, (incipient) actors and (incipient) driving forces—at various points in time as well as the changes between the situations.[7]

Sources

As maintained above, the driving forces to be uncovered must emerge out of a careful reading of the available sources and not out of a number of determining factors decided on beforehand. In relation to the present conceptual framework, this means that evidence of the expression of all elements, relations, actors, and driving forces must exist if the historical analysis of the website is not to be considered unsubstantiated speculation. The task is, therefore, to identify what may not be immediately apparent (elements, relations, actors, driving forces) through what can immediately be seen (the available sources).

As is the case in any other historical study, the website historian must make an effort to provide as many relevant sources as possible to answer the research question. And if the broad approach to website history outlined above is adopted by integrating the relevant entities which surrounded the mediated textual artefact, it becomes evident that archived versions of the website in question do not constitute sufficient source material, which is why other source types must be included.

These sources can be created as well as found by the website historian. For instance, the writing of the history of dr.dk, 1996–2006 will be based on both created sources such as retrospective research interviews and a variety of found sources, including memos, minutes of meetings, reports, policy documents, correspondences, organizational charts, job descriptions, dummies, biographies, diaries, memorabilia, legal texts, statistics, photos, radio/televised interviews and soft- and hardware.

However, although the archived website does not necessarily constitute the main source, in many cases it is an important source, both as the fundamental creator of cohesion in the history we are writing and as a source which may encompass valuable information about the creation of the website. And as a source the archived website stands apart from other well-known document types since it is both created and found. The archived website is based on web material which is actually found on the web, but it is created in the sense that the bits and pieces that stem from the web are collected and preserved based on a number of subjective choices and unpredictable coincidences with regard to software, strategy and purpose, the archiving process, etc. In this sense the archived website can be considered a subjective re-construction on the basis of what was once on the web. Thus, considerations of the website as a document type must also be an integrated part of website history since the archived website makes extraordinary demands on the website historian's ability to evaluate source reliability.[8] Nevertheless, the archived website is not the only source in the writing of website history, nor is it sufficient.

THE CONCEPTUAL FRAMEWORK AT WORK

On January 5, 1996, DR launched the organization's website, DR Online, located at the web address www.dr.dk (cf. figure 5).[9] But what were the forces behind the creation of dr.dk? The following contribution is by no means an exhaustive answer to this question, first, because of the limited space, second, because the search for sources has not yet ended, and third, because the main goal is to illustrate how the above outlined analytical framework can be used in a concrete analysis (also please note that all of the outlined concepts are not used in the following). For these reasons, only the period from January to December 1994 will be discussed in detail and the activities in 1995 are only treated briefly. The following analysis is mainly based on four important sources which are all internal memos and correspondence. Since DR was promised anonymity to protect individuals, the authors of the sources have been left out in the following analysis.[10]

The Impetus

Taking the available sources as our point of departure, our history starts two years before the launch of dr.dk, in the spring of 1994. It soon becomes apparent that the impetus for the creation of the website dr.dk, which will be created two years later, does not only come from inside DR.

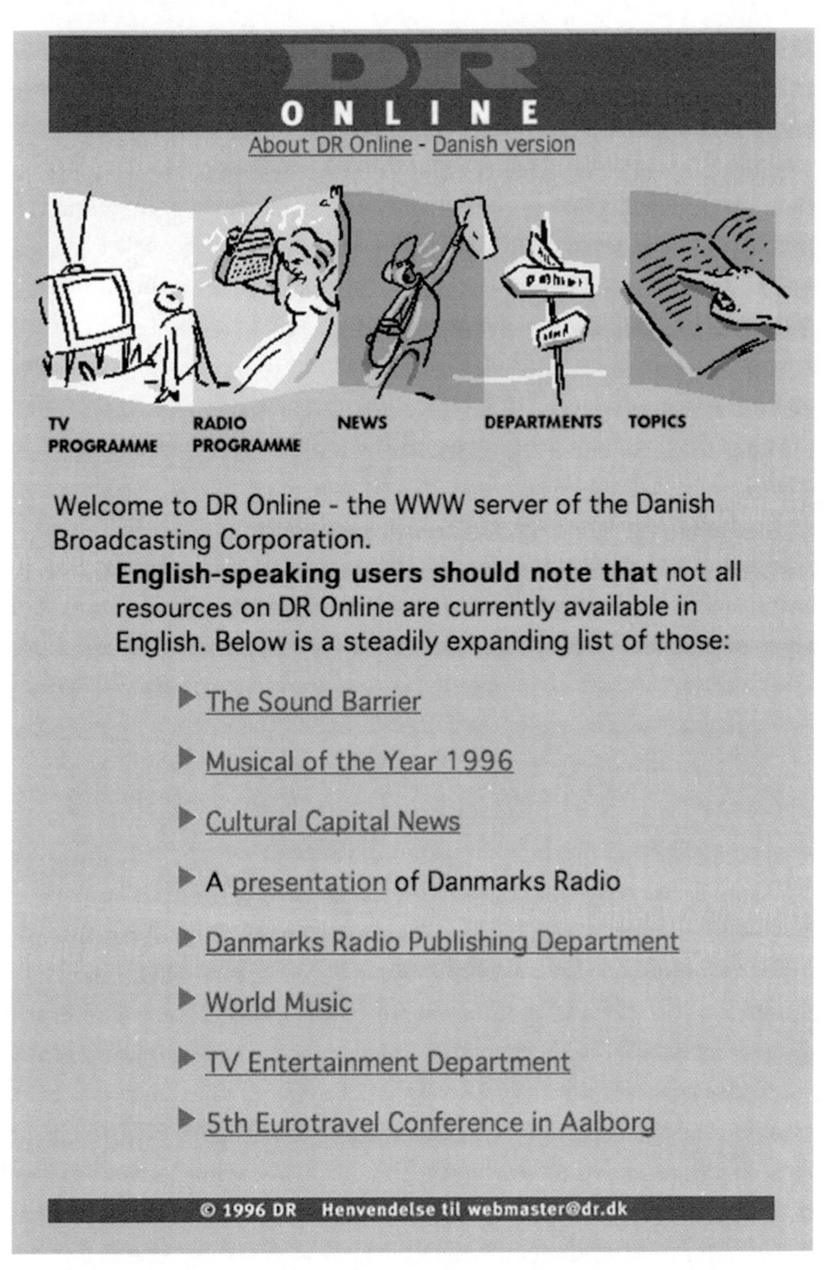

Figure 5: The English front page of dr.dk, November 1996

The initial impetus is in fact an article which appeared in the printed magazine *Emil* (April 1994).[11] Under the heading "Internet Radio," Kenneth Hansen, who studied Danish and Informatics at the University of Copenhagen, gives a brief account of this phenomenon, which is probably new to most of his readers—the internet—and how it may be used for broadcasting radio. However, there is still a long way to go: radio programs have to be downloaded, and half an hour of radio equals 15MB (filling approximately 10 floppy disks), which is why it would take quite a while to download it over a modem. As the author commented: it would be less expensive to mail radio programs recorded on cassette tapes. The website dr.dk is not waiting just around the corner and it is not even mentioned as a possibility by Hansen. But the possible use of the internet by a radio broadcaster has hit the agenda.

The second impetus is not known outside DR and, to a very large extent, not even inside DR. In 1994 it is likely that the majority of people working in DR did not even know what a domain name was, let alone could have imagined the importance of securing dr.dk. But one person, a technician in the television production department, did realise the significance and bought the domain name dr.dk which he ran on a computer at DR that he configured as an email server. This semi-private initiative is probably only known by very few people, but it was in fact the first step towards what would become the website dr.dk two years later (the domain name was later given to DR).

The Radio Program *Harddisken*

The first evidence that individuals within DR had started to think about a website is related to the above-mentioned 'private' email server with the domain name dr.dk. One of the high-profile and trendsetting radio programs about 'new media' in 1994—*Harddisken*, the "Harddisk"—had noticed that their listeners were surprised that this particular radio program did not have an email address. But when some of them emailed the email address dr.dk, the email was actually received by the technician in the television production department (source 1). At that time *Harddisken* did not have any official internet access within DR at all, which is why the research for the weekly feature "News from the Net" was undertaken by a freelancer using his own private internet connection. To avoid this confusion and to put the radio program on a par with the expectations of the listeners, in January 1994 the television technician set up the email address harddisken@dr.dk, which forwarded the email to the private email account of the freelancer, the only person in the editorial team who had access to the internet. This was the first way of making contact with DR via email, and very quickly

Harddisken began receiving up to five emails—per week—which was double the number of paper letters received (source 1, source 3).

The genie was out of the bottle and the staff of *Harddisken* soon started to come up with new ideas of their own. They wrote a letter to the management of the radio department entitled "Considerations and Wishes Concerning *Harddisken* and New Technology" (July 14, 1994, source 1). The considerations in this letter probably provide the first glimpse of how the staff envisioned DR's presence on the internet. Besides the formulation of the fundamental demand—internet access—the editorial team had the following in mind: a bulletin board system for comments from and discussions with and between listeners and program employees, a program overview, and a playlist. And they also tried to look into what, at the time, was considered a very distant future:

> If one looks a bit into the future we imagine: a) that radio programs can be downloaded as written text from a server, b) that one can find supplementary information or stories that were not aired, and c) that one can discuss the content with the producers and other listeners. Looking further into the future we imagine the possibility of making the sound accessible on a harddisk so that the listeners can contact DR's server, search through the programs, and download those of interest. This will also overcome some of the traditional weaknesses of radio, in that the listeners will be able to listen to the program whenever they want to and they will be able to fast forward the program as they please (source 1).

All in all the seeds of what was later to become the website dr.dk are already there.

Three Months Later, Television Enters the Scene: A Different Story

Apparently, the radio program *Harddisken* took the first steps toward using the internet. But in the late summer of 1994 DR's televison department seems to have discovered this new media type. However, their approach is in many ways different from that of *Harddisken*. The point of departure for the considerations about the internet with respect to television is a joint initiative between three subdivisions in the television department: TV-U, TV-IT and TV-ARC (Television Development (TV-U), Television IT (TV-IT), and Television Archiving and Research (TV-ARC)). Approved by television management, the three subdivisions initiated *Projekt Internet* [Project Internet], the primary goal of which was to clarify DR's future use of the internet as an electronic means of information. In the letter "Information on Project INTERNET" (October 10, 1994, source 2), the television group positions the project as part of a strategic paper from DR's executive board entitled *DR 1995–2005*. However, when considering DR's future internet presence, *DR 1995–2005* proves to be a rather poor source since electronic data transmission is barely mentioned—and with good reason, the internet was quite exotic when the paper was prepared. The aim of *Projekt Internet*, therefore, comes across as rather broad and imprecise, namely, "to initiate and support the use of computer networks in DR TV" (source 2).

And whereas the main concern of *Harddisken* was to get in contact with the listeners, the television initiative focuses almost exclusively on the internet as a powerful and fast supplementary tool that can help television journalists in their research process and that should ultimately improve the quality of the television programs.

Apparently, the viewers may one day communicate through emails, if they wished to comment on the programs (but DR television did not imagine that they could email the viewers); the distribution of program information is also briefly mentioned but with no precise considerations as to how this should be done. Thus, the idea of communication with the world outside DR in a broader sense or of something like a website is not yet part of the mindset.

Furthermore, *Projekt Internet* is not initiated on the basis of hands-on experiences with the new medium but on the basis of the documented experiences of other major media institutions as well as analyses and predictions in official reports on the use of 'data transmission' in society at large, including in relation to the Mediator project.[12]

All in all, the document "Information on Project Internet" leaves the reader with the impression that the television department wants to use the internet as a research tool for journalists and that in reality their visions for how the internet may be used are very general.

Two Months Later: *Harddisken* Again

Towards the end of 1994 the first document to discuss WWW more in depth is produced, and again the editorial team of the radio program *Harddisken* is the initiator. In the memo "Proposal for Homepage for Denmarks Radio" (December 7, 1994, source 3), a rather new phenomenon is introduced: the "superstructure on the internet [called] *World-Wide-Web (WWW)*, a network where one can transmit *images* and *sound* from the whole world." However, WWW is by no means the only and most obvious option for internet presence at that time, which is why the author has to explain that another "way of communicating via computer is by creating a so-called BBS, Bulletin Board System, an electronic bulletin board which can be contacted by modem and which does *not* require an Internet connection."

The "homepage" is indubitably understood as DR's communication platform and forum for its listeners and viewers, and the following functions are among those suggested: a debate forum, more detailed program listings than can be found in the printed press—for instance, with small sound bites and/or still images, information on the program editors, and a 'can we see the manuscript?' function—however, genuine transmission of sound and video clips is not considered an option yet. Although *Harddisken* imagines a variety of uses as well as content and navigation structures, it is still not quite clear how the homepage should be realized.

And finally, the author emphasizes two important issues: first, that DR is a public service institution, and therefore DR must provide its customers with a proper service; second, that it is urgent to get started quickly due to the threat of the rival national broadcaster, TV2, gaining a web presence first. Also some international media either are already on www in 1994 (*The Economist, The Times,* MTV) or have an email address (the national Swedish broadcaster).

This document is a continuation of the first letter from *Harddisken* and it is the first document in which something that is close to a genuine website for DR as a whole is presented (in the document called 'homepage'), even though it is neither clear nor clarified whether it should be a website on the WWW or a BBS.

A Voice from the Outside World

Although DR's internet presence has been discussed for some months by a few people inside DR, voices from outside DR can also be heard stating that DR should have a website. On December 28, 1994, a small internet start-up company sends a project proposal to DR which outlines how DR can be present on the internet in the (near) future ("Denmark's Radio on the Internet. Project Description" [December 28, 1994, source 4]). It has not been possible to determine whether this document was sent to DR by request, but it presents a thorough description of some basic features of DR's possible initiatives on the internet; however, to the authors of this project proposal it is still not yet obvious whether it should be based on BBS or WWW, although the balance seems to tip for WWW.

The Driving Forces Behind the Creation of dr.dk

I shall now combine the above short account of the pre-history of dr.dk based on a reading of four of the most important sources with the outlined conceptual framework to make a more systematic analysis of the identification and location of the driving forces behind the creation of dr.dk.

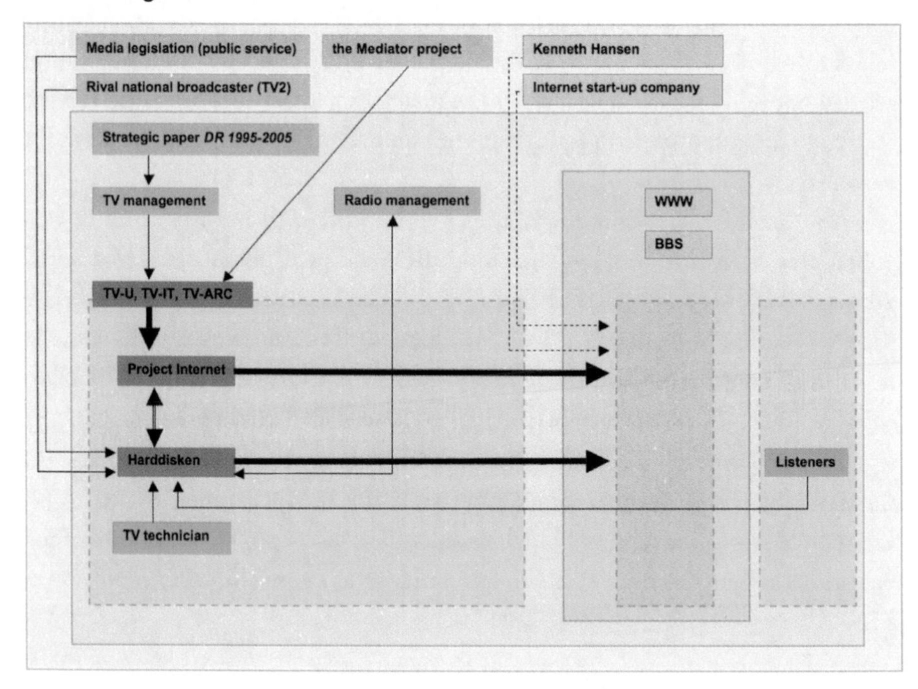

Figure 6: The strategic situation surrounding the creation of dr.dk, January–December 1994

Concerning the period January to December 1994, the strategic situation surrounding the creation of dr.dk can be illustrated as shown in figure 6 (the constituents of 'sender,' 'text,' and 'receiver' have been marked by dotted lines since they have not been constituted yet in the sense that dr.dk remains to be created). The illustration can be explained in the following way.

Two driving forces behind the (future) creation of the mediated textual artefact 'dr.dk' can be singled out. On the one hand, the editorial team of the radio program *Harddisken*, and on the other hand, *Projekt Internet* and the three subdivisions in DR behind it (TV-U, TV-IT and TV-ARC).

To a large extent, *Harddisken* and *Projekt Internet* function as points of convergence and amplifiers for a number of other actors. The editorial team of *Harddisken* is influenced by: (a) an actor on the constituent of 'sender,' namely, the television technician who had the domain name dr.dk, because he facilitates the realization of email correspondence and thereby the building of basic experiences with the new medium); (b) an actor on the constituent of 'receiver,' namely, the radio listeners who formulated a demand that pressed the editorial team to react; (c) two actors in the macro context, namely, the idea of public service as well as the rival broadcasters who legitimize and motivate DR's future internet presence. Each of these actors 'graft' onto *Harddisken* and they all help foster the first ideas of the future dr.dk. However, it is more difficult to judge a relation to an actor in the micro context, namely, the managers of the radio department to whom the letters from *Harddisken* are addressed—either they are a bit reluctant since they have to be contacted twice about the wish to get started, or they do not have the resources which *Harddisken* asks for or the authority to decide.

The initiators of *Projekt Internet* are influenced by an actor in the macro context—the Mediator project—as well as an actor in the micro context—the internal strategy paper *DR 1995–2005* from the executive board—and both legitimize their project. They are also influenced by another actor in the micro context, namely, the managers of the television department who approve and encourage their initiative. All in all these actors fuel and support *Projekt Internet*.

In addition to this, the strategic situation reveals two elements in the macro context that can be considered incipient actors, namely, the article on "Internet Radio" by Kenneth Hansen and the project description from the small internet start-up company; however, they do not have any institutional anchoring in DR, which is why they have great difficulties in becoming actors although one of them actually contacts DR. And finally, two elements articulate the constituent 'medium'—WWW and BBS—and in relation to the creation of dr.dk they are not related to any of the other entities yet, but they are both articulated as elements.

Between the two main driving forces an opposition can be identified (though not a conflict) because each of the forces has a different fundamental rationale:

Harddisken bases its ideas in practical experiences with the new medium (email, BBS, WWW, etc.), and they have a variety of concrete suggestions as to content and navigation, while *Projekt Internet* is based on general visions and on an understanding of the internet as a research tool for journalists. This opposition between the two centres of gravitation can be summarized in the following way: *Harddisken*'s editorial team *has a need that must be fulfilled,* and the television-based *Projekt Internet takes over a number of visions to fill in.*

In the period January to December 1994, it is unresolved whether both or only one of the driving forces will continue. Within the limits of this chapter it is not possible to present a thorough analysis of the following period up until January 1996 when dr.dk was actually launched, but so as to illustrate how the diachronic perspective can be integrated in the analysis, I shall briefly outline some of the relevant events. All the available source material for this period has not yet been studied, which is why the following is only a tentative outline.

In the first months of 1995 it is decided that DR shall be present on the internet, at least on a trial basis (the trial covers the first six months after the launch in January 1996, and starting in the summer of 1996 dr.dk becomes a permanent activity). A project team is formed with participants from radio, television, and the communication department, and a project manager is engaged. This person is associated with the television department, and his only task is to create the website dr.dk, which he does by cooperating with the passionate souls in DR's many production units who were enthusiastic about the new medium and who were eager to contribute, even though they had to work in their spare time. The project manager began to build on this enthusiasm by visiting every corner of DR with a road show where he taught the program producers new words such as the 'at sign,' 'modem,' and 'html.' However, dr.dk is still a small development project within DR, and it has to compete for resources (e.g., manpower, technology) with the two traditional media, radio and television, the staff of which regard the new upcoming web platform with a mixture of reluctance and forbearance.

In addition, in some of the project team's first internal documents about the future dr.dk, one finds a distinction between two separate parts of the website: an external part for the public where, among other things, one can find information about the radio and television programs, about DR, and read news; and an internal part for the employees of DR, where they can find links to research resources.

During 1995, however, the internal part more or less disappears from the documents, and the external part sets the agenda (the internal part is not abandoned, but it never becomes as important as the external part; later on, the intranet 'Inline' takes over many of the proposed features from the internal part).

An outline of the strategic situation reveals that it is characterized by heterogeneous driving forces on different levels: the project team, which becomes a leading driving force on a general level; the project manager, whose staying power and enthusiasm are invaluable; and the individuals in the different production units, who for various reasons want to be part of this new development. However, these driving forces are partly restrained and slowed down by the two strong older brothers, radio and television.

Three diachronic driving forces can be identified between 1994 and 1995. First, the executive board decided that DR should be present on the internet for a trial period, and they established a development project with a project team and a project manager. Second, the project team and in particular the project manager succeeded in making progress and in selling the idea to the passionate souls in DR. And, third, the project team succeeded in mediating the opposition between the two driving forces from the first period—*Harddisken* and *Projekt Internet*—in a constructive way that gave them both leverage in the future work. The radio and the television departments both took part in the project team (however, the editorial team of *Harddisken* is not represented as such), and the two different rationales from 1994 were represented in the distinction between an external and an internal part of dr.dk in the internal documents (the external part is very close to the ideas of *Harddisken,* while the internal part is the continuation of *Projekt Internet*). However, this mediation is performed with varying emphasis and in relation to different production areas: with respect to the content and structure of the website, *Harddisken*'s general ideas have the upper hand, while regarding the organisational power, television takes the lead. For instance, evidence of the first is that in the press release from the launch of dr.dk in January 1996, only one program is mentioned for its web-specific content, namely, *Harddisken*; and evidence of the latter is that the project manager works for the television department, and from the very start the television management has approved the project while *Harddisken* is a bottom-up initiative.

CRITICAL DISCUSSION

As the short analysis of the early history of dr.dk has illustrated, the analytical grid makes it possible to systematically frame a historical study of the forces behind the creation and the development of a specific website. However, using the conceptual framework may give rise to critical remarks. First, a number of general historiographical questions shall be raised; second, the applicability of the conceptual framework in relation to other objects of study within web history needs to be addressed.

General Scope and Operationalization

As mentioned above the conceptual framework is general in scope, which can be considered a strength as well as a weakness because the general scope makes it possible to apply the framework to a number of very different cases and to modify it to fit a variety of individual demands and analytical aims. But it can also be a weakness as the concepts may turn out to be too general and vague and it may be difficult to explain the criteria for their operationalization in a concrete analysis. If we have a collection of sources in front of us such as minutes of meetings, retrospective interviews, and the website as it looked ten years ago, the question of how to identify an 'element' or a 'relation' in these documents arises. Also, there is the consideration of how to associate an element to a specific constituent (sender, medium…).

The vagueness of the concepts ensures the flexibility of the framework by preventing a too limited understanding of, for instance, 'element,' 'relation,' 'sender,' or 'medium.' But the flip side of this is that to a certain extent the web historian is obliged to give these somewhat open categories a more precise meaning to fit the analytical aims of a specific case, and if this additional modification of the analytical grid is not done adequately, then the final analysis may turn out to be too vague and imprecise. To avoid such vagueness, studies must be supplemented with elaborations as to what is understood by, for instance, 'element,' 'relation,' 'sender,' 'medium,' etc. in each specific analytical context. The analytical grid may as such be considered a meta-theory which has to be supplemented with whatever detailed theories are necessary to explain specific parts of it.

So when using the analytical grid, questions such as the following may be raised: How to identify an element or a relation in a source? How, for instance, can a written text in a source be transformed into an element which is positioned relative to another element in a specific way? Questions like these revolve around a general problem within historiography, namely, how to interpret the sources. The sources are in themselves nothing but silent traces which must be brought alive through the historian's interpretation through which certain actions and characteristics are attributed to the individuals, objects, and places of the past. Throughout this process the historian will be guided, inevitably, by a conscious (or unconscious) perspective and pre-understanding that will shape the kinds of questions asked and how the answers are interpreted.

In this sense interpretation of the sources is always guided, and the sources are made to express something that, to a certain extent, is a function of the historian's approach. So if the historian sets out to uncover the driving forces behind a website and how it developed, the conceptual framework outlined above offers a systematic approach that can guide the interpretation and the understanding of the sources. The analytical grid gives the web historian an idea of what to look for in the sources, and it can thus serve as a meta-theoretical framework for more detailed textual and pragmatic analyses of the sources in which the elements, actors, or driving forces may be uncovered either as something mentioned in the text or as part of such extra-textual communicative circumstances as who communicates with whom and with what purpose. But as is the case in any historiography the use of a theoretical vocabulary to frame the analysis must be accompanied by methodological considerations such as: Is the vocabulary adequate? What are the premises for using it? What does it uncover and what does it leave in the dark?

When it comes to the website historian's concrete interpretative work of identifying certain entities in the sources as being specific elements, positions, and relations in the analytical grid, repetition and pattern recognition can play a useful role. If, for instance, a certain entity—an individual, a group, an idea—seems to play a role as actor or driving force in a large number of sources, then that suggests that this entity actually had this role at this specific point in time and in that specific situation and therefore can be understood as a specific actor. However, the opposite is not necessarily the case. If only a small number of sources point in one direction, we cannot conclude that the entity in question did not play the role as actor/driving force—maybe we just have not found enough relevant sources. Identification of the strategic situation and its development is always based on the available sources, and sources found later may give rise to other analytical results. In this sense any historical study is always provisional. Also in this respect methodological considerations must account for our procedures: Why these sources? How and where were they found? Under what circumstances were they created (for instance, retrospective interviews)?

Vertical and Horizontal Scalability

The possible reach of the conceptual framework in relation to other objects of study within web history can be approached by focusing on its eventual vertical and horizontal scalability. To what extent is it possible to scale the analytical grid either vertically in order to write the history of other web entities than a website or horizontally in order to write the history of other types of websites than that of a major national public service broadcaster?

If the analytical grid is scaled vertically, it may be applied to other analytical strata of the web such as the web as a whole, a web sphere, a webpage, or an individual web element on a webpage (cf. Introduction of this volume).

A historical analysis of the web as a whole could, for instance, set out to identify some of the actors and driving forces in the creation of the communicative infrastructure of the web or in the development of web institutions such as The World Wide Web Consortium (W3C), ICANN, or others. A historical analysis of the web activities considered part of a web sphere—for instance, a political event such as the U.S. elections of 2000, 2002, and 2004 (cf. Schneider and Foot, 2006)—could set out to identify and explain the variety of 'senders' and 'receivers' and their roles in the creation and development of that specific web sphere.

If the individual webpage is at the centre of our historical study we could, for instance, account for the development of web design on individual webpages (cf. Engholm, 2002, 2003). And, finally, we could use the analytical grid in writing the history of how specific web elements such as advertisements/banner ads have developed on the web (cf. Li & Zhunag, 2007, or Knox, 2009).

As suggested above, the analytical grid may also be 'displaced' horizontally in order to write the history of other types of websites such as, for instance, commercial websites such as amazon.com, or ebay.com, or social networking websites such as facebook.com, myspace.com, or wikipedia.org. With a view to illustrating how the analytical grid may be used in identifying the driving forces in the history of a social network website, here is a short outline of how a history of facebook.com might look if based on the analytical grid (the following is a rough illustrative sketch and is by no means exhaustive).

At the start of 2004 two groups of elements can be found on the constituent 'sender.' The founders of thefacebook.com, as facebook was called at that time, all played multiple roles: Mark Zuckerberg—"Founder, Master and Commander, Enemy of the State"; Eduardo Saverin—"Business Stuff, Corporate Stuff, Brazilian Affairs"; Dustin Moskovitz—"No Longer Expendable Programmer, Paid Assassin"; Andrew McCollum—"Graphic Art, General Rockstar," and Chris Hughes—"The Secret Weapon" (these 'titles' are the ones used on the webpage 'About' on thefacebook.com Feb 12, 2004, cf. archive.org, accessed March 10, 2009). Another group of elements consisting of the produsers of thefacebook.com (the term 'produser' is borrowed from Bruns, 2008), that is the individuals registered as users who were related to each other in a number of networks. At the time of the launch of thefacebook.com they were exclusively students from Harvard (a harvard.edu email address was required, cf. boyd & Ellison, 2007). Each of these groups of elements contributes specific kinds of text to the website (the constituent 'text'): the founders provide the layout and the design and specific texts such as 'About,' 'FAQ,' 'Term,' and 'Privacy,' as well as specify the textual ways in which the other group of senders—the produsers—may provide information about themselves that will be shared over the network. Most of this content is only accessible to registered users. The microcontext for the founders is their dorm rooms, and an important element in the macrocontext is the competing social networking websites already on the web that had contributed to the shaping of this website genre (e.g., LiveJournal.com or Friendster, cf. boyd & Ellison, 2007) as well as a number of other communities that had other aims than networking with friends (Napster, Wikipedia, SecondLife, LinkedIn, MySpace, and last.fm, just to mention a few). Elements on the constituents 'media' and 'receiver' could be, respectively, the possibility offered by the internet to create social networks, and the prod*users* are now in the role of elements on the constituent 'receiver.'

In the phases following the initial period, a number of changes in the elements can be identified (this short outline does not allow for a more detailed subdivision of phases). The two initial groups of elements on the constituent 'sender' were transformed and supplemented with other important groups of elements in the following years. First, as Facebook grew, it was transformed into a company, which led to the advent of two elements in the microcontext, namely, a board and a number of investors, just as the five founders were supplemented with numerous employees (app. 700 in 2008). Second, the number and the character of the possible produsers were gradually enlarged and changed to include students from other universities in the Boston area, then to students at all universities in the United States and Canada, later to high school students (September 2005), employees in a number of private companies (e.g., Apple and Microsoft, May 2006), and finally in September 2006, Facebook was opened to anyone aged 13 and over who had a valid email address that could be used for registration. Third, two other important groups of elements came to play a role as senders: companies that wanted to use facebook.com for advertising via banner ads (October 2007) and software developers, who were allowed to create applications (May 2007). These two new sender elements created a number of new types of textual elements on the website: the company 'facebook' still delimited the layout and interactive possibilities, while the produsers gave information about themselves and their networks, companies provided advertisements, and software developers made new applications. The former elements in the microcontext—the dorm rooms—were replaced by headquarters in Palo Alto, California (June 2004) and Dublin, Ireland. And as elements in the macrocontext, one may identify the ongoing development of a number of social networking websites as well as a number of other internet companies wanting to take over facebook.com (sometimes in the form of rumours, sometimes based on actual negotiations).

An in-depth historical analysis of facebook.com should then carefully study the available sources with a view to accounting for how these elements did interact (conflicts, balance of power, etc.) as actors and driving forces at different points in time in order to explain the forces behind the creation and development of the website facebook.com.

CONCLUSION

We have just shown that it is possible to formulate a conceptual vocabulary to frame historical studies of individual websites.

The use of the conceptual framework in a short analysis of the early history of dr.dk as well as the critical discussion has indicated, first, that it can help us understand and explain the actors, the acts, the relations, and the stage in the play entitled 'development of a website,' second, that it may be extended to the historical study of other web strata as well as other types of websites.

However, a number of questions still remain: The theoretical premises and implications need to be discussed in more depth just as the analytical productivity needs to be challenged, for instance, in studies based on a large number of different sources.

NOTES

1. The following considerations can probably also be used in writing the history of other media, such as newspapers, radio, and television.
2. It is very difficult to estimate the actual size of dr.dk, but according to the national Danish internet archive Netarkivet.dk, dr.dk takes up approximately a quarter (3TB) of the whole Danish domain .dk (12TB) (by comparison, in 2005 an average Danish website was 12 MB; cf. Andersen, 2006, p. 8). In addition, dr.dk is extremely wide-ranging and deep, and it has been among the three most-visited Danish websites for several years.
3. This argument is discussed in greater detail in Brügger, 2002. Here the constituent 'text' is split up into three constituents ('reference,' 'content,' and 'code'), whereas sender and receiver are termed 'contacter' and 'contacted.' However, in the present context the changes have been made to improve clarity. Although the constituents resemble a communication model, they should be understood as a schematic representation of the possible analytical objects of media studies rather than as a model of communication, seeking to explain the elements and phases in a communicative process (cf. Brügger, 2002, pp. 36–40).
4. In Brügger (2009a) it is maintained that the website is a coherent textual unit that unfolds in the medium 'internet' in one or more interrelated browser windows, the coherence of which is based on semantic, formal, and physically performative interrelations.
5. An idea very close to the distinction between actant and actor in the work of Greimas.
6. This conception of a strategic situation has affinities with ANT/STS (cf. Latour, 1987, 2005; Law & Hassard, 1999). First, the focus is on (power) relations between actors, which both constitute and are constituted by a network, and, second, actors are supposed to be both human and non-human entities. However, in the present chapter, these general insights are not used in an analysis of science, technology, and society but in a study of media.
7. On a general level the outlined analytical grid shares some similarities with Brian Winston's theoretical refinement of communication history in relation to Fernand Braudel's conception of technological progress as a conflict between historical breaks and accelerators or, as Winston puts it, "the balance of forces pushing and inhibiting the technologies" (Winston, 1998, p. 2, cf. also p. 15).
8. For a discussion of the specific nature of the archived website and of how to analyse different versions of archived websites, see Brügger (2008). A presentation of web archiving can be found in Brügger (2009b), and web archiving is discussed in Brown (2006), Brügger (2005), Masanès (2006), and Schneider & Foot (2006).

9. Figure 5 is an archived version of the English front page from November 1996. Judging from printed versions of the Danish front page at the time of the launch in January 1996, the English front page probably looked very much like figure 5 in January 1996.
10. The sources are: (1) letter from *Harddisken* to [name] entitled "Overvejelser og ønsker angående Harddisken og den ny teknologi" [Considerations and Wishes Concerning *Harddisken* and New Technology], July 14, 1994, 3 pages; (2) letter from [name] and [name] to the television management entitled "Information om projekt INTERNET" [Information on the INTERNET Project], October 10, 1994, 5 pages; (3) letter from *Harddisken* to [name] entitled "Forslag til homepage for Danmarks Radio" [Proposal for a Homepage for Denmarks Radio], December 7, 1994, 4 pages; (4) [name of Internet start-up company]: "Danmarks Radio på Internettet. Projektbeskrivelse" [Denmarks Radio on the Internet. Project Description], December 28, 1994, 4 pages + appendix. All translations from the Danish are my own.
11. The magazine *Emil* was published by DR's radio program 1 (after June 1993 it was issued as a supplement to the newspaper *Weekendavisen*).
12. Mediator (1994) was an EU-supported initiative to explore the impact of future data transmission across institutions and national borders within such areas as the home, the workplace, the public sector, research, and education. One of the subprojects of Mediator was called Navigator, which focused on how the new data media could be used in news production.

REFERENCES

Andersen, B. (2006). DK-domænet i ord og tal. *netarkivet.dk*, Århus. Retrieved October 2008, from http://netarkivet.dk/publikationer/DFrevy.pdf

boyd, d.m. & Ellison, N.B. (2007). Social network sites: Definition, history, and scholarship. *Journal of Computer-mediated Communication*, 13(1).

Brown, A. (2006). *Archiving websites: A practical guide for information management professionals*. London: Facet Publishing.

Brügger, N. (2002). Theoretical reflections on media and media history. In N. Brügger & S. Kolstrup (Eds.), *Media history: Theories, methods, analysis* (pp. 33–66). Århus: Aarhus University Press.

Brügger, N. (2005). *Archiving websites: General considerations and strategies*. Aarhus: The Centre for Internet Research.

Brügger, N. (2008) The archived website and website philology. A new type of historical document?. *Nordicom Review*, 29/2, Göteborg.

Brügger, N. (2009a). Website history and the website as an object of study. *New Media & Society*, 11(2), 115–132.

Brügger, N. (2009b). Web archiving—between past, present and future. In M. Consalvo & C. Ess (Eds.), *The Blackwell Companion to internet studies* (pp. 24–42). London: Blackwell.

Bruns, A. (2008). *Blogs, Wikipedia, Second Life, and beyond: From production to produsage*. New York: Peter Lang.

Engholm, I. (2002). Digital style history: The development of graphic design on the internet. *Digital Creativity*, 13(4), 193–211.

Engholm, I. (2003). *WWW's designhistorie: Website udviklingen i et genre- og stilperspektiv* (WWW's design history: The website development from the perspective of genre and style). Copenhagen: The IT University Press.

Knox, J.S. (2009). Punctuating the home page: Image as language in an online newspaper. *Discourse & Communication,* 3(2), 145–172.

Küng, L. (2008). *Strategic management in the media: From theory to practice.* London: Sage.

Küng-Shankleman, L. (2000). *Inside the BBC and CNN: Managing media organisations.* London: Routledge.

Latour, B. (1987). *Science in action.* Cambridge, MA: Harvard University Press.

Latour, B. (2005). *Reassembling the social: An introduction to Actor-Network-Theory.* Oxford: Oxford University Press.

Law, J. & Hassard, J. (Eds.) (1999). *Actor Network Theory and after.* Oxford: Blackwell.

Li, X. & Zhunag, L. (2007). Cultural values in internet advertising: A longitudinal study of the banner ads of the top U.S. web sites. *Southwestern Mass Communication Journal,* 23(1), 57–72.

Masanès, J. (Ed.) (2006). *Web archiving.* Berlin: Springer.

Schneider, S.M. & Foot, K.A. (2006). *Web campaigning.* Cambridge, MA: The MIT Press.

Tosh, J. (1984). *The pursuit of history.* Harlow: Pearson.

Winston, B. (2000) [1998]. *Media technology and society—A history: From the telegraph to the internet.* London: Routledge.

CHAPTER 2

Object-Oriented Web Historiography

Kirsten Foot & Steven Schneider

INTRODUCTION

In the fall of 2001, we worked with the U.S. Library of Congress and the Internet Archive identifying and collecting tens of thousands of websites, organizing these materials into the September 11 Web Archive (viewable at both http://september11.archive.org and http://lcweb2.loc.gov/diglib/lcwa/html/sept11/sept11-overview.html), and completing several scholarly articles and chapters based, in part, on analyses of archived materials. One of us recently attended a meeting with some of the staff of the National September 11 Memorial & Museum—the organization that is building a memorial and a museum at the site of the World Trade Center in New York. The purpose of the meeting was to explore ways of explaining the role of the web in the aftermath of the 2001 terrorist attacks in the United States. The September 11 Museum staff were interested in exploring how they might tell this part of their story. Although budget constraints prevented us from moving the project forward, the discussion illustrated for us some of the central issues of web historiography that we hope to highlight in this chapter. To start with, of course, the Museum staff wanted to understand what, how much, and how many: What had been archived? How much material had been collected? How many times had we collected each site? This led to a discussion of why: Why had we collected web materials? Why had we made the choices we did? Why did we start? Why did we stop? And finally, we turned to the notion of: How did the archiving software work? How were archived pages displayed? And how did our technologies and techniques influence what was available to be shared with others? The answers to these very practical questions begin to provide a helpful point of departure for this chapter.

As the chapters in this book illustrate, there are a variety of approaches to conducting web history, from the building of archives and developing tools for historical analyses, to retrospective studies of web production, to analyses of web-mediated practices such as the use of webcams. We hope the ideas presented in this chapter have some value for practitioners of each of these approaches. We have three aims in this chapter: (1) to propose an "object-oriented" approach to researching and writing web history; (2) to suggest ways that such an object-oriented approach makes developmental analyses of the web more rigorous, and (3) to encourage social researchers to be proactive in building web archives, alongside archivists and librarians—as well as interfaces and tools for analysis to facilitate robust web historiography (and in this we concur with several other contributors to this volume). We come to this subject of web historiography as social researchers—we have novices' appreciation of and enthusiasm for history but do not share the kinds of expertise cultivated by historians and archivists. Whether it is an advantage to be free from the constraints of practice imposed by those professions or a disadvantage to be unaware of the epistemological and methodological controversies of these fields is a judgment yet to be rendered. We write as practitioners with experience initiating and completing studies of the web based largely on analyses of objects contained in archives of web objects we collected on our own and with other institutions, contemporaneously or prospectively during our studies. Our patterns of practice as web archivists, and our motivations to archive, have evolved since we began archiving in 1999, and we welcome the opportunity this chapter has afforded to reflect and critique, in a serious way, on what we have done and what our choices have meant for the types of web histories we have been able to write.

Our initial foray into web studies was born of an opportunity—we were both hired in 1999, independently, to work on a large, well-funded research project seeking to assess the impact of the web on the U.S. federal elections in 2000. We quickly concluded that a reasonable starting point would be to analyze the development of election-oriented websites over the course of several months leading up to the election. We recognized the impossibility of personally observing, on a daily basis, all or even a substantial percentage of election-related websites and determined that some sort of archive or collection of sites would be necessary to complete a post-hoc developmental analysis. The principal investigator we worked with was fond of doing experiments, so we pitched our idea for an archive as the mother of all stimuli for experiments on citizens' reactions to online election phenomena. He bought our pitch, so we jumped into the challenge of prospective archiving to enable the retrospective analyses we wanted to be able to conduct.

Our experiences in that initial study of the 2000 election connected us to a small and eclectic group of scholars, librarians, and archivists who had similarly recognized the challenge—and desirability—of archiving web materials for future analysis and assessment. In the summer of 2001, we contacted staff at the U.S. Library of Congress (LC), who were at the time starting to create collections of born-digital materials thought likely to have historical significance. We discovered that the LC, not surprisingly, approached web archiving from the perspective of librarians: expert selection of materials with no attention to (or even recognition of) sampling strategies; cataloging basic meta-data about an entire collection, with no consideration (or even recognition) of the need for site-level cataloging, or tools to facilitate analysis. We sketched a plan for our newly emergent research group, WebArchivist.org, to work with the LC, and later the Internet Archive, to create web archives for social researchers as well as the U.S. Congress and the public at large. When terrorist attacks occurred in the United States a few weeks later, the three organizations, joined by the Pew Internet & American Life Project, quickly agreed to immediately implement our sketchy plan for collaboration. The result was the September 11th Web Archive, consisting of daily captures between September 11 and December 1, 2001, of all the URLs that the librarians at LC, the researchers in our group, and web users around the world added to the Internet Archive's seed list for this collection. We conducted a preliminary categorization of 2,500 sites, including creation of meta-data describing the type of producer and some assessment of content—this was a big advance for LC as it moved beyond the collection-level cataloging it had done for previous collections. We also conducted detailed analysis of 250 sites by examining the kinds of activities in which site visitors could engage. We developed two distinct interfaces to the archive to facilitate interaction with collected materials. Eventually, other researchers joined us in producing a compilation of reports on a range of post-9/11 web phenomena captured in the archive, and at least three journal articles were published based on materials from the archive. We have gone into our origins as web archivists and web historians in some detail because the observations that follow have been deeply colored and shaped by our experiences.

We refer to the September 11 Web Archive project throughout this chapter, because it was a formative experience for us, and because it has been the basis for several web histories to date—thus, it is a useful case study in the emergent field of web historiography. In contrast to our earlier study of the 2000 election, we had no particular research questions in mind on September 11, 2001, as we launched into helping build an archive of whatever was happening on the web related to the terrorist attacks. But our election research convinced us that a contemporaneously collected archive would document and preserve some of the ways that the web served as a significant surface for social, political, and cultural activity in the wake of the attacks. Over time, we and other researchers formulated a variety of questions, and retrospective analyses were conducted on a range of topics, including the rise of do-it-yourself journalism, governmental web responses, visual imagery, religious organizations' web actions, and shifts in online personal expression regarding the attacks during the autumn of 2001 (see Rainie, Schneider, & Foot, 2002). Throughout this chapter we refer to the experience of building the September 11 Web Archive, conducting our own studies based on materials drawn from it and other archives, and interacting with other scholars about researching and writing histories of the post-9/11 web.

THE COMPLEXITY OF OBJECT

In proposing an object-orientation to research and writing web history, we first offer a two-pronged definition of object summed up as object as motive and object as artifact. We then turn to some implications of this dual notion of object for the practice of web historiography. We identify potential motivating aims for this kind of scholarship, suggest levels and units of analysis that we consider especially important to think through, and consider the relationship between these motives, units of analysis, and the practices of doing web historiography. Developmental analyses of any aspect of the web, whether engaged in contemporaneously or retrospectively, entail dynamics within and between the (co)producers of web artifacts, production practices and techniques, and web artifacts themselves—as well as between the researcher(s) and the phenomena under investigation. These dynamics make it difficult but very important for scholars to identify and situate their object(s) of analysis historically, theoretically, and as methodologically constructed.

"Object" is used within fields as diverse as object-oriented programming and museology to reference discrete units and distinguish them from each other, as well as to define and demarcate those things that are being described or analyzed. Similarly, we suggest that the practice of researching and writing web history is enhanced when it is artifact-aware—that is, grounded in and shaped by a researcher's interactions and experiences with web artifacts as well as the artifacts themselves. This notion of object is consistent with the German term *Objekt*. A second, more abstract definition of object is "the end toward which effort or action is directed." This definition is resonant with the German philosophical term *Gegenstand*, which entails the concept of the embeddedness-in-activity of objects—both material and immaterial—that serve as motivating but largely unattainable horizons. This is the concept of object employed by Lev Vygotsky (1978) and the activity theorists who succeeded him, and it is the one with which we begin our discussion. In the next sections of this chapter, we explore these two aspects of "object" in greater depth.

AN ACTIVITY THEORY PERSPECTIVE ON OBJECT AS MOTIVE

The activity theory notion of object is richly complex, but English-language elucidations of this essential concept are scanty (see Foot, 2002; Kaptelinen, 2005; Leont'ev, 1978; Miettinen, 1998, 2005; Tuunainen, 2001). The activity theory concept of object can be difficult to grasp, in part because the German and Russian terms in which it developed are not easily translatable into English. Stated briefly, an object (*Gegenstand*) may be understood in the framework of activity theory as a collectively constructed entity, in material and/or ideal form, through which the meeting of a particular human need is pursued. To elaborate, activity theorists (Engeström, 1990, 1999; Engeström & Escalante, 1996; Lektorsky, 1984; Leont'ev, 1978) argue that the process of object formation arises from a state of need on the part of one or more actors. The need state, which is usually unconscious and thus not clearly definable, precipitates a set of "search actions" (Engeström, 1999, p. 381), during which any number of potential artifacts may be encountered. In most cases, it is only when search actions result in an encounter between the need and an artifact that the need begins to be experienced consciously. A social researcher as subject(s) orients toward one of these artifacts, such as a website, through actions mediated by both personal experiences of the researcher and reifications of cultural-historical experience, and a "motive" arises out of the encounter of the need state and the artifact. Thus a social researcher's motive for doing web historiography can be understood as the interpolation of any one of a number of potential needs, such as the need to understand, the need to create knowledge, or the need to achieve fame and fortune with any kind of web artifact or virtual phenomenon.

Since the relationship between object (*Gegenstand*) and motive is dialectical in the activity theory framework, in that motive energizes object-oriented activity, and the conjoining of object and need state evokes motive, it is essential to maintain a clear analytical distinction between the two concepts (Foot, 2002). For our purposes in this chapter, it is sufficient to note that this notion of object-orientation is useful for understanding and advancing web historiography as a particular kind of activity that always involves researchers' own need states, which give rise to particular motives that may differ among researchers studying same pool of web artifacts or may vary over time for a single researcher. The activity-theoretic distinction between web artifacts and the motivating/orienting horizon of the object (*Gegenstand*) of the activity of web historiography is also very useful. In this perspective, the object-as-motive (*Gegenstand*) embedded in the activity of researching and writing web history can be understood as activity-context dependent and socio-culturally formed and thus historically evolving.

Drawing together these insights from activity theory, we view one prong of object-orientation in web historiography as the researcher's motive for researching and writing web history, that is, the horizon that is being reached for in the activity of web historiography. Others have written on the motives that give rise to the creation of archives. In an essay on representations of the 9/11 attacks in internet news stories, Brown et al. (2003) suggest that there is an "archival impulse" that prompts attempts to capture and preserve instances of media performance. Similarly, social theorist Jacques Derrida (1995) argues that archiving is an ancient practice, reflecting a deeply embedded human drive, and that archives are shaped by individual psyches and sociocultural formations that are remarkably persistent over time. Derrida argues that in spite of this, and simultaneously, humans in general (and elites in particular) are intrinsically motivated to manage social memory in ways that are often distortive and sometimes destructive of memory, and ultimately of society. He terms this malevolent motive "archive fever." Whether due to this notion of archive fever or not, in our experience web archives are few and far between. While we are not suggesting that our fellow scholars are afflicted by a strain of archive fever (although Derrida might), the use, much less the creation, of web archives by social researchers is relatively rare.

Certainly there are instances when people (ourselves included), whether expert archivists or not, build web archives to preserve web phenomena that they (or their institutions) find meaningful. Other motives for web archiving are more interactional in nature, such as preserving what others would rather erase or expunge. For example, the Internet Archive collection (accessible at http://www.archive.org) includes pages produced by U.S. government agencies and industry groups which were removed from the web in the weeks after the attacks because they were deemed to be interesting to potential terrorists (Soraghan, 2001; Toner, 2001). Similarly, the archive preserves a site that may have been produced by the Taleban (http://www.taleban.com) in Afghanistan that was subsequently defaced (Smetannikov, 2001) and later removed from the web altogether (Di Justo, 2001). But in general, at least in North America, far more of the web is over-written, erased, or deleted than captured, due to the dominant ideology of perpetual technological innovation and the widespread cultural impulses to revise or forget (web) history. Perhaps ironically, the more macro cultural-historical context of technological determinist ideology and historical amnesia in which we work as scholars is part of what motivates us to archive and engage in web historiography. It is our way of being countercultural as Americans, and as social researchers working in fields where ahistorical, point-in-time studies are the norm.

Going beyond our own motives for researching and writing web history, we suggest several other plausible motives for doing web historiography, for example, the need to make sense of socio-cultural-political relations and events or to try to understand development and evolution on the web at different levels and/or over time. Some social researchers may be motivated to retrospectively trace the emergence of a web phenomenon in order to get a read on its trajectory.

Illustration: Our motives for retrospective analyses of web memorializing.

Although neither of us lost anyone we knew in the 9/11 attacks, we, along with many, experienced a persistent sense of collective loss and grief along with empathy for those who had lost loved ones. Somewhat consciously, and perhaps more so subconsciously, we desired to validate the losses through our research expertise in the realm of technology and society. As academics who get paid to do research as well as to teach, we write to live, and we want to invest our research and writing in subjects that hold significance for us and others. We also happen to both be drawn to the emergent, to web phenomena that are nascent and unfolding. And so, nearly two years later, when the topic of memorializing on the web came up in a conversation with another colleague, Barbara Warnick, one of us (we can no longer recall which one) said "someone should study this," and the other said "we should." In that conversation the need states we have just described encountered the artifacts of memorial websites, and gave rise to the object-as-motivating-horizon of understanding and rendering web-based memorializing more visible by researching it in its historical context on the post-9/11 web (see Foot, Warnick, & Schneider, 2005, to view the outcome of our collective research activity).

OBJECT AS ARTIFACT

The second perspective on which we focus conceptualizes object as artifact. The sense of object as artifact is derived from the conceptualizations of object offered by a diverse set of professionals, including object-oriented programmers, museologists, art curators and social theorists. When thinking of object as artifact, we suggest that consideration be given to the demarcation of objects, the properties of objects, and the process by which objects become artifacts. We examine each in turn.

The first consideration with respect to object as artifact is to demarcate the object, a process in which boundaries around objects are determined. Web objects can be considered along a continuum ranging from "bits" to "experiences" (Arms, Adkins, Ammen, & Hayes, 2001), and each step along this continuum can be considered as an object. Few historians are likely to engage the web at the bit level—though those interested in the use of the web to distribute hidden messages would certainly be the exception (Provos & Honeyman, 2001). Some historians studying the web may consider page elements as artifacts—for example, examining patterns of images found within a set of web pages (Dougherty, 2003;

Sillaman, 2000). More commonly, either web pages—groups of elements assembled by a producer and displayed upon request to a server—or websites—groups of pages sharing a common portion of their URL—are treated as objects for the purposes of analysis. Alternatively, some historians might be interested in examining links between pages as objects. At another level altogether, some historians may wish to examine the experience of the web, without particular attention to sites or pages or elements, and define a browsing session as an object. In many respects, the challenges facing historians and archivists as they grapple with this aspect of object as artifact are similar to those actively engaged with curating objects for museum exhibits or art exhibitions. Within museology, "object" refers both to the "specimens" of artistic and cultural activity and the "re-presented" or "staged" associations between specimens crafted by historians and museologists (Preziosi, 2006). In short, we need to bring to the surface the underlying assumptions made when examining web objects as artifacts and to recognize the role of the researcher in constructing the object as artifact.

At the same time, it is important to establish some common framework of web objects in order to give the artifact aspect of the concept some meaning. To that end, we turn to a consideration of the term "object" as it is conceptualized in the field of object-oriented programming—the inspiration for the title of our chapter. In this field, "object" is considered as a discrete unit that has "the same power as the whole" (Kay, 1993). Objects, within this perspective, are sets of programming instructions that stand on their own, can be re-used in multiple applications, and carry with them a set of properties or inheritances that affect the way they "behave" or are interpreted by other objects within different contexts. Objects, wrote Alan Kay, a pioneer in object-oriented programming "are a kind of mapping whose values are its behaviors" (Kay, 1993). The "values" of objects are expressed in the properties assumed to be inherent in them. Encapsulation highlights the independent standing of an object and allows objects to be used or referenced by other objects under terms established and enforced by the object itself, even if not visible to other objects. Polymorphism suggests that some characteristics of objects are dependent on the context in which they are encountered. Inheritance assures that an object defined in relationship to another object possesses the traits and characteristics associated with those other objects. In short, from the perspective of object-oriented programming, objects as artifacts are stable, dependable, and predictable.

Finally, we should reflect on the process by which an object becomes an artifact. From the perspective of the historian, considering an object as an artifact requires that we consider the tension between the object as "performed" and the object as archived (Schimmel & Ferguson, 1998). Jacques Derrida (1995) is particularly attuned to this distinction and draws our attention to the impact of the archival act on the impression of the object being viewed. While archived web objects may not obviously suffer the same degradation in meaning and expressive capacity as objects representing performance art do upon being re-presented in artifactual form (Greenstein, 1998), the challenges associated with re-presenting web artifacts remain significant and should be acknowledged by historians. For example, consider the re-presentation of a page from a web archive: the page is likely to consist of several elements (html code, images, etc.) assembled together and presented as a single artifact; however, each of the elements, archived individually over a period of time, may have changed during the archiving process, thus potentially rendering the archived object differently than it would have been performed (Arms et al., 2001). Furthermore, the viewing of web objects as artifacts may take place using browsers and displays significantly different than the technology available at the time the object was archived. In short, objects as artifacts are less fixed and more fluid (Levy, 2001) than some might have initially perceived.

Illustration: The website as artifact encompassing potential action.

As we began our analysis of the post-September 11 web through an intensive observation of pages archived in the immediate aftermath of the terrorist attacks, we began to notice that visitors to sites had been invited to participate in a variety of specific actions, some online and some offline. We further noticed that some of these potential actions were found within sites produced by types of actors not often associated with the actions observed, especially in 2001. For example, some national governments created pages within their sites, providing the opportunity for unrestricted text to be entered on memorial pages. We decided to expand our notion of online structure potentiating action (see Schneider & Foot, 2002) by focusing on the artifacts representing potential actions.

Our subsequent analysis (Schneider & Foot, 2003) focused on the types of online and offline actions in which visitors to websites in the September 11 Web Archive could have potentially engaged, and examined the distribution of actions potentiated across different types of site producers. This analysis illustrates the use of a website as an object, as we ascribed the potential action observed to the site encompassing the action. Our analysis further abstracts the observed objects by tallying actions across sites produced by types of actors (i.e., government agencies, educational institutions, etc.) as an estimate of how different types of site producers developed the web during the observed time period.

THE PRACTICE OF OBJECT-ORIENTED HISTORIOGRAPHY

Having proposed a two-pronged notion of object as motive and artifact, we now turn to the implications of this notion for the practice of object-oriented web historiography. We begin our discussion of object-oriented practice by considering different approaches to the web as a focus of study. If some aspect of web is to be the focus of an historical analysis, it is helpful to examine the underlying practices associated with a variety of approaches that are taken by social researchers and that can be useful in web historiography. We have elsewhere (Schneider & Foot, 2004, 2005) identified three sets of approaches that have been employed in web studies and examine each to highlight their relevance to distinctly historical research. We conclude with a focus on the techniques associated with scholarly web archiving.

Approaches that employ discursive or rhetorical analyses of websites—treating artifacts (sites, pages, elements) essentially as texts and drawing conclusions based on an analysis of content—allow historians to focus on the emergence of specific communicative phenomena that occur on the web. However, this approach is likely to downplay or ignore the impact of structuring elements and fail to provide an opportunity to assess the role of links among pages and between sites. Structural/feature analysis methods—which tend to use individual websites as their unit of analysis and focus on the structure of sites, such as the number of pages, hierarchical ordering of pages, or on the features found on the pages within the site—draw historians attention to aspects of web production and development but tend to understate the contribution that content may make to the overall experience of web users. We term a third set of approaches that analyzes multi-actor, cross-site action on the web as sociocultural analyses of the web (see several examples in Beaulieu & Park, 2003). Lindlof and Shatzer (1998) pointed in this direction in their article calling for new strategies of media ethnography in "virtual space." Hine (2000) presented one of the first good examples of sociocultural analysis of cross-site action on the web. By appropriating the term "sociocultural" to describe this set of approaches, we seek to highlight the attention paid in this genre of web studies to the hyperlinked context(s) and situatedness of websites—and to the aims, strategies, and identity-construction processes of website producers—as sites are produced, maintained, and/or mediated through links. This approach affords the historian an opportunity to take a broader view of the development of websites in historically evolving contexts.

Through a consideration of these approaches, whether employed concurrently with or retrospectively to evolving web phenomena, the differentiated attention and focus on object become clear. We have found that a multi-method approach that balances the analysis of evolving content with an analysis of changing structure and includes an assessment of artifacts within the context of the web provides the most illuminating and comprehensive outcomes in retrospective developmental analyses.

As part of the practice of object-oriented web historiography, we find it necessary to involve ourselves as researchers in the collection of artifacts. This may be temporary: as the practice of web archiving becomes standardized over the next decade or so, the robustness of archiving may become sufficient to support scholarly work. The first generation of web archiving—illustrated by the Internet Archive—has not provided us with archives that are robust enough to write some kinds of scholarly web histories. We suggest that historians become familiar with, and seek to influence, web archiving at several levels. To that end, we turn to a discussion of the techniques associated with web archiving. Our view of web archiving encompasses a wide range of activities, from the identification of objects to archive, through the stages associated with "getting bits on disk" as well as associating meta-data about the objects with the objects, and finally to representing archived objects in a web browser. Web archiving began in earnest in the mid-1990s, as sufficient technologies were developed and the recognition of rapidly disappearing content was acknowledged. The impetus behind web archiving activity was clear: The web was doubling in size every three to six months from 1993 to 1996, and it appeared that it had the potential to become a significant platform on which a wide variety of social, political, scientific, and cultural phenomena would play out. Individuals at different types of institutions, such as libraries and archives, whose mission included the preservation of cultural and historical artifacts and materials, recognized the challenge that digital materials presented.

The use of web archives by scholars, including historians, obligates us to identify the specific techniques used by both archivists and researchers when examining web artifacts. Archivists engage in a number of discrete processes on the path from conceptualizing an archive to making archived objects and associated meta-data available to others (Brügger, 2005). The historian employing web archives to support historical claims must account for the potential impact of these processes on their conclusions. This is especially important when claims are being made about the web itself (as opposed to claims being made about individuals or organizations whose materials happened to be on the web). Together with Paul Wouters, we have explicated our view of these practices elsewhere (Schneider, Foot, & Wouters, 2009) and here will only briefly touch on those relevant to historians and other research-oriented users of archives. We intend to draw attention to the techniques associated with making web objects accessible as artifacts for future analysis.

One set of processes involves management tasks associated with the selection and representation of artifacts in archives. *Identification* includes the steps necessary to make known to an archiving system those web objects to be considered for inclusion. Obviously, any archiving process involves the selection of some web artifacts and the exclusion of others. There are innumerable techniques associated with this process. Experts can identify websites of interest, often from a published directory. Query results from a search engine can be processed using a fixed set of rules to select artifacts of interest. In any case, it should be expected that historians using archives specify the underlying assumptions by which objects were identified. A separate but related process, *curation,* involves creating a set of rules and procedures necessary to collect the desired objects. These rules might specify, for example, the instructions to be given concerning whether to follow or to ignore links to other artifacts. This process, often a highly technical and specialized procedure, is frequently opaque to researchers and other users of archives. However, these procedures determine the specific artifacts that are included in archives, and, more importantly, those that are excluded. Making curatorial processes visible allows historians and others who develop evidence from archives to document, as necessary, those factors which may influence the observed results. Finally, *representation* is the process of retrieving archived artifacts from a collection and presenting them in a web browser. Considering this step as a process associated with web archiving will draw attention to the fact that rendering archived artifacts involves affirmative choices that affect the ways in which the rendering is performed. Historians and other researchers need to be aware of how the practice of representation is shaping their perspective on archived artifacts.

We have also suggested that there are three distinct processes of associating and presenting meta-data, or data about the artifacts, with the objects themselves. We distinguish these meta-data processes from each other on the basis of the techniques utilized. *Indexing* refers to the process of generating meta-data about collected artifacts or groups of objects algorithmically, while *categorization* is the process of generating meta-data through human observation and analysis. Indexing can involve developing meta-data from one of the available sources of information about archived artifacts, including the artifact itself, log files from crawling programs, and data developed externally to the archiving project. Categorization of meta-data can be applied to different types of artifacts (such as page elements, pages, or sites). The process of *interpretation* provides meta-data about collected artifacts, derived through the processes of categorization and indexing, to support sense-making activities such as discovery and search and to facilitate selected representation of collected artifacts. This process may include the design and implementation of an interface to a web archive, allowing users to select archived artifacts for examination or analysis. Providing full-text search of both meta-data and archived artifacts is an interpretation technique especially well suited to presenting unstructured data generated through annotation of artifacts, as well as providing access to archived artifacts containing text matching submitted queries.

Our aim in elucidating the techniques associated with web archiving is to increase their transparency and enhance the ability of social researchers to assess archives and interfaces that they confront while writing web histories, as it will often be the case that historians will address archives through an interpretative interface that masks categorization, indexing, and representation processes. In addition, we hope to provide a framework that will encourage social researchers to actively participate in constructing archives and designing interfaces that support their ongoing efforts to write web histories.

Illustration: The web archive interface as an interpretative frame.

Our work creating two distinct interfaces to the September 11 Web Archive illustrates the importance of interpretative frames to archived collections. The interface visible at http://september11.archive.org focused on the type of actions in which visitors to websites in the collection could engage. This interface incorporated specific meta-data fields for 250 analyzed sites; these meta-data were collected as part of our research process. The interface links to web pages, archived on a specific date, that were identified as potentiating the actions identified.

A second interface, visible at http://www.loc.gov/minerva/collect/sept11, allows visitors to browse 2,313 websites for which other fields of meta-data were collected. Visitors to the archive can browse sites by selecting the first letter in the producer name, producer type, producer country, language, and the presence/absence of content associated with bioterrorism, September 11, and the Afghan War. Successively selecting meta-data fields narrows the search parameters to matching sites; clicking on the producer name provides access to a web archive record hosted by the U.S. Library of Congress. The web archive record links to a listing of dates on which the website was archived for this collection.

These interfaces provide dramatically different access to the same underlying archive. The first interface described would be valuable to those visitors particularly interested in the specific actions about which meta-data was collected, and would direct visitors to specific objects that potentiated these actions. The second interface described would serve visitors interested in framing their own questions about objects in the archive and may enable them to develop samples of sites for further analysis.

CONCLUSION

In summary, we have proposed an object-oriented approach to web historiography based on a two-prong definition of object, involving object as motive and artifact. We have sought to demonstrate that such an object orientation serves to make web historiography more transparent and more rigorous. We hope this object orientation will evoke greater reflexivity among social researchers regarding their motives for doing web historiography, the ways in which web elements are made into and selected as artifact-objects of research, and the full range of techniques employed in web historiography. This proposal of object-orientation in web historiography will also, we hope, cause social researchers to become more proactive in (co)constructing the archives and interfaces in/through which they seek to conduct their scholarship.

So what does this kind of object orientation imply for historians of the web in contrast to historians of other media? Scholars who desire to research and write web histories face significant challenges. The affordances of the web render object orientation very important. Some characteristics of the web that are generally agreed upon by new media scholars include its scale and scope, its pervasiveness and ubiquity, the immeasurable and evolving volume of its content and its ephemerality, and the widespread access to the means of production that an unprecedented number of non-professional users have. In view of these affordances and a general lack of awareness of the connection of current web phenomena with the past, web phenomena must be understood as (already) history that is relevant to human activities in the present and the future.

William Thomas (2004) cites literary scholar Espen Aarseth's (1997, p. 62) insightful characterization of cybertexts as non-linear, dynamic, explorative, configurative narratives and argues that the "ergotic" nature of web artifacts, combined with the selective and iterative nature of web archiving, holds significant implications for historians:

> For historians the first stage in such textual developments for narrative have [sic] already been expressed in a wide range of digital archives. While these archives might appear for many users as undifferentiated collections of evidence, they represent something much more interpreted. Digital archives are often themselves an interpretative model open for reading and inquiry, and the objects within them, whether marked-up texts or hypermedia maps, derive from a complex series of authored stages.

A commitment to object-oriented historiography can shape our work as researchers in how we archive, how we identify/articulate our motives and the aims of our research, and in how we conduct our research practice in view of the research questions we pursue.

REFERENCES

Aarseth, E. (1997). *Cybertext: Perspectives on ergotic literature*. Baltimore and London: Johns Hopkins University Press.

Arms, W., Adkins, R., Ammen, C., & Hayes, A. (April 15, 2001). Collecting and preserving the web: The MINERVA prototype. *RLG DigiNews, 5*.

Beaulieu, A., & Park, H. W. (Eds.). (2003). *Journal of Computer-Mediated Communication, 8*(4) (special issue Internet networks: The form and the feel), retrieved September 1, 2009, from http://jcmc.indiana.edu/vol8/issue4/.

Brown, M., Fuzesi, L., Kitch, K., & Spivey, C. (2003). Internet news representations of September 11: Archival impulse in the age of information. In S. Chermak, F. Y. Bailey & M. Brown (Eds.), *Media representations of September 11* (pp. 103–116). Westport, CT: Praeger.

Brügger, N. (2005). *Archiving websites: General considerations and strategies*. from http://www.cfi.au.dk/publikationer/archiving/.pdf.

Derrida, J. (1995). Archive fever: A Freudian impression. *Diacritics, 25*(2), 9–63.

Di Justo, P. (2001). Does official Taliban site exist? [Electronic Version]. *Wired*. Retrieved January 25, 2009 from http://www.wired.com/politics/law/news/2001/10/47956.

Dougherty, M. (2003). Images of September 11th on the web. In L. Rainie, S. M. Schneider & K. Foot (Eds.), *One year later: September 11 and the internet* (pp. 62–65). Washington, DC: Pew Internet and American Life Project.

Engeström, Y. (1990). *Learning, working and imagining: Twelve studies in activity theory*. Helsinki: Orienta-Konsultit.

Engeström, Y. (1999). Innovative learning in work teams: Analyzing cycles of knowledge creation in practice. In Y. Engeström, R. Miettinen & R.-L. Punamaki (Eds.), *Perspectives on activity theory* (pp. 377–406). New York: Cambridge University Press.

Engeström, Y., & Escalante, V. (1996). Mundane tool or object of affection? The rise and fall of the postal buddy. In B. Nardi (Ed.), *Context and consciousness: Activity theory and human-computer interaction* (pp. 325–374). Cambridge, MA: MIT Press.

Foot, K. A. (2002). Pursuing an evolving object: Object formation and identification in a conflict monitoring network. *Mind, Culture and Activity, 9*(2), 132–149.

Foot, K. A., Warnick, B., & Schneider, S. M. (2005). Web-based memorializing after September 11: Toward a conceptual framework. *Journal of Computer Mediated Communication, 11*(1), retrieved September 1, 2009, from http://jcmc.indiana.edu/vol11/issue1/foot.html.

Greenstein, M. A. (1998). The act and the object. *Art Journal, 57*(3), 85-88.

Hine, C. (2000). *Virtual ethnography*. Thousand Oaks, CA: Sage.

Kaptelinen, V. (2005). The object of activity: Making sense of the sensemaker. *Mind, Culture, and Activity, 12*(1), 4-18.

Kay, A. (1993). The early history of smalltalk: 1960–66—Early OOP and other formative ideas of the sixties. Retrieved September 4, 2008, from http://www.smalltalk.org/smalltalk/TheEarlyHistoryOfSmalltalk_I.html

Lektorsky, V. A. (1984). *Subject, object, cognition*. Moscow: Progress.

Leont'ev, A. N. (1978). *Activity, consciousness, and personality*. Englewood Cliffs: Prentice-Hall.

Levy, D. (2001). *Scrolling forward*. New York: Arcade Publishing.

Lindlof, T. R., & Shatzer, M. J. (1998). Media ethnography in virtual space: Strategies, limits, and possibilities. *Journal of Broadcasting and Electronic Media, 42*(2), 170–189.

Miettinen, R. (1998). Object construction and networks in research work: The case of research on cellulose degrading enzymes. *Social Studies of Science, 38*, 423–463.

Miettinen, R. (2005). Object of activity and individual motivation. *Mind, Culture, and Activity, 12*(1), 52-69

Preziosi, D. (2006). Art, art history, and museology. *Museum Anthropology, 20*(2), 5–6.

Provos, N., & Honeyman, P. (2001). *CITI technical report 01-11: Detecting steganographic content on the internet* (PDF). Ann Arbor: Center for Information Technology Integration, University of Michigan.

Rainie, L., Schneider, S. M., & Foot, K. A. (Eds.). (2002). *One year later: September 11 and the internet*. Washington, DC: Pew Internet and American Life Project.

Schimmel, P., & Ferguson, R. (1998). *Out of actions: Between performances and the object, 1949–1979*. New York: Thames and Hudson.

Schneider, S. M., & Foot, K. A. (2002). Online structure for political action: Exploring presidential Web sites from the 2000 American election. *Javnost (The Public), 9*(2), 43–60.

Schneider, S. M., & Foot, K. A. (2003). Crisis communication & new media: The web after September 11. In P. N. Howard & S. Jones (Eds.), *Society online: The internet in context* (pp. 137–154). London: Sage.

Schneider, S. M., & Foot, K. A. (2004). The web as an object of study. *New media & society, 6*(1), 114–122.

Schneider, S. M., & Foot, K. A. (2005). Web sphere analysis: An approach to studying online action. In C. Hine (Ed.), *Virtual methods: Issues in social research on the internet* (pp. 157–170). Oxford: Berg.

Schneider, S. M., Foot, K., & Wouters, P. (2009). Web archiving as e-research. In N. Jankowski (Ed.), *E-research: Transformation in scholarly practice* (pp. 205–221). London: Routledge.

Sillaman, L. (2000). *The digital campaign trail: Candidate images on campaign websites*. Unpublished Master's thesis, University of Pennsylvania, Philadelphia, PA.

Smetannikov, M. (2001). Taleban site defaced after terrorist attacks [Electronic Version]. *Extreme Tech*. Retrieved February 20, 2009 from http://www.extremetech.com/article2/0,2845,120263,00.asp.

Soraghan, M. (2001). In wake of terrorist attacks, industry Web sites remove information [Electronic Version]. *Denver Post*. Retrieved January 25, 2009 from http://www.accessmylibrary.com/coms2/summary_0286-7842438_ITM.

Thomas, W. G. (2004). Computing and the historical imagination. In S. Schreibman, R. Siemens & J. Unsworth (Eds.), *A companion to digital humanities*. Oxford: Blackwell.

Toner, R. (2001). Reconsidering security, U.S. clamps down on agency Web sites [Electronic Version]. *New York Times on the web*. Retrieved January 25, 2009 from http://query.nytimes.com/gst/fullpage.html?res=9C06E5D81F31F93BA15753C1A9679C8B63.

Tuunainen, J. (2001). Constructing objects and transforming experimental systems. *Perspectives on Science, 9*(1), 78–105.

Vygotsky, L. (1978). *Mind in society: The development of higher psychological processes*. Cambridge, MA: Harvard University Press.

PART TWO

Web Cultures

CHAPTER 3

Evolution of U.S. White Nationalism on the Web

Alexander Halavais

For many, organized racism in the United States evokes images of groups of isolated and less educated people, most likely in the southeastern part of the country, gathering in a barn to plan violence against their neighbors. It is perhaps because of that stereotyping that the rise of a technologically sophisticated group of racists, of various stripes, went relatively unnoticed. The "new racism" makes substantial use of the internet to persuade audiences and recruit new adherents to their cause. What follows is an attempt to trace a brief history of white nationalist organizations on the web, attending particularly to the use of the web by groups in the United States, and the evolution of their web presence since 1997.

In many ways, the United States is a unique case, given its history of slavery and institutionalized racism. But understanding the ways in which the web has been employed by these groups provides a model for its use in other countries and in other contexts by other movements. Perhaps more than this, what we have seen in the last decade or more is the growing globalization of extremist groups. There is some awareness of this in the parallel and related case of terrorism, with techniques, training, and resources being shared among groups that are otherwise ideologically incommensurate. This extends to organizations and media outlets that may not be actively engaged in large-scale violence but serve as a support structure and breeding ground for those who are. When, at the end of this chapter, I make an argument for archiving, monitoring, and studying the work of these groups, it extends to all extremist groups on the web, no matter where they are geographically concentrated.

The argument presented here is that, first, there is no clean line between speech and organizations that are extremist and those that are not. Many of the sites under examination explicitly condemn violence, even as they are linked to (and from) groups that embrace it. Second, it is argued that looking at these sites as a whole, and their evolution over time, provides an example of a new kind of rhetoric, one that is distributed and relies on collaborative construction over time and through hyperlinks.

THE NEW NETWORKS

As Tucker (2001) notes, the "striking thing about the networked structure of the new terrorism is that it differs little from the structure of the old terrorism," and such networked architecture can be found in all manner of resistance groups throughout the course of history. The idea of cells and of distributed control has long been important to such groups, as it provides a smaller target for surveillance and government action. In the most extreme cases, participants are organized under a banner but without any single point of control. Decapitation attacks that might have been effective against hierarchical criminal organizations are relatively useless within networks of individuals that share the same ideology but do not rely upon coordination or planned cooperation.

The growing ubiquity of networked telecommunications, and particularly the development of the internet and global popularization of the World Wide Web during the 1990s, provided networked communication for organizations that were already organized as networks. Over the last decade there has been mounting evidence of a "new terrorism" that makes use of these global networks to reach its ends. The growing recognition of this new use of technology to create global networks appears most popularly in Arquilla and Ronfeldt's (2001) *Networks and Netwars,* which examined the use of networks by a number of insurgent organizations. They were not, of course, alone in recognizing the shift toward globalization of organized crime. Unlike the growth of global corporations, transnational crime relied on loose networks of affiliation, reputation, and interaction. Because so much of globalization was occurring "from below," with connections across national borders by individuals and small groups, tracking changes in these relationships and combating networked threats, were all seen as particularly difficult for police organizations accustomed to more traditional, hierarchical organizations.

Arquilla and Ronfeldt took the Zapatista movement to be iconic of this new form of networked warfare. The movement grew out of an armed rebellion in Chiapas, Mexico, and made use of the internet to spread its message and to gain support from a global community. United behind the anonymous Subcomandante Marcos, the movement made use of frequent internet announcements spread through the internet to present its case to Mexico and to the world and managed to maintain a significant resistance both in terms of political perception and real progress on a number issues. Perhaps equally importantly, the movement acted as a catalyst, linking together global grassroots organizations into loose global affiliations (Garrido & Halavais, 2003). The networked nature of grassroots organizations generally seems to encourage porous organizational boundaries.

There is rarely a clean line between networks of violent extremist groups and other groups outside of the normal range of public participation but not engaged in violence. Many have recognized that global networks of terror exist within a larger network of transnational organized crime, and this crime network, in turn, is connected in various ways to other sorts of organizations. It is not surprising, perhaps, that members of these organizations have engaged in loose affiliations, creating weak ties with compatriots and even with those who share only the slimmest plots of common ground.

THE NEW AMERICAN RACISM

Particularly over the last twenty years, groups that explicitly endorse violent action against non-whites have largely (though certainly not completely) faded from the public stage. In their place has come a movement that goes under the banner of white nationalism. It does not openly support violence as a means for change and generally differentiates itself from white supremacists by arguing not for the superiority of European-Americans but rather the advantages of promoting white culture and racial segregation. One prominent apologist for the movement defines it as the "idea that Whites may need to create a separate nation as a means of defending themselves" (Yggdragsil, 1996).

This claim springs from an argument that heredity leads to significant differences among races and that these differences are leading to a subjugation of white culture. The characteristics of this group defy the stereotypes of racist groups in the United States, appealing to "more heterogeneous, ideologically complex, technologically advanced, market savvy men *and* women from diverse social, political, geographical, and educational backgrounds" (Schafer, 2002). Those who believe in "racialism" can find support in popular books like *The Bell Curve*

(Herrnstein & Murray, 1996), and those seeking support for the idea that European-American culture in the United States is under assault need turn only to Samuel Huntington, who notes that "a plausible reaction to the demographic changes underway in the United States could be the rise of an anti-Hispanic, anti-black, and anti-immigrant movement composed largely of white, working- and middle-class males, protesting their job losses to immigrants and foreign countries, the perversion of their culture, and the displacement of their language" (Huntington, 2004). While neither of these opinions endorses segregation or white nationalism explicitly, they suggest that the opinions held by such groups are potentially legitimate reactions to current policy. As a result of this legitimization, white nationalists find themselves within a nexus of issues on the right, particularly with regard to immigration issues.

The connections stretch further than this. White nationalists generally reject both the subjugation of non-whites and violence against them, but the line between non-violent and violent racism is not always as clear as it might be. White nationalists see themselves as threatened by those of other races and may encourage ambiguous forms of self-defense. The rhetoric and arguments for the value of a white identity are often shared among white nationalists, neo-confederates, and supremacists. On the other end of spectrum, the views of white nationalists find their way into the rhetoric of mainstream conservative politics in the United States (Blumenthal, 2008). This should not be read as equating white nationalism either with the Aryan Brotherhood on one end or with the Republican Party on the other. Instead, it is useful to understand these as existing on a broad discursive continuum and that ideas and individuals find connections across this continuum. Groups like the Council of Conservative Citizens and National Alliance—which are identified by both the Anti-Defamation League and the Southern Poverty Law Center as hate groups—occupy a powerful position within the political wing of a racist ideological movement.

GHOSTS OF WEB EVOLUTION

Although there is a great deal of hyperbole present in popular media portraying the internet as a training environment for future criminals and terrorists, there seems to be little evidence that it is being deployed in such an organized fashion (Stenersen, 2008). Rather, it is used just as other individuals and organizations use the web—as a repository of information, a channel of persuasion, and a medium for sociability. These three functions have always been a part of what organizations attempted to do online, and over time white nationalist sites have become more sophisticated in the ways in which they present their message. Moreover, it appears that there is a growing recognition of these texts as meaningful in the collective sense as well as on an individual or site-by-site basis.

A number of projects have described the content of contemporaneous white supremacist websites, and their persuasiveness, without examining explicitly how these have changed over time. Several of these are fairly exploratory, cataloging the presence of particular features, type of content, or means of communication. Ray & Marsh (2001) examined the sites included here as well as more extreme sites by the Church of the World Creator and the National Alliance during a two-week period in 2000. Having cataloged the materials and appeals at these sites, and particularly those intended for children, they found that with the exception of music, these sites posed a fairly insignificant recruitment risk. Gerstenfeld, Grant & Chiang (2003) examined a larger collection of sites and coded for the presence and absence of particular features, including links to other extremist organizations and racist imagery, and Schafer (2002) undertook a similar catalog of features and arguments for a sample of hate sites. More recent work has focused less on the content of these sites and more on the connections between them (Chau & Xu, 2007; Zhou, Reid, Qin, Chen, & Lai, 2005).

We must look for what changed over a ten-year period, from 1997 to 2006, both in terms of the content of the most widely read sites and the ways in which they interact. As Quentin Jones (1997) suggests, there is something archeological to many investigations of online activity, because the investigator is left only with the material remains of the communities that made use of the collective virtual spaces. These traces are even thinner as we rely on archived versions of those pages. Many large, static pages have been recorded by the Internet Archive and are available through its Wayback Machine, but these archives generally only cover a small part of each site, over only a small part of its existence. The incomplete nature of the Archive, particularly in the early days of the web, is well known (Edwards, 2004), and the problems of collection are particularly acute for extremist sites. Several of the sites make use of dynamic content from fairly early on in the web's history, for everything from discussion boards to games to link counters. As a result, significant parts of the sites are not recorded by the Archive. Likewise, any part of a site protected by passwords is excluded from analysis.

Using the Internet Archive introduces further difficulties. Links are not always to copies of sites recorded contemporaneously. Because it is impossible to take a "snapshot" of the web at any one time, it is sometimes necessary to rely on a copy of a page that may have been collected many months from when the initial link was made. Of course, this is true with the live web as well—links disappear and change in content over time—but it is particularly acute in the case of an archived copy of the site. An added difficulty comes, particularly in the early days, when small sites were forced to change their URLs because a service provider found that white supremacist sites violated their content restrictions.

As a result, it is necessary to cobble together a picture of the development of white nationalism online by digging backwards from the conditions today and carefully unearthing connections between the sites and conversations about them. It is easy to concentrate on what was lost, but what is present provides enough of a picture to be able to generalize and characterize the development over time. To that end, a collection of sites was assembled, starting with the four with the highest traffic as measured by Alexa at the time of writing, namely: Stormfront, White Aryan Resistance (WAR), Crosstar, and the European-American Rights Organization. Each of these has archived pages running back at least to 1997, with the exception of the European-American Rights Organization, which appears in the Internet Archive beginning in 2001. These most certainly do not represent all white nationalist, anti-Semitic, Christian Identity, or similar groups on the web. They do, however, command a significant amount of traffic and provide windows on a large and growing community of white nationalist sites online.

A careful reading of the last archived copy of each site for each year starting in 1997 was undertaken. Particular attention was paid to interpretations of the movement by authors and textual mentions of other sites. The analysis was not confined to the initial four core sites: outward links were followed to other sites (when available in the archive) to provide a better idea of the hyperlinked neighborhood these sites existed in and how this neighborhood changed over time.

WHITE SUPREMACIST SITES ON THE EARLY WEB

Racist organizations were early users of computer networking. Before the popularity of the internet and the World Wide Web, Bulletin Board Systems (BBS) were employed by many racist organizations to share information and plan activities (Stills, 1989). A BBS could be reached via telephone, and often the telephone numbers and passwords were distributed at meetings and parties, so that the users could be assumed to already be active in various pro-white activities. Some of these systems, including the White Aryan Resistance and Stormfront, continue today, having made the transition to the World Wide Web. Stormfront's originator, Dan Black, was involved in the Ku Klux Klan and other supremacist groups through the seventies and eighties. While serving a prison sentence for a failed white coup in Dominica, he gained an education in computing and put this education to use in campaigning for former Knights of the KKK Grand Wizard David Duke's bid for the U.S. Senate (Kim, 2005). Creating a website based on a BBS started in 1995, Dan Black says he established the site as

> a resource for a movement which we call white nationalism. Our purpose is to provide an alternative news media with news and information and online forums for those who are part of our movement or for those who are interested in learning more about white nationalism. (quoted in Swain, 2002, p. 31)

It became the most significant hub of communications not only for the white nationalists but for white supremacist organizations of all stripes. At the time of writing, Alexa indicates that it remains the most visited of such sites. Drawing on its roots as a BBS, Stormfront has always been a sociable site and has encouraged conversation and posting in an online forum.

The second highest amount of traffic for white nationalist sites today, according to Alexa, goes to the website for the "White Aryan Resistance Hate Page." WAR was created by Tom Metzger, a former organizer for the Knights of the Ku Klux Klan who frequently appeared on talk shows through the 1980s as a representative of white pride and created his own programs on public access cable television (Turner, 1986). WAR represents a vision of the white nationalism movement, perhaps because of these television appearances, that we are more familiar with. It does not reject the label "racism," clearly feels the white race is superior, and encourages threats and violence against ethnic minorities. Metzger first created a BBS with the help of compatriots in the 1980s and frequently shared media space with a broad range of racist groups. He referred to this network as a kind of "electronic village square" (Anti-Defamation League, 2002). However, when WAR made the shift to the web, it became more insular.

The WAR website started in 1996, with the earliest capture on the Internet Archive dating from 1997, when it was little more than a collection of scanned advertisements for supremacist and neo-Nazi materials in traditional media: books, videos, and audio cassettes. By the following year, the site looked a bit more like many of the other successful white nationalist sites on the web and included brief position statements on a range of issues. In terms of content, it was sparse, without substantial discussions of ideology that appeared on other sites. Instead, it relied heavily on galleries of racist cartoons that appeared several years into the site's operations.

Unlike the Stormfront and WAR sites, which were extensions of BBSs, the Crosstab site was created by Richard Barrett and the Nationalist Movement, after a decade of publishing the right-wing newspaper *All The Way*. It hosted the newspaper and publicized Barrett's efforts to gain public attention for his cause.

The language on each of these sites reflected a working through of the appropriate strategy for resistance and social change. Metzger and WAR endorsed the "lone wolf" approach as he and others called it: the idea that symbolism was empty, and that for direct action to be effective it had to be accomplished by individuals with strong convictions working alone. The *Turner Diaries* (Macdonald, 1980), a novel by National Alliance leader and former professor William Luther Pierce that is seen as a template for the Oklahoma City bombings and other attacks, decries "the fainthearts and hobbyists—the 'talkers'"—in favor of provocative and open propaganda as the way of attracting the most militant and disciplined members. It was this view, particularly in the form articulated by Louis Beam (1992), that animated Metzger's position. His own difficulty with the courts led him to encourage others in the movement to not speak with authorities or negotiate for reduced sentences when captured. The site is replete with racist epithets and cartoons reinforcing negative stereotypes. There is very little here for those who want to meet with and talk to likeminded individuals; the era of the electronic village square had dwindled.

Richard Barrett's Nationalist Movement takes a very different tack and has come into conflict with WAR because of it (Anti-Defamation League, 2005). Barrett's focus is not on private violence but on generating publicity for the movement and exercising its rights, he has led the fight for equal application of the First Amendment to racist groups, taking one case, in 1992, all the way to the U.S. Supreme Court. While he may not subscribe to Metzger's lone wolf approach, his rallies often do not attract many white nationalists. They do bring out anti-racist protesters in large numbers and television cameras to cover the event. The Crosstar site reflects this, to some degree, highlighting past and planned events. It does, however, provide access to a forum for user discussions.

There is an obvious tendency to think of these sites as units and consider their content from that perspective. In practice, a web surfer may spend only a few moments on a site before moving on. It is therefore important to understand the context of these sites on the web at large and how hyperlinks connect them with other sites in the movement.

As noted above, the Stormfront site of 1996 was fairly sophisticated, incorporating a number of social elements. This extended to links to potential compatriots. A number of Bulletin Board Systems (including Stormfront's own) are listed, as are email lists and Usenet newsgroups, but the links to external websites, both on the front page and throughout the site, outnumber these. In total, 38 sites are linked. Several of these are incidental: Black's homepage, for example, or a hit counter, or a Jefferson biography at the Virginia Tech University website. Others sketch out the major resources available for those interested in racism online.

Particularly notable, however, is that not a single one of the sites linked by Stormfront during its first year appears to reciprocate and link back. This is not surprising in some cases. The EFF Blue Ribbon free speech campaign, for example, was linked by a wide variety of sites on the web without reciprocating those links. In the case of the sites dedicated to Pat Buchanan and David Duke, there were clearly political reasons not to return the link to Stormfront. By 1997, the number of sites linked from Stormfront climbed to 49—bolstered by the addition of websites for various chapters of the Ku Klux Klan—and several of these sites now linked back to Stormfront. This open linking of many other sites in the movement (though not, notably, the National Movement site, which has attempted to distance itself from the former Knights of the KKK members in leadership positions at Stormfront and WAR) establishes Stormfront as a portal for white nationalism from early on, a position it continues to hold. Because it provides these outbound links and collects links more easily from the rest of the web, it represents the public face of the movement on the web. At the time of writing, a search on Google for white nationalism yields Stormfront as the second hit after the Wikipedia article. These outbound hyperlinks suggest a certain current or directionality that continues to develop within the white nationalist web, even as the link directory has faded away.

In 1999, an article in *Time* magazine pegged the number of internet hate sites at 254 (up 55% from the previous year according to Grace, Monroe & Roche, 1999), but the real growth would occur through the early 2000s. During this period some of these sites became more interlinked rather than a collection of unlike parts.

FROM ISSUE TO COMMUNITY

One of the consistent arguments among many white nationalists is that they are no different than other cultural identity groups, including those who support African Americans or Latinos, and should be afforded the same sort of respect. Shared across almost all white supremacist and white nationalist sites is the fear that the "white race" is facing extinction at the hands of other races, through either population dynamics, the hybridization of races, race violence against whites, secret control of government and economy (i.e., the "Zionist Occupied Government," or ZOG), or some combination of these things.

As noted, the first step toward networking for many of these organizations was not the web, but Bulletin Board Systems, email lists, and, to some degree, Usenet. The experience with these media appeared to influence Stormfront throughout its many years. In 1997, they added the Aryan Dating Page. In some ways this is an iconic view of the social nature of the site. As Les Back (2001) suggests, the construction of white nationality works best in cyberspace, where individual differences can be easily ignored. The Dating Pages extended the process of making real-world connections, acting as a kind of early Facebook for white nationalists.

Over the period of study, Stormfront added a greater degree of support for women in the movement. Initially, women's issues (which, naturally, differ somewhat from the issues expressed by women in mainstream organizations) were addressed in passing, but by 1999, there was a prominent link to a "women's page." That year also saw the inclusion of a "kid's page," managed by 10-year-old Derek Black, that included rants against Pokemon, and downloads of a White

Power version of Doom (in which monsters are replaced with stereotypical depictions of ethnic minorities) among other material.

The Crosstar (Nationalist Movement) site followed suit in many ways during the 2000s. They continued to offer polls on their site ("Should Gore bow out now?" "Is Rumsfeld putting Israel before America?"), but also dedicated a section of the site labeled "interactivity" to a forum, chat room, an area for online petitions, and an apparent link to the alt.national newsgroup. Unfortunately, because of extensive use of Java applets, much of the functionality of the site has not been well preserved in the Internet Archive, but much of the structure and content can still be gleaned. The focus remained on using their site as a way of presenting white nationalist views to the media. As with other sites, photos depicted heroes and "martyrs" of the movement, but the gallery at Crosstar also showed a chart of website traffic over a period of several years, with indications of events occurring at the time (e.g., "Daily Show," "Serb Support," "Manville Attack").

Even WAR, a site that included a prominent link entitled "Don't Talk," took modest steps toward community. The evolution of the site was slow, but by 2004, they had included a section for photographs of white-power meetings (with faces obscured), provided email accounts at the resist.com domain for a fee, and offered an email list for those who wanted to learn more about the organization. But even at this late date, the site was almost entirely without outbound links. The exception was a link to "Better than Auschwitz," a site that centered on forums, chat, and filesharing. Unfortunately, these forums were not captured by the archive, and the site is no longer active, having apparently merged with the main WAR site in 2005. According to the Anti-Defamation League (2001), the site provided information on effective bomb construction and ways of luring targets and doing substantial physical damage. Even lone wolves apparently desired online interaction and instruction.

The European-American Rights Organization (whitecivilrights.com) was initially announced in 1999 by David Duke, then under the banner of the National Organization for European-American Rights (NOFEAR). Especially in the earliest versions of this site, Duke's own mark on it is clear, with a prominent advertisement for his book on the front page. Beyond this, it reflects the sort of sites it sought to emulate—the contemporaneous National Association for the Advancement of Colored People (NAACP), or the National Council of La Raza (NCLR), for example. This extended so far as to offer a section on "teaching tolerance," offering pointers on reducing anti-white bias and stereotypes and providing "great White men" to serve as role models for young white children. Though it is easy to read this as mere satire, it appears to be in earnest. The site was marked by a lack of community-oriented features but clearly represented the intended face for white nationalism—a mirror of existing identity groups with an effort to change attitudes and policies through political pressure.

These websites represented some of the most visited sites relating to white nationalism and presenting the image of the movement to the public, but something else was happening just below the surface. Beginning in the early 2000s, there was an increase in the variety of sites, and as time went on, the connections among these sites increased significantly. Many of these new sites presented the thoughts and arguments of their authors, which were then linked heavily by the growing online white national community. Sites created by those with academic standing or similar mainstream credibility, no matter how slim, were heavily linked by the community. Websites like those written by Arthur Butz at Northwestern University or Kevin MacDonald at California State University, for example, generally did not link back out to sites like Stormfront, though they likely received substantial visits from those who passed through that portal and ones like it. White nationalist sites that existed on the margin of academia did tend to link to and support one another but rarely made the jump to more openly racist sites.

However, that did not extend further. Stormfront, for example, continued to be a gateway to the breadth of white nationalist sites, and its link pages grew exponentially. By 2003, the site provided links to over 400 different sites, with a mechanism for rating the quality of each site. This included a fairly broad set of sites, from those supporting the Ku Klux Klan, neo-Nazis, and the Church of the World Creator, to a small section marked "The Other Side," and linking to sites like the Simon Wiesenthal Center, Nizkor, and the Center for the Study of Hate and Extremism, each of which enjoyed healthy traffic by users. Still, this represented only a fraction of the white supremacist sites available on the web, and the number increased quickly. Note that it is difficult to conclude much from this—the extreme growth of the web meant that the number of sites on any topic increased rapidly. By 2004, the Stormfront site went entirely to the discussion board format, a focus it retains through today, and many of the links to outside sources were no longer as prominent in the archives.

PATHS TO VIOLENCE

If these two approaches to white nationalism—community and identity—grew during this period, what of the activity that is most troubling: organized violence? With good reason, those who have watched white power organizations are skeptical of the aims of those who claim to seek separation and protection from rather than power over non-whites. Many members of these organizations who disavow violence publicly appear to support it privately. While they argue that they have no desire to express superiority, they continue to promote stereotypes, demean non-whites, and spread what can only be considered willful ignorance when it comes to the history of the Holocaust or the relationship of descent to social behavior. Moreover, those racists who have traditionally encouraged violence have clearly indicated the need to go underground and to act independently.

Legitimization, at least among those on the right wing in the United States, of views that had been largely relegated to back rooms after the civil rights movement provides a kind of soft support for the violent acts that continue to be perpetrated. Few anti-abortionists support the killing of doctors, and few environmental activists support Earth Liberation Front arsonists, but those who engage in direct action in these areas recognize that their ideology is representative of a group, and there is support for their cause, if not their means. They can be nurtured by a community that shares their values. There has traditionally not been that sort of mainstream acceptance of white nationalism in the United States, at least not publicly. The idea that one could, for example, walk into a Target store and purchase a shirt emblazoned with the coded number 88 (an international sign for "Heil Hitler") would have seemed unlikely in the 1990s and somehow less surprising in the early years of this decade (Teaching Tolerance, 2002). Particularly through the internet, the social stigma of talking about race, and differences among races, has diminished. What does this mean for racial violence and inequality?

We have understood for some time that broadcast media present a message that passes through layers of social networks as it diffuses through a population, and those intermediary opinion leaders represent an important source of influence (Katz & Lazarsfeld, 2006). Networked media have increased the importance of these informal channels. To borrow from Marshall McLuhan, the social network is the message. In this context, recruitment takes on a greater importance. It may not be necessary to recruit members to actively participate beyond acting as a conduit for the message—as part of an ever-increasing broadcasting footprint on the network. The internet provides a number of advantages to spreading a message that may be abhorrent to a large portion of the population. The anonymous nature of the net allows individuals to explore material that they would otherwise find dangerous to their social standing. Such anonymity provides for the potential for increased in-group identification and for prejudice to be more easily applied because it is necessitated by the lack of social context (Glaser & Kahn, 2005).

The "kinder, gentler" racism on these sites also provides a sort of welcoming center for the curious. In tone and style, many appeal, by design, to standards of scholarly discourse and citation. A student seeking information on the Holocaust, the Civil War, Segregation, or Civil Rights is likely to find herself on a white nationalist website without even realizing it. Stormfront, for example, hosts a page at www.martinlutherking.org that is explicitly directed toward students (at present, the banner at the top of the page reads "Attention Students: Try our MLK Pop Quiz"), and until recently was the first link on a Google search for the civil rights leader. It remains on the first page of responses on Google and Microsoft Bing, among others. Students who have not been taught to seek out and evaluate multiple alternatives may find themselves learning about the civil rights leader from a source that is anything but balanced.

Recruiting need not mean taking someone who has never been interested in the issue of race and turning her into a racist. Instead, it may be a process by which people are welcomed into a community and gradually come to share a common set of understandings. The open nature of some of these online communities gives people a chance to try on an identity, and a network of potential friends, without a significant commitment. That trial period provides an opportunity to bring new people into the fold.

Once someone is a part of a community like that on Stormfront, it can easily become the new normal. Exploration of more extreme ideas seems natural. Although most members of the mainstream public will never encounter *The Turner Diaries*, David Duke's rhetoric, the antiquated racist evolutionary material found on some of these sites, or similar ideas, these are part of the common discourse on Stormfront and the web that surrounds it. And while more extreme sites tend not to link to Stormfront, the pattern created early on of a fairly ecumenical portal to white power has not diminished over time; though as the number of sites approaching the topic grows larger and more diffuse, the role of portals becomes less vital.

ENGAGING THE WIDER CONVERSATION

These days, the term "white nationalist" is often eschewed by those on the far right who, nonetheless, are proposing policy that is consanguineous with the stated aims of white nationalist groups (e.g., Brimelow, 2006). The explosion of blogging, especially with the U.S invasion of Iraq, brings a diverse set of voices to the web. Many blogs express positions found on the extreme right, positions that would rarely be exposed on mainstream media, and if they were, would be contextualized or framed by mainstream views.

How would someone find their way to the Stormfront site, aside from the search engines? To provide one example, consider an article about Stormfront founder Don Black meeting presidential candidate Ron Paul that was published on the popular political blog Little Green Footballs (Johnson, 2007). Little Green Footballs generally takes a conservative perspective, as do its readers, but it is a popular mainstream blog. The pro-Israel positions it often adopts—it was ranked the best overall blog by the *Jerusalem Post* ("2005 Jewish and Israeli Blog Awards," 2006)—would not endear it to many white nationalists. There can be little question that the connection of Paul to the movement is being exposed as an embarrassment to the candidate. Nonetheless, Stormfront is linked to the article, and some segment of the blog's readership will visit the Stormfront site for the first time. Similar kinds of links, by supporters and non-supporters, appear on blogs across the web.

The availability of these ideas means that they enter into the common discussion of many on the right. In a blog post entitled "Why I am not a white nationalist," the author states early on, "I am not a white nationalist, but I do read white-nationalist blogs, and I'm not afraid to link to them," and then proceeds to do just that before providing an argument against white nationalism as a political project (Moldbug, 2007). The entire web is a slippery slope, but this willingness to engage and discuss, this openness to treating the arguments of white nationalists as worthy of serious contemplation, is new.

The result, however, may not be entirely what white nationalists expected. An examination of the discussions on Stormfront today reveals more of the expected racist rhetoric, along with surprising suggestions. One poster wonders if it wouldn't be better practice to be more accepting of those in the Jewish faith. Another decries the use of demeaning language as ineffective to the cause. It is easy to see much of this as mere posturing, an attempt to gain mainstream credibility by applying a sheen of political correctness to the same racist ideas some members had when they were in the Klan. On the other hand, over time they may have found that they played this role too well. There are still barely veiled calls to violent action, but the discussion that occurs on the boards themselves has created a tenuous bridge from the extreme right wing on the web to those who are part of the mainstream right, and that bridge provides for some traffic in both directions.

POLICY MOVING FORWARD

The United States remains unusual within the developed world for permitting speech that advocates race-based discrimination and violence. The neo-Nazi rhetoric found on many of these sites, and many of the sites they link to, is considered illegal in much of Europe and in other parts of the world. Perhaps because of this, a large number of the sites in this web neighborhood discuss white nationalism in contexts other than that of the United States. As racism grew on the web through the end of the 1990s, many in the United States questioned whether permitting hate speech was in the best interests of the country, and if not, what the alternatives might be.

At the core of many of these questions is whether the danger is great enough to justify forcibly silencing opinion. In fact, the open communication of white supremacist materials has in some cases made it easier to pursue the publishers for contributing to bias crimes. Tom Metzger has demonstrated, however, that even a very sizable civil judgment against the publisher of such material is not enough to keep them from continuing to encourage race-based violence. Nonetheless, a reasonable argument can be made that it is better to permit pernicious discussion to occur in open venues, where it is more available to surveillance by both the public and the police, than have it be forced underground and no longer easily accessed or countered. Filtering and annotation systems may yet provide some counterbalancing effect on the influence of such sites.

The web is sometimes considered an echo chamber in which people only hear ideas that they are most likely to agree with. The growth of sites online that are related to white nationalism, including anti-immigration sites and those dealing with questions of affirmative action, is certainly troubling from the perspective of a liberal democracy dedicated to liberty and justice. Whether the stridently racist messages of the old racism or the more intricate arguments of the new racism, continued exposure to these messages without a counterbalance affects viewers' thinking (Lee & Leets, 2002). On the other hand, if a diversity of ideas is being presented on the web, and if the authors of those ideas are coming more frequently into contact and discussion, this may lead to more deliberative approaches to working through racial grievances. For this to occur effectively, analytical tools and mediation must be provided that promote ways of understanding and evaluating material found online. Certainly this should be supplied by schools and other educational contexts: media literacy and critical thinking skills are the best antidote to racism (Messner & Apple, 2003).

But there is also a place for such material on the web itself. There are a number of organizations that chart the existence of hate groups, and their activities, and have been effective in linking instances of violence to online sites. These organizations have also provided a valuable service by informing the police, teachers, and students about these materials (Perry, 2000). But especially as these networks grow, they must be met with networks in response. Further work needs to be done to debunk the foundational ideas they profess and provide web surfers the tools for thinking through faulty reasoning.

As the size and complexity of this discussion continue to grow, better ways of archiving, tracing, and summarizing that growth are necessary. The discussion above provides a bit of a view of the development of this particular corner of the web but was hampered by archives that were incomplete and a lack of tools necessary for tracking the large-scale evolution of web structure. The Internet Archive is an invaluable resource, and its creators should be lauded for having the foresight to begin collecting documents from the web before others recognized the importance of that project. It is imperative that we continue from this step, toward the ultimate aim of being able to surf the web directly as if we were viewing it during another time period. We also must continue to develop tools that provide us with clear indications of how the linkage structure within these neighborhoods evolves over time and provide clues as to when changes happen not just on a single site, but across a range of interconnected sites.

It is too easy to compare the web of a decade ago, where only a few dozen racist sites existed, to today's web, where comments on YouTube devolve to racist remarks and political blogs justify segregation, and decide that the web has a negative effect on racial understanding. The reality is more complicated and confusing. More people are talking openly about race and expressing their deeply held opinions in forums where they feel anonymous enough to allow this to happen. This increase in conversation may ultimately prove to be detrimental to the slow progress toward racial equality the United States has enjoyed, but there is some hope that problems laid bare are more easily solved than those unobserved. We must actively observe these conversations and find ways of engaging the participants.

BEYOND WHITE NATIONALISM

An investigation of white nationalist websites in the United States is determined in large part by the context and histories of the organizations and individuals involved, but within this micro-history we can see the playing out of a rhetorical structure for the web that may be employed by other marginalized groups on the web that are seeking to make their ideas heard and win new adherents. Barriers to interaction are lowered on the web, and organizations may find a flexible network of people from around the globe that share objectives or methods. In many cases, involvement is likely to be less intensive, as people may engage anonymously, and the discussion may never lead to significant collective action. Nonetheless, the ability to engage in discourse and play with identities and allegiances provides a new opportunity for those who wish to recruit individuals to more violent forms of action.

Those eager to draw people toward more extreme beliefs have found less extreme venues on the web to be likely recruiting grounds. Although it is unlikely that the average searcher or browser on the web will encounter the most violent extremist sites directly, sites like Stormfront provide a portal to more extreme material. While we may think of a web "portal" as nothing more than a directory or collection of links, it is clear that Stormfront and similar sites perform additional roles. More than just a front door, these sites are the welcome mat for a large, more private "dark web." They provide a source of basic information, fully indexed by the search engines and easily found. In many cases, they also provide a space in which people who may find themselves outcasts in their everyday lives can find an accepting, if anonymous, community. De Koster and Houtman (2008), in looking at right-wing extremists in the Netherlands, question whether this relation between offline stigmatization and online community may form the mechanism for organization among a range of marginalized groups around the world.

Although such a welcoming area may shape a space for recruiting curious people into networks that espouse more violent acts, if the reaction of the most violent of these groups is any indication, they also serve to dull violence and encourage discussion, in some cases. The natural tendency of those wishing to diminish the influence of extreme opinions is to isolate communities on the margin that espouse what are seen as deplorable ideas. However, the fact that the most violent groups seek to isolate themselves and avoid discussion suggests that a strategy of purposeful inclusion provides an opportunity for these portals to act as a space for elucidation and discussion rather than solely recruitment by extremists.

One of the advantages to examining white nationalist websites over a period of several years is that it provides an indication of how an ecosystem of related organizations and web structures has evolved. Although it is not as visible to the mainstream as other forms of interaction, some of these sites were at the forefront of employing social technologies on the web to build distributed communities. Identifying centers of extremist speech, particularly when tied to violent acts, remains an important part of the modern policing function, but a better understanding of the continuity between mainstream websites, marginal portals, and the dark web they connect to is important to discovering the environment in which individuals become interested in, and acculturated into, extremist groups. It is also important to recognize that these portals have the potential to reflect back to the rest of the web and represent a recruiting ground not only for extremist organizations, but for those hoping to educate, inform, and invite disaffected citizens into broader discourse and debate.

REFERENCES

2005 Jewish and Israeli Blog Awards. (2006, February 2). *Jerusalem Post.* Retrieved August 1, 2008, from http://info.jpost.com/C005/BlogCentral/JIB.2005/index.html.

Anti-Defamation League. (2001). The consequences of right-wing extremism on the internet. *Extremism in America*. Retrieved August 1, 2008, from http://web.archive.org/web/20041206045147/www.adl.org/internet/extremism_rw/guiding.asp.

Anti-Defamation League. (2002). Tom Metzger and White Aryan Resistance (WAR). *Extremism in America*. Retrieved July 22, 2008, from http://www.adl.org/learn/Ext_US/Metzger.asp.

Anti-Defamation League. (2005). Richard Barrett. *Extremism in America*. Retrieved August 1, 2008, from http://www.adl.org/learn/ext_us/barrett.asp?xpicked=2&item=barrett.

Arquilla, J., & Ronfeldt, D. F. (2001). *Networks and netwars*. Santa Monica, CA: RAND Corporation.

Back, L. (2001, January). White fortresses in cyberspace. *UNESCO Courier*, 44.

Beam, L. (1992). Leaderless resistance. *The Seditionist, 12*. Retrieved August 1, 2008, from http://www.louisbeam.com/leaderless.htm.

Blumenthal, M. (2008, January 18). Mike Huckabee's white supremacist links. *The Nation, Campaign 08 Blog*. Retrieved August 1, 2008, from http://www.thenation.com/blogs/campaignmatters?pid=272545.

Brimelow, P. (2006, July 24). Is VDARE.COM "White Nationalist"? *VDARE.com*. Retrieved August 1, 2008, from http://www.vdare.com/pb/060724_vdare.htm.

Chau, M., & Xu, J. (2007). Mining communities and their relationships in blogs: a study of online hate groups. *International Journal of Human-Computer Studies, 65*(1), 57–70.

De Koster, W. & Houtman, D. (2008). "Stormfront is like a second home to me." *Information, Communication & Society, 11*(8), 1155–1176.

Edwards, E. (2004). Ephemeral to enduring: The Internet Archive and its role in preserving digital media. *Information Technology and Libraries, 23*(1), 3–8.

Garrido, M., & Halavais, A. (2003). Mapping networks of support for the Zapatista movement. In M. McCaughey & M. D. Ayers (Eds.), *Cyberactivism* (pp. 165–84). New York: Routledge.

Gerstenfeld, P. B., Grant, D. R., & Chiang, C. (2003). Hate online: a content analysis of extremist internet sites. *Analyses of Social Issues and Public Policy*, 3(1), 29–44.

Glaser, J., & Kahn, K. (2005). Prejudice, discrimination, and the internet. In Y. Amichai-Hamburger (Ed.), *The social net: Human behavior in cyberspace* (pp. 247–75). Oxford: Oxford University Press.

Grace, J., Monroe, S., & Roche, T. (1999, March 8). Trading white sheets for pinstripes. *Time, 153*(9), 30–1.

Herrnstein, R. J., & Murray, C. (1996). *The bell curve: Intelligence and class structure in American life*. New York: Free Press.

Huntington, S. (2004). The Hispanic challenge. *Foreign Policy, 141* (May/April), 30–45.

Johnson, C. (2007, December 20). Ron Paul's photo-op with Stormfront. *Little Green Footballs*. Retrieved August 1, 2008, from http://littlegreenfootballs.com/weblog/?entry=28353&only&rss.

Jones, Q. (1997). Virtual-communities, virtual settlements, and cyber-archaeology: A theoretical outline. *Journal of Computer Mediated Communication*, 3(3). Retrieved August 1, 2008, from http://jcmc.indiana.edu/vol3/issue3/jones.html.

Katz, E., & Lazarsfeld, P. F. (2006). *Personal influence: The part played by people in the flow of mass communications* (2nd ed., p. 400). New Brunswick, NJ: Transaction Publishers.

Kim, T. (2005). Electronic storm. *Southern Poverty Law Center Intelligence Report.* Retrieved August 1, 2008, from http://www.splcenter.org/intel/intelreport/article.jsp?aid=551.

Lee, E., & Leets, L. (2002). Persuasive storytelling by hate groups online: Examining its effects on adolescents. *American Behavioral Scientist, 45*(6), 927–957.

Macdonald, A. (1980). *The Turner Diaries.* Washington, DC: National Alliance.

Messner, B. A., & Apple, A. L. (2003). *Fighting cyberhate: A guide to critical evaluation of hate on the internet.* Lilly Endowment, Inc.

Moldbug, M. (2007, November 22). Why I am not a white nationalist. *Unqualified Reservations.* Retrieved August 1, 2008, from http://unqualified-reservations.blogspot.com/2007/11/why-i-am-not-white-nationalist.html.

Perry, B. (2000). Button-down terror: The metamorphosis of the hate movement. *Sociological Focus, 33*(2), 113–31.

Ray, B., & Marsh, G. E. (2001). Recruitment by extremist groups on the internet. *First Monday, 6*(2). Retrieved August 1, 2008, from http://firstmonday.org/htbin/cgiwrap/bin/ojs/index.php/fm/article/viewArticle/834

Schafer, J. A. (2002). Spinning the web of hate: Web-based hate propagation by extremist organizations. *Journal of Criminal Justice and Popular Culture, 9*(2), 69–88.

Stenersen, A. (2008). The internet: A virtual training camp? *Terrorism and Political Violence, 20*(2), 215.

Stills, P. (1989, December). Dark contagion: Bigotry and violence online. *PC/Computing, 2*(12), 144–50.

Swain, C. M. (2002). *The new white nationalism in America.* Cambridge: Cambridge University Press.

Teaching Tolerance. (2002, August 26). TARGET: Retailing white supremacy. *Hate in the News.* Retrieved August 1, 2008, from http://www.tolerance.org/news/article_hate.jsp?id=603.

Tucker, D. (2001). What is new about the new terrorism and how dangerous is it? *Terrorism and Political Violence, 13*(3), 1.

Turner, W. (1986, October 7). Extremist finds cable TV is forum for right-wing views. *The New York Times,* 23.

Yggdragsil. (1996). What is white nationalism? Retrieved August 1, 2008, from http://www.whitenationalism.com/wn/wn-06.htm.

Zhou, Y., Reid, E., Qin, J., Chen, H., & Lai, G. (2005). US domestic extremist groups on the web: Link and content analysis. *IEEE Intelligent Systems, 20*(5), 44–51.

CHAPTER 4

A History of Allah.com

Albrecht Hofheinz

Figure 1: Allah.com in May 2005

Perhaps the most prestigious web address that Muslims might imagine is Allah.com. Who stands behind it, and what can we learn from looking at its history? This chapter provides a 'thick description' (Geertz, 1973) of the development of the site. As a historian of Islam, my primary interest here has been to find out what attitudes and changes in attitudes among contemporary Muslims the site may reflect and help us to understand. The analysis, I hope, will demonstrate that proponents of 'traditional' content on the internet not only use the new medium to extend existing practices and ideas but thereby also contribute to structural changes that may be interpreted as 'modern.'

I cannot begin here to engage the vast literature on Islam and modernity. Space only permits me to outline my own use of the term. A key mark of 'modernity' is a conscious distancing from an inherited/past order of things (often conceptualized as 'tradition') in favor of an order devised according to transparent standards that are deemed valid not because they already exist and are therefore socially self-evident, but because they conform to norms of rationality and are open to public review and criticism in which every rational individual can partake. 'Rationality' here need not necessarily mean 'secular' rationality. In this definition, it is not only the contrast to tradition, but the subjugation of 'traditional' order to conscious criticism according to newly evaluated standards that is central. Equally so it is the agency of individuals as participants in a public sphere ideally open to all, the idea that 'everyone' has the right to a voice, and therefore the obligation to understand, in order to participate in judgment and control. Therefore, processes of individualization are an integral structural part of modernization, even if the individuals concerned do not necessarily embrace 'modernist' views. It is this latter aspect that the present case can serve as an example of.

My sources are largely drawn from the Internet Archive Wayback Machine (http://www.archive.org) as well as 'live' interaction and private archiving, chiefly in 2003, 2005 and 2008, supplemented by common web tools such as whois or Alexa. I did *not* resort to interviewing, in an attempt to demonstrate the extent (and thereby also the limits) of what can be found out using only online sources. Methodologically, I contend that web texts (understood broadly as what in internet speak is often called 'content') are not fundamentally different from other texts. The internet may pose particular challenges such as defining a text that is part of a hyperlinked reality and that includes multimedia elements, or of establishing and retrieving 'versions,' given not only the rapid and frequent changes that many sites undergo but, even more so, the intrinsically dynamic character of interactive sites. But texts have at all times been fluid and embedded in intertextual relations and extratextual contexts. Further, texts have always been mere traces surrounded by what has been lost or suppressed; not only the Internet Archive but any archive has always had many more holes than content. For all these reasons, the interpretation of texts must necessarily be an open-ended, dialectical hermeneutic exercise. A hermeneutic approach can therefore go a long way to helping us understand web texts, just like any other text.

THE SURFACE: SUFIS QUICK TO EMBRACE A NEW MEDIUM?

The address Allah.com was registered relatively early (in 1995) and for many years (1998–2006) appeared to be linked to the Sammāniyya *ṭarīqa*, a Sudanese religious brotherhood. More specifically, it referred to the spiritual authority of Shaykh Ḥasan al-Fātiḥ Qarīballāh (d. 2005), a famous Sammānī shaykh from Omdurman. Originally, it called itself "the site (electronic mosque) of Allah on the Internet"; later, this was toned down to the "First Site of Prophet Muhammad on the Internet."[1] Its mission was to promote "The Leading Islamic Revival Plan," as follows:

1. FREE www.JesusMuhammad.com all Global Preaching and Beautiful Dialog and Guidance where Bishops, Priests and Rabbis are embracing Islam
2. FREE Download area contains over 70 Free books and children collection and growing children
3. FREE Muhadith Training Across the Globe—www.Ghumari.org (Please sign-on). No longer the youth will abandon the Prophet Saying
4. FREE Intensive Islamic Training—(Please sign-on for the opportunity near you) with FREE Multimedia Presentations (NEW)
5. FREE Islamic University for traditional Islamic Scholars
6. FREE Ihsan Spiritual Training—www.Qadiria.com (Please sign-on) Over 1000 Centers in Africa and growing worldwide
7. FREE Muslim Marriage with local Imams/Mosque Interface[2]

In other words: it promised to be a hub for dialogue-oriented Muslims with an education programme ranging from children's books to general Islamic training to university-level studies to spiritual guidance in the Sufi tradition, and how to complement one's faith with a good Muslim marriage.

In earlier incarnations, the site advertised an "Interactive Islamic College (tm)" with a relatively short (i.e., excellent) authority link (*sanad*) of only "35 teachers between you and the Prophet" and which contained for download "all the documents that every Muslim needs."[3] It was also characterised by outspoken antipathy towards the Wahhābīs.

So was this a modern incarnation of an ancient Sufi brotherhood from the Sudan? Was the Sammāniyya under Shaykh Ḥasan Qarīballāh (who was well known for his anti-Wahhābī stance, and who was regarded as a 'modern' shaykh in the Sudan due to his adoption of 'modern' bureaucratic and technical means of organisation and communication) particularly quick to embrace the new medium of the internet to spread their message to the world?

A look behind the scenes reveals a somewhat different yet no less interesting picture.

A SOFTWARE GURU

The domain Allah.com was registered, together with Muhammad.com, on June 28, 1995, by Ahmad Darwish (ldapguru@yahoo.com) of Linuxvision, Sharjah (United Arab Emirates). Less than two months later, on August 17, 1995, Ahmad Darwish registered mosque.com, completing a trinity of sites that for many years mirrored each other. His "Mosque of the Internet" was first hosted by New Mexico Internet Access, Inc., in Albuquerque, under the address http://www.nmia.com/~mosque. Ahmad Darwish's email address reflected the fact that he was a specialist in LDAP, or "Lightweight Directory Access Protocol," an internet protocol that email programs can use to look up contact information from a server. Together with one Ahmad Abdel-Hamid, Ahmad Darwish (who seems to have been an employee or consultant of Sun Microsystems at some point and who in most of his computer-related business refers to himself as "Alan Darwish" (Alan Darwish < Main < TWiki, 2004) (Ahmad Darwish's profile, 2005) founded the company Linuxvision sometime in the 1990s. Linuxvision was apparently first registered in the Cayman Islands for tax reasons and later moved to St. Louis (but with a phone number in Dallas, TX), and had representatives in Chicago, Dallas, Cairo (Ahmad Abdel-Hamid), Karachi, Bombay, and Dubai. Linuxvision struggled to find a market niche as the internet boom took off, betting on offering Arabic support under the Linux platform. In 1999, it developed the "Sheba Linux Arabization Server Package" which added some Arabic support to Linux's GNOME user environment. Sheba was doomed, however, due to a lack of development in a constantly changing open-source world. A number of other projects—e.g., to establish a *ḥalāl* offshore bank or "the largest Muslim public offering company"—also never got off the ground.[4] In late 2001, the people behind Linuxvision established a new company, iShebaPlanet, which focused on tailor-made software solutions, especially medical software for Palm Pilot handhelds.[5] iShebaPlanet was headquartered in Dubai with a branch in Egypt. The new enterprise was not especially successful either. In 2002, it was taken over by CompuEx, an Arab-Pakistani-American computer retailer, later eReseller, and then Internet Service Provider in Houston (est. 1984), who continued to sport the logo for a few years before dropping it again in favour of "Linuxvision," which continued to be run by Ahmad "Alan" Darwish out of his home in Oak Park, IL, where he offered support for the company's three Linux-based Java applications until 2006.[6] During 2007, Linuxvision.com mirrored Allah.com, then in 2008, after Ahmad's return to the Middle East (see below), it was transformed into the "Royal Global Hajj Gateway," Mecca, the "First Global Enerprise [*sic*] EGov Solution with the new advanced Royal visionary gov2gov, B2B, end2end architecture." Again, however, this project faltered before being realized. The domain 'hajjgateway.com,' registered by Ahmad Darwish on February 18, 2008, was not taken into active use; instead, linuxvision.com was redesigned as HajjGateway. By March 2009, there was no active content behind the Flash façade (which was maintained at http://salicehajic.startlogic.com/darwish/hajjgateway/, the site of Samir Alicehajic, a Muslim software engineer from Croatia, that also mirrored Allah.com). Curiously, Ahmad Darwish submitted spurious contact information to the Whois record for hajjgateway.com, including a telephone number belonging to the National Transportation Safety Board in Washington, DC. By May 2009, both linuxvision.com and salicehajic.startlogic.com had disappeared from the net.

What drove this software broker to open the "first mosque on the internet"? Aḥmad [b.] [Kāmil] (al-)Darwīsh (as he refers to himself in Arabic) was born in Cairo in 1952. He studied for a while at al-Azhar University, but like many other Egyptians left his home country in 1976 to seek his fortune in the oil-rich Gulf, where he found work as a computer specialist. It is from Sharjah that we have his first picture, showing him as a devotee of *al-Sharīf al-Muḥaddith* ʿAbdallāh al-Ghumārī—more on him below.

Figure 2: Ahmad Darwish with al-Ghumārī, Sharjah 1977

In 1981, Ahmad Darwish moved to the United States "after receiving an invitation from the Muhammad Ali Foundation";[7] he settled in the Chicago area (where Muhammad Ali lived). Later, he married an English woman from Bristol whom he presents as a "poet ... from an established English family" (Darwish, 1998) who converted to Islam (Anne Khadija Stephens); they have a daughter, Noor. In the 1990s, the family lived in Freemont, California, for a while before moving back to Chicago. In 2006, after a quarter century in the United States, Ahmad abandoned his Linux software business. He returned to live in Cairo (and Jakarta) and began to present himself more outspokenly as a religious authority in his own right, as "Shaykh Ahmad Darwish."[8] His path from systems engineer to turbaned Shaykh can be retraced through an archaeology of Allah.com.

Figure 3: Shaykh Ahmad Darwish, 2008

REVIVAL THROUGH TRADITION

The first noted religious influence on Ahmad Darwish came from ʿAbdallāh b. Muḥammad b. Ṣiddīq al-Ghumārī (1910 or 1914–1993), a Moroccan of Prophetic (Idrīsī) descent from a well-established family of religious leaders in Tangiers.[9] His father was a leading Moroccan scholar of *ḥadīth* (Prophetic Tradition) at the turn of the twentieth century who headed a (Shādhilī?) lodge there named al-Zāwiya al-Ṣiddīqiyya. After early studies in Tangiers and Fes, ʿAbdallāh left for Cairo in 1930 to become a professor (*ʿālim*) at the Azhar, Sunni Islam's most renowned university, where he specialized in the sciences of *ḥadīth*. He worked for religious revival and had links to the Muslim Brotherhood and several related organisations. Because of these links, he was imprisoned from 1959 to 1969 like many other Muslim Brothers. He resumed his teaching and writing after his release, and at the end of his life returned to Tangiers, where he died.

ʿAbdallāh al-Ghumāri was as staunch a supporter of Islamic revival based in the Prophetic Tradition as he remained committed to his Mālikī and Ṣūfī heritage. In other words, unlike more radical reformers, he did not use Prophetic Tradition to question the juridical and ethical/mystical heritage that centuries of scholastic learning had produced. He therefore opposed the fundamentalist Wahhābi and Salafī version of Sunni thought as epitomized by the influential Shaykh Nāṣir al-Dīn al-Albāni (1914–1999), and was ready to cooperate with the Zaydī *imām* of Yemen on a revision of the Zaydī Shīʿite Encyclopaedia. He thus represented a 'traditionalist' version of Islamic thinking, one that was more accommodating of Islam's legal and mystical heritage than the Wahhābī/Salafī trend.

It remains unclear how Ahmad Darwish first came to know ʿAbdallāh al-Ghumārī (whether in Cairo or in Sharjah) and what the precise nature of their relationship was, but when al-Ghumārī visited the United States in 1981—apparently at the invitation of Warith Deen Muhammad (1933–2008), who steered the black separatist Nation of Islam towards a more 'orthodox' understanding of the faith—Ahmad Darwish was actively involved by helping, i.e., to put together a collection of "8381" *ḥadīths* that were given to Warith Muhammad in a clear effort to support his "Sunni" reform.[10] Whether or not it was al-Ghumārī who inspired Ahmad Darwish to turn to Prophetic Tradition as a key source of Islamic revival, Ahmad came to share his basic outlook: one that defends the centuries-old heritage of Islamic scholarship against 'fundamentalists' attacking the scholastic law schools as well as mystical interpretations of Islam.

ISLAMICWARE: COMPUTERIZING THE HERITAGE (1986)

In 1986, Ahmad Darwish first seems to have put together his computer skills and his impulse to spread Islam by making English translations of the Koran available in electronic form. This included not only the well-known translations by Yusuf Ali and Muhammad Marmaduke Pickthall but also a translation by himself and his wife, dubbed a "contemporary" version. To market this product, he established "IslamicWare," which by December 1992 advertised version 7 of "The Holy Koran & Explanation" on a number of CDs ranging from 80–280 USD. The "first electronic Islamic encyclopedia" was also announced to be due shortly (Hashem, 1992).

ONTO THE HIGHWAY: THE MOSQUE OF THE INTERNET (1995)

Meanwhile, however, the world of electronic information changed. In 1993 the Mosaic browser was released, an easy-to-use interface to the newly conceptualized World Wide Web. By late 1994, the public at large became increasingly aware of the internet. Ahmad Darwish was quick to react. In 1995, he registered his first three domains to create what he advertised as "the first Islamic site on the web ever" (http://web.archive.org/web/20000229112458/http://www.Allah.com/). This is clearly an exaggeration—Islamicity, for example, started earlier, in February 1995—but nevertheless, Ahmad Darwish was among the pioneers of Muslim domains on the internet.

The principal raison d'être of Ahmad's "Mosque of the Internet" was the distribution of his Koran translation, which he advertised as "the most authentic translation in modern English."[11] Ahmad also offered "for the first time in the West, the explanation of Al Fatihah (...) available to non-Arabic speaking readers," as well as a Life of the Prophet Muhammad that he and his wife had written and for which in December 1996 they received high praise from "Prof. Hasan al Fatih Qaribullah, former Dean, Umm Durman Islamic University": "This work—in my opinion, and I am sure that other Muslim scholars will agree—is far more authentic than the work of Martin Lings and definitely more authentic than Haykal's work, moreover it will touch the hearts of all ages…"[12]

Building on these translations and texts, Ahmad developed the idea of making other texts on Islam available for download, first in the form of "the largest DB 18mb on Islam," and then as a "trademarked" "Interactive Islamic College" (IIC) that promised instruction in the Koran, Hadith, jurisprudence, Arabic, Seerah [the life of the Prophet], Islamic economy, as well as sections on "interfaith" and a "children Islamic corner." The college was to be the Islamic part of the 'information superhighway' that was the catchword of the day: "The Prophet said: 'Any one that takes a highway for knowledge, Allah will make an easy highway to Paradise for that person.'" "Enrol Now," Ahmad urged Muslim surfers, there are "Limited Openings." The courses offered "are only available in this Interactive Islamic College and are the highest in quality and written by traditional professors of Islam with 310 years of combined expertise, each having written over 100 books, headed by Prof. Hasan Muhammad El-Fatih Qaribullah." The 310 man-years of combined experience—an idea borrowed from the software industry—later grew to 4,000 years.

Ahmad went as far as to call his site "The Azhar Mosque of the Internet," and proudly presented his credentials from leading North American media sites:

> PC Computing Nov 1996 selected this site together with 6 Islamic sites to be amongst top 1001 sites in the whole WWW. Two out of these 6 sites have unfortunately shut and the remaining sites have 10 times less visitors than the Mosque.
>
> The Azhar Mosque of the Internet has been elected by the Discovery Channel; AOL linked to the Mosque on 9/27/1996; CNN has linked to the Mosque under: http://www.cnn.com/EVENTS/world_of_faith/links.html and http://www.zdnet.com/ pccomp/cdron/951228/webmap/religion.html PC top 1001[13]

The texts actually offered on the IIC amounted to an eclectic collection: the translation by Ahmad and his wife of the Koran and of a selection of Prophetic Traditions (the beginnings of a project to translate 8,266 *ḥadīth*s common to both al-Bukhārī's and Muslim's canonic collections); the couple's Life of the Prophet; a translation of Ghazālī's "Pure Faith Defined" (*Qawāʿid al-ʿaqāʾid*, a chapter from *The Revival of the Religious Sciences* of the famous twelfth-century renewer of Sunni thought); a text by David Benjamin Keldani (1867–1940s), a former Uniate-Chaldean bishop from western Iran who converted to Islam in

1904 and later wrote the book *Muhammad in the Bible* under the name ʿAbd al-Aḥad Dāwūd; plus texts for children and texts denouncing other groups that had originated from within Islam but are generally regarded as heterodox by mainstream Muslims: the Aḥmadiyya-Qādiyāniyya, the Ismāʿīliyya, the Bahāʾīs, and Rashad Khalifa's "United Submitters International."

For free downloading without registration, the site offered a selection of "cool Islamic references"—mainly pamphlets and introductions to "what every Muslim in the West should know."[14]

Instruction was at first not to be free but was to be provided as a subscription service. This did not pay, however. In 1996, the site was "visited by over 39,000 visitors and only 29 cotributed [*sic*]." Ahmad asked for checks to be mailed to him in anticipation of the large Islamic library to come:

> We have over 80 mb of Koran almost ready to download, in shaAllah, just need more coding!
>
> We will have Islamic oreintaion/introduction to Islam in the InterActive Islamic College, please let us know what you like to learn, even if you do not attend. Though if you subscribe will be a great help to you and your beloved.
>
> *$ If you and your family can help to support the Mosque & College will be greatly rewarded and appreciated.*[15]

Ahmad also explored other possibilities for income generation. Faithful to the metaphor of the "Mosque on the Internet," he constructed—playing for the first time with the new Java Script—an area called "Souks around the Mosque," which was meant as a space to advertise the businesses of fellow Muslims. His first entry there was CompuEx.com of Houston, the company that had taken over his Linuxvision. But except for two other entries, the Souk remained empty, and the area quietly lapsed. Later, Ahmad abandoned the idea that he could derive an income from his site and advertised it as being free.

NEW SPONSORS: THE "OFFICIAL SAMMANIA SITE" (1998)

After the 'real' Azhar came online in late 1997/early 1998, Ahmad included a link to it on his site, which was still called the "Azhar Mosque of the Internet."[16] But only a few months later, reference to the Azhar was dropped (from the text as well as from the title) when, between September and December 1998, Ahmad

Darwish turned his "Azhar Mosque of the Internet" into the "Official Sammania Site." Ahmad had been in contact with the Sammānī *shaykh* Ḥasan al-Fātiḥ before (cf. the latter's praise for Ahmad's Seerah of December 1996). It is unclear how this contact first came about—perhaps during a visit by Ḥasan al-Fātiḥ to the United States in 1993. In September 1998, however, Ḥasan al-Fātiḥ attended the "2nd International Islamic Unity Conference" in Washington, DC, which was organized by the traditionalist "Islamic Supreme Council of America" (dominated by Hishām Ḳabbānī's Naqshbandiyya-Ḥaqqāniyya). Most likely it was there that Ahmad Darwish introduced Ḥasan al-Fātiḥ to his internet site and received permission from the *shaykh* to turn this Mosque into the "Official Sammania Site."

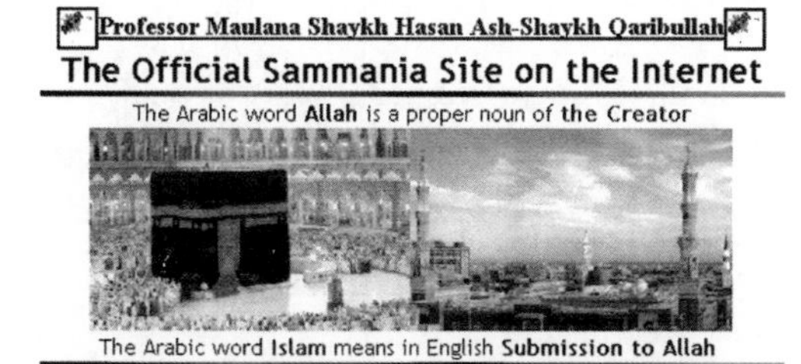

Figure 4: Mosque.com, December 1998

Not much else changed at first. An email address for Shaykh Ḥasan was added, and in addition to his being "Chancellor of Um Durman Islamic University," he now became "Chancellor of The Sammania Mosque of the Internet."[17] The Life of the Prophet, which previously had been advertised as a work by Ahmad and his wife, was now attributed to Ahmad's *shaykhs*, Ḥasan Qarīballah and ʿAbdallāh al-Ghumārī as well, as follows:

> The Millenium Biography of Muhammad, by Grand Shaykh, Professor Hasan Qaribullah, Dean of Umm Durman Islamic University and Sammania Grand Shaykh; Grand Muhaddith Master Abdullah Ben Sadek; Shaykha Anne Khadijah Darwish; Shaykh

Ahmad Darwish (Shaykh Qaribulla USA Personal Secretary, The Founder of the Mosque of the Internet); Reviewed in part by Former manager of Muhammad Ali.[18]

Last not least, a lengthy introduction to the "Sammaniyyah Heritage Path" was published that began in quite 'traditional' style:

- Spiritual ancestry (Qadiria, Khalwatia, Naqshbandia, T. al-Anfas, T. al-Muwafaqa)
- Prerequisites and principles
- Base [*al-asās*]
- Four Cornerstones (reducing food, talk, sleep, socializing)
- Awrad
- Zikr
- Chain of Affiliation (Silsilah / Sanad)
- Symbol (waist wrap)
- Initiation Pledge [*al-bay a*]
- Seasons [annual & weekly rituals]

It ended, however, with a detailed "socio-scientific, educational and academic program" that reads like operationalized "objectives" in places and that mentions "task committees" responsible for specific tasks—a form of presentation and organization that cast traditional elements of Sufi education into a model clearly derived from modern academic institutions:

- Socio-scientific, educational and academic program
- Lectures for brothers, youth, sisters
- Academic tutorials
- Social activities (visits, zakat, marriages, ...)
- Training activities (da'wa, practice of the 4 cornerstones)
- Spiritual activities (dhikr, ziyara, rabita)
- Islamic Sciences (publication & outreach)
- Construction (mosques, secondary schools, ...)
- Medical services (Koranic healing; free med. serv. during Ramadan)
- The ultimate goal: from loving Allah ... to being loved by Him
- At-Tariqah as-Sammaniyyah Task Committees[19]

It remains unclear, however, what impact the Sammania page ever had. A year after its start, by October 1999, it was obvious that Shaykh Hasan did not use his email address, and although two distinct buttons for him and for "Shaykh

Ahmad" remained on the site for a year, both pointed to Ahmad's address only (which for a while was guru@hypersurf.com). In August 2000, the "Shaykh Ahmad" button was removed and only "Shaykh Hasan" remained—with an email address linking to Ahmad. After the site allegedly was hacked in late 2001, the "Shaykh Hasan" link was never restored. But a letter from one convert was published in June 2002, seeking clarification on a matter pertaining to the daily office (*awrād*), and stating that "I wish to proceed to the Divine Face in a short time" and "I plan to go to the Shaykh within the next month as I have been looking for a true teacher on the path for a long time."

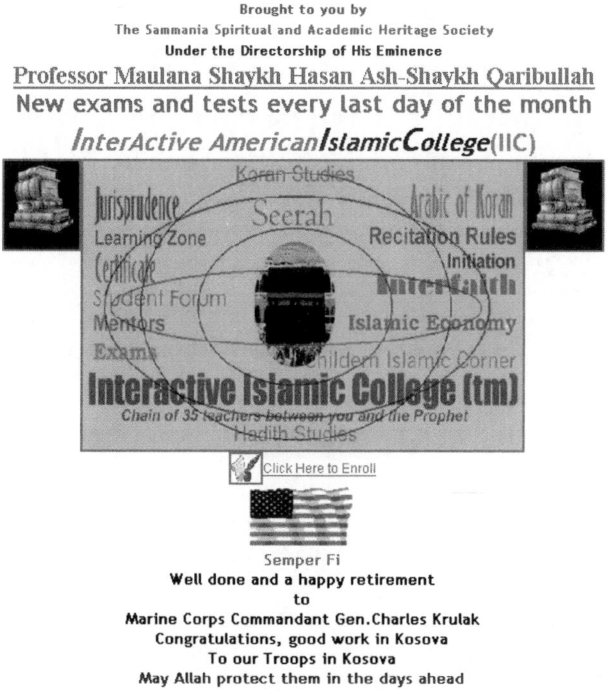

Figure 5: Mosque.com, Oct. 1999

SEMPER FI: AMERICANIZING THE MOSQUE (1999)

Again, however, Ahmad felt the need to react to outside developments. A radical overhaul of the site was implemented in late March/early April 1999, when NATO attacked Serbia over Kosovo. The U.S. flag came up flying on the home page, which now was styled "The Free Islamic University." Allah.com and Muhammad.com (registered years ago) were now prominently promoted alongside Mosque.com, and Shaykh Ḥasan became director of the "Sammania Spiritual and Academic Heritage Club." The designation "Official Sammania Site on the Internet" was dropped, on the other hand. Further down on the site, cash donations were solicited for "your Muslim Brethren in Kosova." The "Interactive Islamic College" turned into the "InterActive American Islamic College," and by October 5, 1999, Ahmad had added the motto of the U.S. Marine Corps to the U.S. flag waving on his site: "Semper fi," in honour of their "good work in Kosovo."

OPEN SOURCE? HARVESTING MUHADDITH.ORG— TO SURPASS HARVARD! (1999)

By Oct. 1999—a year after the Sammania started to 'sponsor' the site—the material offered by the IIC began to be upgraded. This was made possible "thanks to Linuxvision of the Cayman Islands," who by April 1999 had started to be featured on the site—Ahmad Darwish's company that produced the Sheba Arabization package for Linux based servers. Substantial numbers of Arabic texts were made available—and most of these were taken from Al Muhaddith, one of the earliest examples of online Islamic libraries that began in 1988 and opened its website (muhaddith.org) on November 30, 1998, offering "Search software, Islamic books and prayer times program for free download by all Muslims with online search. 119+ classic Islamic books, 300 MB+: Holy Quran, Hadith, Fiqh, Dictionary and other Islam books." A few months later, texts from Al Muhaddith started to appear on mosque.com, and subsequently, Ahmad Darwish's eulogy of his own site reached new heights:

> Starting 1/1/2000 the Mosque.com [the first Islamic site on the web ever] has surpassed Harvard and Oxford in the Islamic Scholarly Leadership. We register 10 Islamic Studies Students a day added to our daily over 1500 visitors actively in study, soon we will challenge the students of both Harvard and Oxford on-line. Stay Tuned.
>
> Mosque's World Largest & Free Islamic and Arabic Collection: 85 References (some made of 20 volumes) Containing 2000,000 lines or 150,000 WebScreen-Pages[TM]. Enjoy, and give credit to al-Muhaddith and Sheba Arabic Systems of Linuxvision.com of Cayman Islands.[20]

Respectability for the enterprise was sought by the Sammānī Shaykh "[o]ffering an Accredited Ijaza and an optional Accredited MBA. You will receive a traditional sealed Ijaza In Islamic Studies signed with the seal of Prof. Shaykh Hasan Qaribullah."[21] The "accredited MBA" linked to the (commercial) Heriot-Watt University MBA, of which mosque.com became a distributor for a few months (by May 2000, this reference had been removed).

Registered visitors (registration was free) gained access to an impressive collection of both English and Arabic texts that, as explained, was taken from muhaddith.org—a source which, however, was only vaguely acknowledged. Where muhaddith.org works through a dedicated reading application that users need to download, the idea behind the IIC was that registered users study the texts online. Visitors were urged to read continuously and attentively—if one clicked the "Next" button too quickly, the following message popped up: "Please do not keep clicking, read in full then click! Please take your time reading, otherwise it will cost us bandwidth, we are sure you understand. Please go back and complete the reading of this WebPageScreen [tm]—Thanks." Alternatively, however, a download option was still offered for offline html browsing.

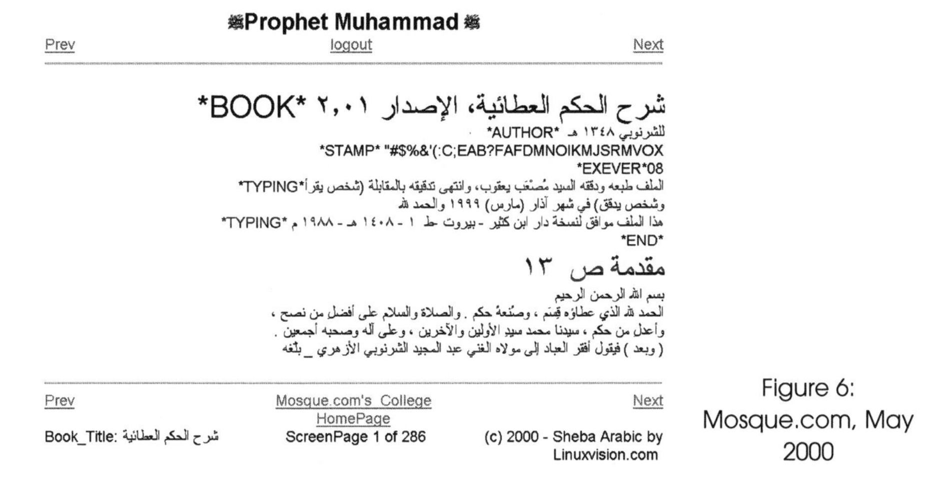

Figure 6: Mosque.com, May 2000

REACHING OUT TO SPREAD 'WHAT EVERY MUSLIM NEEDS'

Beyond the classical texts harvested from Al Muhaddith, Ahmad worked to expand his downloadable library of "all the documents that every Muslim needs":

1. Islamic Brief (islam.doc a.k.a What is Islam?) **What everyone should know about Islam. Get it in Spanish Que es Islam? Se Hable Espanole?**

2. Holy Koran (koran.doc) by the mosque.com's founders Grand Shaykh Hassan and Shaykh Ahmad—in a word document format
    A. **The best Koran program on the Planet** (koran.zip) with all world languages by yildun.com
    B. **TOPICS OF THE HOLY KORAN IN A NUTSHELL (KoranGlossary.doc)**
3. Prophetic Quotes (hadith.doc) 171 subjects (1950 Prophetic Quotes.)
4. Life of Prophet Muhammad (seerah.doc) "Prophetic Bibliography: IN LOVE OF PROPHET MUHAMMAD" (first time in English)
5. Principles of Faith (faith.doc) by Algazel (al-Ghazali)
6. Civil Islamic Law (fiqh.doc) Coming Soon!
7. Praise Prophet Muhammad (dalail.doc) Burdah Poem and Dalail: Seven parts, one per day, a must have for peronal spiritual growth.
8. Daily Contemplation (owrad.doc) of Sammania Qadiria—Coming Soon!
9. Reviving the Science of Religion (ihya.doc) by Algazel (al-Ghazali) Coming Soon!
10. Sex and Reproduction (halaljoy.doc)[22]

In the same spirit of reaching out and engaging "every Muslim," he attempted to initiate *daʿwa* (the call to Islam) by email:

> Enjoy and let others know about the college (...) Please share in the reward of spreading Islamic knowlege by telling your local Imam and the Islamic school near you about the Mosque of the Internet's InterActive Islamic College. Be as creative as you can, the more you guide people the greater the reward. Here is some ideas: Make a flyer, send email to the friends and local community members. Write to the teachers of local Islamic and American schools. Tell Sunday school teachers. Write local news papers and journals.[23]

Internet technology was to help every believer to familiarize him/herself with the basic texts and to take responsibility for spreading this knowledge and belief to as many people as possible. The net was a tool to empower believers to become agents of 'basic Islam,' made accessible through "the best Koran program on the Planet" and summarized in short "Islamic Briefs."

The education programme was not immune, however, from the vagaries of the dot-com economy. The IIC had been Linuxvision's first and major project by which the company hoped to make a name for itself in the Arabization market. When that failed, and after the 2001 site hack/hard disk crash, the IIC was much less vigorously pursued and vanished from the web sometime between August and December 2001.

"WAHABI TERROR"—WE HAVE KNOWN IT SINCE THE NINETEENTH CENTURY!

Another prominent feature of the site had long been its anti-Wahhābism, in the form of polemics against what it presented as the Saʿūdī/Najdī "Taimia cult" (a derogatory reference to the paramount significance for contemporary Wahhābism and Salafism of the purist scholar Ibn Taymiyya [1263–1328]):

> Can Saudi Prince & Family, Sultan of Brunai and all Muslim Millionaires and Billionaires in the west or the east spare another 2.5% annually of private wealth for Allah and His Prophet to put Islam right on the map as the Prophet's friends did? Also Can the 19 children of Ibn-AbdulWahab spare 2.5% annually from their $28 Billion dollars trust? Hay, sometime the market goes down and you loose 2.5% anyway.
>
> You know what with Allah will never go down![24]

Ṣaudis do not give money for true Islam—instead, they are trying to steal money from credulous Muslims, fraudulently using good names such as that of Shaykh Ḥasan:

> WARNING WARNING WARNING WARNING WARNING WARNING The Saudis: Dr. Ali Naseef, Muhammad Al-Ghamdi and Tayyib of Canada are running unauthorized and fraudulent activities stealing $100s of dollars from Sufi sites throughout the world. They are very creative sometime they pose as they are from Mecca helping Mosques and mention good clean names like Prof Hasan al-Fateh etc or sell you domain name or banner on Allah.com. They ask people to transfer money to them via Western Union etc. If they contact you please call your local police immediately to trap them, or Email the FBI Cyber Squad. sandiego@fbi.gov Or call FBI Federal Bureau of Investigation J. Edgar Hoover Building 935 Pennsylvania Avenue, N.W. Washington, D.C. 20535-0001 (202) 324-3000. **Allah.com, Muhammad.com, and Mosque.com neither request nor accept any money**[25]

It is thus no wonder that the Wahhābīs turned out to be terrorists. "The World Trade Center bombings [of 1993] were not the first of the Wahabi Terror"—Wahhābī terror had already been condemned by Sunnī scholars in the nineteenth century:

> At the beginning of the Arabian kingdom, thousands of Muslims were massacred by the deviant Wahabis, and the scholars of Ahlesuna (The people who follow the Prophetic ways) wrote their opposing response in their books. The Mufti of Mecca wrote, "...they marched with big armies...lay siege the Muslim area...killed the people, men, women, and children. They also looted the Muslims belongings. Only a few escaped their barbar-

ianism." Ref.: Mufti of Mecca, Ahmad Zayni Dahlan al-Makki ash-Shafi'i in his work "Fitnatu-l-Wahhabiyyah."[26]

9/11—A CHANCE FOR TRADITIONAL ISLAM?

After the attacks of September 11, 2001, Ahmad Darwish added new texts to his site to emphasize that Islam is beautiful, and not about terror, that "Americans are NOT anti-Islam [... and] that Muslims are NOT all in support of Bin Laden," that the WTC victims are in Paradise, and that it was time to act: that "all Wahabi schools (that teach fanatical anti-Islamic actions)" should be converted "into traditional Islamic schools (that teach love and compassion and tolerance)." More concretely: "We recommend that the US investigate the $18 billion in American Banks belonging to a Wahabi Fund, controlled by Wahabi's 19 children (Dallas, TX), and invest it in True Islamic schools."[27]

True Islam was "traditional Islam"; it was about love, tolerance, learning, and spiritual development. This was the message of traditional shaykhs. And true to this spirit, "Professor Hasan El Fateh Qaribullah," "spiritual coach" to Mosque.com and "spiritual head of Sammania Spiritual Heritage Society with millions of followers worldwide, some of whom are dignitaries, USA ambassadors and USA army personnel," condemned the WTC attacks and blamed them on the Wahhābīs.

At a time when U.S. authorities were obsessed with scrutinizing the financial affairs of Muslim institutions, Ahmad Darwish decided to make his Koran translation (now presented outright as undertaken by Shaykh Ḥasan!) available for free:

> It is in this spirit [of love] that Professor Hasan is freely giving away, this first clear translation of the meaning of the Koran to visitors of the above mentioned sites. He neither accepts nor asks financial support and his spiritual order has no bank account.[28]

This was in clear contrast to 1997, when (on June 26) Mosque.com did ask for help: "If you and your family can help to support the Mosque & College will be greatly rewarded and appreciated. (...) Please mail checks to: (...) Fremont."

But although Ahmad Darwish took every opportunity to emphasize that "Prophet Muhammd ain't one of them!"[29] he agreed with widespread feelings in the Muslim world when he spoke out against the U.S. attack on Afghanistan and against U.S. troops in Saudi Arabia:

there is an Islamic tradition served in Koran and Prophetic quotations, that non Muslims should not be in Arabia, Now over 20,000 American troops (some with civil duties) are in Arabia, this is again a very dangerous situation, since fanatic Wahabis will use this to their own advantage, since they pick and choose Islamic traditions.

On the other hand the masses of middle eastern Islamic countries feel that there is a conspiracy in the west to suppress Islamic societies. This is easily understood when one reviews the news of Bosnia, Chechnia in former USSR, and Kashmire where the west allowed a holocaust to continue.

Where is the sound advice of Harvard for the US Administration? I bet you they do not have a clue !!![30]

Figure 7: Mosque.com, January 2003

Overall, however, the site maintained a positive spirit in the hope that 9/11 could be a chance for true, traditional Islam: "America said, 'Bye bye!' to Ben Laden. America says, 'Hello Prophet Muhammad of Allah'." Ahmad Darwish hoped that the alarm over the Saʿūdī involvement in the terror attacks would lead to a backlash against his old enemies, the Wahhābīs: "If you like the wahabies click here—If you dislike the wahabies click here," read two buttons on his site. Clicking on the first one led to a tirade equating the Wahhābīs to Satan:

> **Preaching Islam is the first right due to all people in the west, not blowing them away falsely in the name of Jihad. Duh!** Wahabi: Their Wahabi (Najdi) satan attempting to Kill Prophet Muhammad, then killed many many Muslims, now killing innocent christians besides Muslims who were in WTC

> Not to mention that the Afgani people have suffered a lot because of this war financially and civilian casualties, and that we pray for them all
>
> What will happen if Wahabi, Brotherhood (Ikhwan) and Ben-Ladenism go away?
>
> The best thing will happen, is setting Islam free from these Hijackers of Islam! Where the west will welcome Prophet Muhammad and Islam.
>
> Remember Wahabi started their system by killing many Muslims to take over Arabia, Now they started Killing innocent Christians (WTC). So EVIL DOING is their nature. satan loves najd, Saudi Arabia, the Wahabi head-office, all us know when satan appeared advising the unbelievers of Mecca to Kill prophet Muhammad, in Najd look and feel image)[31]

Clicking on the second button ('I hate the Wahhābīs') led to the solution: opening the way of the "good traditional Shaykhs:"

> We have reached a cross road, either promote pure Islam or pay heavily for a fanatic takeover. Remember fanatics do not co exist with pure Islam.
>
> I mean by promoting pure Islam, the media should open themselves for good traditional Shaykhs—it is strange that "Good Morning America" continue to welcome a priest and a rabbi to chat without a shaykh, while the number of Jews and Muslims in the US is equal![32]

THE MOSQUE SOLD?

In the beginning, Allah.com, Muhammad.com, and Mosque.com had more or less mirrored each other. Later, Mosque.com became the main site; Allah.com was used to showcase the Koran translation, and Muhammad.com the Prophet's biography. By 2003, Ahmad Darwish began to think of a reorganization of his sites.[33]

Then, in summer 2004, a disclaimer came up on mosque.com reading: "This site has nothing to do with www.Allah.com and www.Muhammad.com per sale's agreement." The person having bought it was presented as "Sheik AbdelRasool HANAFI, Lagos, Nigeria," and the sole purpose of the site was now to "be the most authority listing 1911 errros [*sic*] of wahabi in faith, fiqh and manners (akhlaq calling bad names to Imam Azam Abu Hanifa) and money, oil zakat fraud," and in particular "to reveal the truth about the Wahabi group who killed slaughtered my great great family in early 1800 who were living in Mecca and Medina during the siege by Wahhabis." In fact, however, the domain mosque.com continued to be registered until late 2007 in the name of Linuxvision, Sharjah, and has since been transferred to Ahmad Darwish. Even during the "Nigerian" interlude of mid-2004 until July 2006, the language of the site closely resembled that of Ahmad Darwish's other enterprises, so one is left to wonder about the significance of this "sale."

ALLAH.COM VOTES KERRY

Following his celebration of the United States for their support for the Kosovo Muslims and his declaration of national solidarity after the terrorist attacks of 9/11, we have seen how Ahmad Darwish opposed the U.S. invasion of Afghanistan. When U.S. and British troops invaded Iraq in March 2003, however, he remained conspicuously silent. He neither commented on the event, nor removed the button saying: "Please listen to this American anthem that turned into the whole world anthem" (which linked to http://www.paperveins.org/anthem, Virgil Wong's version of the U.S. national anthem celebrating global solidarity with the 9/11 victims). When President Bush stood for re-election a year later, however, Ahmad Darwish warned his fellow Muslims that

> If you are a Muslim and USA citizen who does not vote or votes for Bush on November 2nd Allah will ask YOU for each drop of all the blood of Muslims 330,000 children shed in Bush's wars for Oil in both Afghanistan, Iraq and Palistaine and more to come. All Ulama of Pakistan, Malaysia and Azhar passed such verdict and Shaykh Ahmad Darwish of this Site confirms it. Remember your punishment will start in your tomb and Allah will bankrupt you in both your green dollar and in your Iman in this life. The Prophet will be quit of you here and hereafter and you receive no shafaat or mercy. Amen[34]

When Bush had won, all political overtones disappeared from the site, which refocused on offering the major texts that always had been at its core: the Koran translation, the biography of the Prophet, the translation ("in progress") of Prophetic Traditions. Together with his wife and daughter, Ahmad also began a new translation of parts of Ghazālī's *Revival of the Religious Sciences*.

CUSTODIAN OF PROPHETIC TRADITION ON THE INTERNET

In autumn 2005, Ahmad's sites suffered a technical problem. In October, Allah.com temporarily linked to other sites, and in November, Ahmad began to reconstruct the site, apparently from earlier backups. This reconstruction (which eventually also was reflected on Muhammad.com, and in the course of which mosque.com returned to Ahmad's trinity of sites) took some time; it was not completed until January 2006. The site's temporary instability may not be unrelated to the fact that around this time, Ahmad left the United States and his Linux business there and returned to Cairo. Between August and October 2006, mosque.com and Muhammad.com redirected to alnabee.com, a minor Saudi site promoting the veneration of Prophet Muhammad, recommended by al-Azhar (Mawāqiʿ mufida, 2008). Around November 2006 they reverted to mirroring Allah.com. Not much else changed until May 2007, when a design update was implemented on all three sites and Ahmad built a separate Arabic site.[35] Simultaneously, the site's host was temporarily moved to the Moroccan genious.net. By November 2007, Ahmad had integrated both Arabic and English into "the first and largest sites on Iman, Islam, Ihsan," "Allah.com from Chicago, the land of Muhammad Ali, Da Bulls, Mayor Richard Daley, who votes against any war. Take a virtual tour of Chicago."[36] At the same time, Ahmad also began to experiment with new formats. On Oct. 28, 2007, he uploaded two short 'video' lectures (which besides the audio only contain a few still images) to Google Video. This was followed on March 28, 2008, by a proper video showing al-Ghumārī together with Ahmad Darwish.[37] Apart from these novel formats, he continued to offer and expand on translations (not least, works by Ghazālī), and very clearly positioned himself against the Wahhābī followers of "Ibn Taymia's Monkey Business in Faith": "700 years after the Prophet a man came to change all the faith of Islam and start spiritual and metarial [*sic*] terrorism!"[38] He lined himself up with like-minded defenders of a traditionalist understanding of Islam based on the four established Sunni schools of law (*madhhabs*), of Prophetic piety, and others who oppose what they regard as militant and extremist deviation from the true faith: Ḥasanayn Makhlūf (Egyptian state jurisconsultant [*muftī*] in the 1940s and 1950s), Muḥammad Mitwallī al-Sharʿāwī (1911–1998), Egypt's most famous twentieth-century TV preacher, ʿAbd al-Ḥalīm Maḥmūd (1910–1978, *shaykh al-Azhar* 1973–1978), Aḥmad al-ʿAlawī (1869–1934), Sufi *shaykh* and teacher of several European Traditionalists, René Guénon (1886–1951), the father of contemporary Traditionalism, ʿAbd Allāh al-Ghumārī (*v.s.*), Muḥammad al-Fātiḥ Qarīballāh (1915–1986), Sammānī *shaykh* and father of Ḥasan al-Fātiḥ, Hishām Ḳabbānī (b. 1945), Naqshbandī *shaykh*, particularly active as missionary in the West, here portrayed together with Prince Charles, al-

Ḥabīb ʿAlī al-Jifrī (b. 1971), traditionalist Muslim preacher (see http://www.alhabibali.org), ʿAmr Khālid (b. 1967), contemporary Islam's most influential televangelist (see http://www.amrkhaled.net), and Ḥamza Yūsuf [Hanson] (b. 1960), well-known traditionalist Muslim missionary from California (see http://www.zaytuna.org). Very openly he sought to find a new home in a "traditional" Sunni environment as he announced his hope, to the traditionalist Dār al-Muṣṭafā (http://www.daralmustafa.org) in Yemen, "to move Allah.com administration to your office and to join you. He is waiting to hear from you!"[39]

Figure 8: Screenshot from Allah.com, March 4, 2009

Figure 9: Screenshot from Muhammad.com, March 4, 2009

Sometime in mid-2008, Ahmad Darwish implemented another change in the layout of his site, announced since mid-January 2008, with the aim of making it easier to navigate his "Hadith Data Warehouse."[40] The design, which had become very bloated, reverted to a more streamlined format similar to the one of early 2007 and presented "Shaykh Ahmad Darwish" as the faithful servant (*khādim*) of the two master keepers of proper Islamic tradition, al-Ghumārī (for Prophetic Tradition) and al-Shaʿrāwī (for the Koran).[41] Framed by these two, and visually cast in their mold, he was now founder and "full time research scholar" of *Dār al-Ḥadīth al-Sharīf ʿalā 'l-Internet* ("The House of Prophetic Tradition on the Internet"). His goal there, he proclaimed, was to establish, with computer aid, no less than the definitive text of Prophetic Tradition, in analogy to the establishment of the canonical text of the Koran undertaken under the authority of the third caliph, ʿUthmān (r. 644–656) ("*taghlīq jamʿ al-Ḥadīth al-Sharīf mithla jamʿ al-Qurʾān al-Karīm*")—a goal that he set after looking at the shortcomings of The scholarly Saudi Society for the Prophetic Sunnah (http://www.sunnah.org.sa), a high-profile official institution that became operative in 2005. "Praise be to God who graced me with designing a 'live' effort that is better than the effort of the team of [*Shaykh* Māhir] al-Faḥl, all by myself. This is *Majmaʿ al-aḥādīth*, a work that encompasses 50000 man years of the work of scholars in one vessel, with the help of only two assistants [… The method I followed is one that God] inspired me with (*alhamanīhi* [*sic*]) and that no book or person pointed me to" (Darwīsh, n.d.).

CONCLUSION: THE INDIVIDUAL ON THE "HIGHWAY FOR KNOWLEDGE"

From degreeless software engineer to scholar-*shaykh* and "custodian of Prophetic Tradition," Ahmad Darwish set out to complete the work of the first caliphs by establishing the definitive canon of Islam. Fantastic as this story may sound, it is nevertheless not untypical. The challenge posed to the hegemony of traditional Muslim scholarly elites by people with an educational background in secular fields (medicine, engineering, law, literary studies, accounting, etc.) has been a dominant theme in twentieth-century Islam; many of the most prominent names in contemporary Islamic thought and action exemplify this challenge. Ahmad Darwish is far from being prominent; even his high-profile web address has not helped to secure him a significant audience.[42] But he is typical of a larger trend: that the individual believer—in principle, *every* individual believer—assumes the right and takes responsibility for understanding what Islam means and for spreading this understanding in public. To interpret, in public, what Islam means, to talk in the name of Islam, is no longer the prerogative of scholars who have had a specialized education, or of Sufi *shaykhs* who have been initiated by their masters. It is the right—and some would say, the duty—of every believer. This threatens the power to define Islam that religious specialists have long held: 'I don't need a *shaykh* when I can have direct recourse myself to the basic sources, to Koran and Prophetic Tradition.'[43] At heart, this is a development that advances the emancipation of the individual from traditional authorities.

The internet is an ideal platform for spreading and facilitating such an individualized approach to interpreting Islam, and software guru Ahmad Darwish has indefatigably used it, like other technologies before, to present his version of the faith. He constructed "the First Site of Prophet Muhammad on the Internet," *the* "Mosque of the Internet," "The Free Islamic University," to make available "all the documents that every Muslim needs," the "Largest Give Away Free Islamic Authentic References" (first and foremost, his own Koran translation—with subtitles to facilitate understanding!—and his "Millenium Biography of Muhammad"), and thus spread "The Leading Islamic Revival Plan." This particular case may reflect an inflated self-confidence, but it merely magnifies the growing importance of the role the individual assumes in picking and choosing and remixing and interpreting for him- or herself what Islam means.

Of course, we must not ignore that Ahmad set himself up in the garb and as the successor of Islamic scholars (*ʿulamāʾ*), as the one who serves and completes Tradition. But his own words (in the preface to his *al-Taʿqīb al-matīn ʿalā "Manāhij al-muḥaddithīn"*) betray that behind the ostensible defence of tradition there lingers a new distribution of roles. The scholar—that is now *me*: "Praise be to God—I was the first to establish Islamic sites on the Internet (...) I was Shaykh of the Internet in the early eighties, all in order to serve our traditional heritage (*al-turāth*) [and on my sites can be found] most of what God has made to come about through my hands and those of my wife."

Defining 'myself' as the foremost scholar signals a dialectic whereby traditional learning is not merely reconfirmed as a model but is consciously reconstructed by the individual. Allah.com illustrates this reconstruction of tradition, illustrates how an individual uses new technology to make sense, by way of bricolage, of the world and of his religion, reducing and simplifying the canon for the sake of facilitating access to what "every Muslim needs," and breaking down hierarchies in the process.

The process whereby such individualized reconstruction of meaning displaces the traditional self-evidence of a world-view defined by scholarly experts may be regarded as an important component of modernity. The internet facilitates this process, but it has not caused it. The website Allah.com is embedded in a longer historical dynamic where Muslims at the social periphery of the scholarly establishment questioned the epistemological basis of scholastic tradition and employed a variety of means including the new media of the time to make every individual responsible for understanding a simplified body of knowledge and for implementing it in practice (Hofheinz, 1993). Website history, at bottom, does not begin with the web.

NOTES

1. Allah.com, Nov. 2001 vs. May 2005. Until Aug. 2004, the site called itself "The First Mosque of the Internet—Since 1986." As for other domains based on the name "Allah," "Allah.org" was registered March 19, 1996 by islamicity.com (the oldest and still one of the biggest Islamic portal sites); it contains an interface to RealAudio and text versions of the "Names of Allah" but no other content. "Allah.net" was registered on 21 Nov. 1996 by Bhavesh Sutaria, a Hindu or Jain from the U.K., possibly out of commercial interest; it has never been in active use.
   "Allah.info" was registered in 2002 to refer to "Quran.org," a minor English-language Islamic site based in New Hampshire that is online since 1996. A few country-specific "Allah" domains exist but only outside the Arab world, and none of them is prominent: e.g., allah.ir (a general Persian portal), allah.pk (a Flash Koran), allah.tk (a Pakistani personal site not updated since the 1990s), or allah.jp (the Islamic Circle of Japan).
2. http://web.archive.org/web/20041229171046/http://www.Allah.com/.JesusMuhammad.com, Ghumari.org, and Qadiria.com were maintained by the owner of Allah.com in 2004–05 but lapsed in 2006. The "Free-For-Ever Matrimony Club of Muhammad.com" section (http://66.240.115.245/ad101/Muhammad/marry.html) in mid-2005 contained nine female personal ads lifted from matrimony.org ("Muslim Marriage Link") around Sep. 2004—ads that had since been deleted from the original site. The Matrimony Club never went beyond the "Under construction" stage.
3. http://web.archive.org/web/20020524034244/http://www.Allah.com/.
4. On May 5, 1999, Ahmad Darwish registered the domain grandcaymanbank.com, for a planned *ḥalāl* offshore bank. The project never materialized, and by Dec. 2007, Ahmad Darwish had changed the registration to "US law firm Sidley Austin for Former 3 times WHW. Boxer Muhammad Ali Hon. Herburt [sic] M," with himself as contact person. For a while, Linuxvision also held the domain 1iso.com ("global Muslims and guests web presence covering all Muslims (and the world) languages." This pointed to ishebaplanet.com in 2002, but lapsed shortly afterwards. In spring 2000—just before the New Economy bubble burst—Ahmad Darwish attempted to solicit Muslim software developers to increase his Internet business: "Remember, there are even lots of free software that can help us to outperform them all. If we do not do it, Muslims will be lead for ever and pay 'others' their money, having no control of morality or children safety. IF YOU ARE INTERESTED PLEASE CONTACT ME, IT WILL BE A FAIR EXCHANGE LEADING TO THE LARGEST MUSLIM GLOBAL PUBLIC OFFERING COMPANY IF ALLAH WILLS INSHAALLAH" (ldapguru@yahoo.com, email communication to those registered on mosque.com, May 7, 2000).
5. iShebaPlanet was named after iPlanet (an erstwhile E-commerce solutions platform by Sun) and the "Sheba Linux Arabization Server Package."

6. "Java Six Sigma Server," for "analyzing and controlling organizational quality and performance"; "Java Candlestick Server," a "tool in interpreting trends in share prices, invented by the Japanese some two hundred years ago"; and "Java Medical Server," for patient record management and medical office work (http://www.linuxvision.com, retrieved September 17, 2005). The coding for these applications was largely done through contracted third-party coders (*About Buyer Alan Darwish*, 2007).
7. http://Muhammad.com, retrieved Sep. 1, 2008.
8. The first instance I found of Ahmad Darwish referring to himself as "Shaykh Ahmad" dates from 1998/99, then as a clear subordinate to "Shaykh Hasan" (see p.117). But only after what appears to be the definitive end of his software career and his move to Cairo, he consistently presents himself as a Shaykh, first as "Ahmad Darwish (al-)Shaykh" (used after the reconstruction of his site in January 2006), then from mid-2006 with increasing prominence as "Shaykh Ahmad Darwish."
9. Ahmad Darwish acknowledged the importance of this influence when on July 22, 2004, he established the website Ghumari.org (which lapsed, however, shortly afterwards). Already in October 2003, he appears to have set up the Yahoo! Group *Al-Ghumari • Zawiyyah Siddiqia* (http://uk.groups.yahoo.com/group/Al-Ghumari/). For information on al-Ghumārī, see his works in the reference list, as well as Tamām (n.d.); Mursī (1968); Sadek (n.d.).
10. Apparently based on Abdallāh b. al-Ṣiddīq al-Ghumārī, *Bāb al-taysīr fī radd iʿtibār "al-Jāmiʿ al-Ṣaghīr"*—and thus ultimately on the Egyptian polymath al-Suyūṭī's (1445–1505) summary of Prophetic Traditions for a larger public. In Chicago in 1981, Ahmad received a formal authorization (*ijāza*) from al-Ghumārī to transmit the famous *ḥadīth* scholar and revivalist Muḥammad al-Shawkānī's (1759–1834) list of works he had studied, *Itḥāf al-akābir bi-isnād al-dafātir* (reproduced at http://www.Allah.com/pdf/42VirtuesOfTheProphet.pdf, retrieved March 4, 2009).
11. Interactive Islamic College, June 1, 1997. For a sense of the whole, here is the 1997 version of his translation of the first Surah: "In the Name of Merciful, the Most Merciful Allah, // The praise is for Allah , // the Lord of the worlds. // The Merciful, the Most Merciful. // The Owner of the Day of Repayment. // You we worship and You we rely upon for help. // Guide us to the Straight Path. // The Path of those You have favored, // not those upon whom is the wrath, // nor those who are astray." The translation was subsequently amended, last in 2001 (http://salicehajic.startlogic.com/cgi-bin/mt.cgi?lang=en&cfile=KoranIn Subjects, retrieved March 2, 2009).
12. "Interactive Islamic College," June 1, 1997. Reference is to two of the most well-known twentieth-century Muslim biographies of the Prophet.
13. http://web.archive.org/web/19971221131210/http://mosque.com/.
14. On June 1, 1997, the following texts were offered, in addition to "selections" from IIC: "What Every Muslim in the West Should Know / What Everyone, Should Know About Islam "What is Islam" / Click for your local approximate prayer time +10/-10 mins / Celebrated Azharian said: "Shakespeare said 'BLOOD is no argument'" / Cool Islamic References—refuting the anti Ahle-Sunat wa al-Jamaat. / She said when her marriage broke up: "The only good thing was Islam" / New Trends: selections from American Scientists who embraced Is-

lam / Prophet Muhammad's Birthday Celebration. / Muhammad Yunus, ABC's Person of the Week / Need a Prayer or to pray?", http://web.archive.org/web/19970601012129/http://www.mosque.com/)

15. "Interactive Islamic College," June 1, 1997.
16. http://web.archive.org/web/19980520053205/http://mosque.com/.
17. http://web.archive.org/web/19990128174557/www.mosque.com/sammania.html.
18. http://web.archive.org/web/20021128053622/http://Muhammad.com/.
19. Both outlines are from http://web.archive.org/web/19990128174557/www.mosque.com/sam mania.html.
20. http://web.archive.org/web/20000301142242/http://mosque.com/.
21. http://web.archive.org/web/20000301142242/http://mosque.com/.
22. http://web.archive.org/web/20020528125431/http://www.mosque.com/.
23. Email sent to registered users by mosque@mosque.com, May 5, 2000. In another email (July 8, 2000) with the subject "Vote for the Prophet," we read: "Please go to this site and vote for the Holy Prophet (PBUH). 'http://www.msnbc.com/modules/Millennium_People/Mill P_ReligPhilos.asp' It is an NBC channel site. They are polling for the person who did the best for mankind."
24. http://web.archive.org/web/20000301142242/http://mosque.com/.
25. http://web.archive.org/web/20010802154719/http://www.mosque.com/index.html. The names given cannot easily be identified.
26. http://web.archive.org/web/20020330100947/12.109.24.76/index.html. Aḥmad Zaynī Daḥlān was a leading nineteenth-century Meccan scholar whose anti-Wahhābī stance has served as a reference for many twentieth-century opponents of the Wahhābiyya.
27. http://web.archive.org/web/20020528125431/http://www.mosque.com/.
28. http://web.archive.org/web/20011217234327/12.109.24.76/.
29. http://web.archive.org/web/20011217234327/12.109.24.76/.
30. http://web.archive.org/web/*/http://www.compuex.com/ad101/wahab2.html.
31. http://web.archive.org/web/*/http://www.compuex.com/ad101/wahab.html.
32. http://web.archive.org/web/*/http://www.compuex.com/ad101/wahab2.html.
33. First, a note was added to mosque.com saying: "The Mosque of the Internet: Mosque.com is soley for Non-Muslims to welcome them to Islam NOT to wahabi-colt. Please, if you are already a Muslim, go to the physical mosque near you. To learn ONLY from Allah and His Prophet Muhammad. Please visit www.Allah.com and www.Muhammad.com," http://web.archive.org/web/20030130074537/http://mosque.com/.
34. http://web.archive.org/web/20041020060725/http://www.Allah.com/.
35. First visible on archive.org in June 2007 in a subdirectory of Allah.com ("The First Global Prophetic Sayings DataWarehouse", http://web.archive.org/web/20070626100051/www.genious.net/~darwish/indexar.html), than directly under Muhammad.com ("1st Global Hadith Dataware House", http://web.archive.org/web/20070803005941/http://www.Muhammad. com/).
36. http://web.archive.org/web/20071118061052/http://www.Allah.com/.

37. http://video.google.com/videoplay?docid=2089838931886508940&ei=fPeuSZq4IoySiQK47diwDg&q=%22ahmad+darwish%22;http://video.google.com/videoplay?docid=7836303352863805471&q=source%3A005164779214297107404&hl=en; http://video.google.com/videoplay?docid=8196279095019931285&q=source%3A005164779214297107404&hl=en.

38. http://web.archive.org/web/20071118061052/http://www.Allah.com/. Cf. also his call, in Arabic on the same site, that the Saudi government should stop supporting "suicidal Wahhābī activities and fighting the proponents of the four law schools."

39. http://web.archive.org/web/20071221021306/http://www.Allah.com/. On Traditionalism, see Sedgwick, 2006. In January 2008, Ahmad Darwish expanded his list of men considered authoritative and included many links from Omar Klaus Neusser's Traditionalist Sufi website livingislam.org (establ. 2003).

40. http:// web.archive.org/web/20080127163408/Allah.com/newindex.html.

41. In mid-November 2007, Ahmad Darwish changed his self-description from *khādim al-mawqiʿ wa'l-ḥadīth al-sharīf* ("custodian of the site and of the noble Prophetic Tradition") to *khādim al-ḥāfiẓayn* ("custodian / servant of the two Keepers [of the Koran]"; http://web.archive.org/web/20071108205444/http://www.Allah.com/ vs. http://web.archive.org/web/20071118061052/http://www.Allah.com/). The designation *khādim al-ḥāfiẓayn*, which evokes the official title of the Saudi King, *Khādim al-Ḥaramayn* (Custodian of the Two Holy Places) was later changed to the somewhat less ambitious *khādim al-ḥuffāẓ* (servant of the Keepers) (http://Allah.com, retrieved March 4, 2009).

42. See *Allah.com—Traffic Details from Alexa*. (n.d.). Retrieved regularly over the years, last on March 7, 2009 from http://www.alexa.com/data/details/traffic_details/Allah.com. In August 2008, Shaykh Ahmad undertook his "1st UK tour." This venture into the off-line public caused both Salafis and rival Sufis to denounce him in public (*Shaykh Darwish 1st UK Tour*, 2008). This was related to Ahmad himself polemicizing against Muslims whom he accused of having sold their faith, naming not only several Sufi *shaykh*s prominent in the West but also lashing out at Muslims in the Arab world in general, in an August 2008 interview with an English adept that was leaked on YouTube (*Ahmad Darwish (Allah.com) on Islam & Muslims Part 1 of 6*, 2008). References to the tour were removed from Ahmad's websites after September 2008.

43. Cf. *Ahmad Darwish (Allah.com) on Islam & Muslims (FULL)* (2008), in particular around minute 37.

REFERENCES

About Buyer Alan Darwish. (2007, August 19). Retrieved March 7, 2009 from RentACoder: http://www.rentacoder.com/RentACoder/DotNet/SoftwareBuyers/ShowBuyerInfo.aspx?lngAuthorId=466205

Ahmad Darwish (Allah.com) on Islam & Muslims (FULL). (2008, August 13). Retrieved March 7, 2009 from Google Video: http://video.google.com/videoplay?docid=-6401053654340257536

Ahmad Darwish (Allah.com) on Islam & Muslims Part 1 of 6. (2008, August 12). Retrieved March 7, 2009 from YouTube: http://www.youtube.com/watch?v=2oeRerr3v3I

Ahmad Darwish's profile. (2005). Retrieved March 7, 2009 from Amazon.com: http://www.amazon.com/gp/pdp/profile/A1Q124NJYGQ5YK/ref=cm_cr_dp_pdp

Alan Darwish < Main < TWiki. (2004, May 7). Retrieved March 7, 2009 from TWiki: http://twiki.org/cgi-bin/view/Main/AlanDarwish

al-Ghumārī, ʿA. (n.d.). *Irghām al-mubtadiʾ al-ghabī bi-jawāz al-tawassul bi'l-Nabī.* Retrieved Feb. 27, 2009 from Soufia.org: http://soufia.org/vb/showthread.php?t=466

al-Ghumārī, ʿA. (n.d.). *Murshid al-ḥāʾir li-bayān waḍʿ ḥadīth Jābir.* Retrieved Feb. 27, 2009 from Association of Islamic Charitable Projects: http://www.aicp.org/SupportingDocs/MurshidAlHaairLibayan.pdf

al-Ghumārī, ʿA. (n.d.). *Rutab al-ḥifẓ ʿind al-muḥaddithīn.* Retrieved Feb. 27, 2009 from Mawqi ʿal-Imām Fakhr al-Dīn al-Rāzī: http://www.al-razi.net/website/pages/m13.htm

Darwīsh, A. (n.d.). *al-Taʿqīb al-matīn ʿalā "Manāhij al-muḥaddithīn".* Retrieved March 7, 2009 from Allah.com: http://www.Allah.com/pdf/MuhaddithMethod.pdf

Darwish, A. (1998, January 25). *New Translation and Concordance of the Koran.* Retrieved March 7, 2009 from Azhar Mosque of the Internet: http://web.archive.org/web/19980125045745/www.mosque.com/koran.html

Geertz, C. (1973). Thick description: Toward an interpretive theory of culture. In C. Geertz, *The interpretation of cultures: Selected essays* (pp. 3–30). New York: Basic Books.

Hashem, B. (1992, December). *Islamic Computing Resource Guide.* Retrieved March 2, 2009 from http://www.africa.upenn.edu/Software/Islamic_Computing_11765.html

Haykal, M. Ḥ. (1956). *Ḥayāt Muḥammad.* Cairo: Maktabat al-Nahḍa al-Miṣriyya.

Hofheinz, A. (1993). Der Scheich im Über-Ich, oder Haben Muslime ein Gewissen? Zum Prozeß der Verinnerlichung schriftislamischer Normen in Suakin im frühen 19. Jahrhundert. (S. Faath, & H. Mattes, Eds.) *Wuqûf, 7–8,* pp. 461–481.

Lings, M. (1983). *Muhammad: His life based on the earliest sources.* London: Islamic Texts Society / Allen & Unwin.

Mawāqiʿ mufīda. (2008). Retrieved March 7, 2009 from Mawqiʿ al-Azhar al-Taʿlīmī—al-Idāra al-ʿĀmma li'l-Computer al-Taʿlīmī: http://www.alazhar.gov.eg/Links.aspx

Mursī, A. b. (1388/1968). Taʿrīf bi'l-muḥaddith ʿAbdallāh b. al-Ṣiddīq. In A. b. Mursī, *al-Kanz al-Thamīn.*

Sadek, A. b. (n.d.). *Al Mahdi, Jesus and Moshaikh [the Anti-Christ].* Retrieved Feb. 27, 2009 from mosque.com: http://web.archive.org/web/20010205070900/www.mosque.com/mahdi.html

Sedgwick, M. (2006, July 21). *Traditionalists.org: A website for the study of Traditionalism and the Traditionalists.* Retrieved March 7, 2009 from http://www.traditionalists.org/

Shaykh Darwish 1st UK Tour. (2008, August 28). Retrieved March 7, 2009 from Marifah: Knowledge and Realisation: http://www.marifah.net/forums/index.php?showtopic=2971

Tamām, A. (n.d.). *ʿAbdallāh al-Ghumārī ... baqīyat al-salaf al-ṣāliḥ (fī dhikrā wafātihi: 19 min Shaʿbān 1413 h.* Retrieved Feb. 27, 2009 from IslamOnline: http://www.islamonline.net/Arabic/history/1422/11/article05.shtml

CHAPTER 5

Historicizing Webcam Culture: The Telefetish as Virtual Object

Ken Hillis

Consider the vast number of networked webcameras on the web: Webcam-Search.com and ChatCamCity.com, to cite only two of the million-plus Google hits on "webcam," provide links to thousands of cameras in dozens of categories, including animal cams, indoor cams, college cams, surveillance cams, UFO cams, volcano cams, and cams on all seven continents and in fifty U.S. states. Cams are deployed worldwide to watch and surveil traffic, public spaces, bank interiors, parking structures, shopping malls, the weather, nature views, major tourist attractions, and the unsuspecting baby sitter back home. Millions use them to make personal pc-to-pc phone calls and to "meet" others in video chat rooms organized by conversation topic and special interest—everything from alternative medicine and spirituality, to online therapy, poker playing, and for-pay "live sex chat." Increasingly, events are broadcast through webcams, including the political rallies of politicians competing in the 2008 U.S. presidential campaign, the Nobel Prize ceremonies in Stockholm, and Sunday morning services of churches across the planet. The sheer variety of webcam uses now on display suggests the need to speak of webcam *cultures*.

The networking of webcameras began with Cambridge University computer scientists Quentin Stafford-Fraser and Paul Jardetzky. In 1991, they wrote software that allowed them to watch, via a Local Area Network, the collective coffee pot located elsewhere in the building where they worked. The CoffeeCam was connected to the internet in 1993 as the first live, twenty-four hour webcam site (Stafford-Fraser, 1995). When taken down in 2001, it had amassed more than 2.4 million visitors.[1] Jennifer Kaye Ringley's JenniCam is possibly the best known twenty-four hour personal webcam site from the late 1990s. Ringley launched her cam in 1996 when she was an undergraduate at Dickinson College in Carlisle, Pennsylvania. At its peak, her site attracted an estimated four million hits per day and remained popular until Ringley closed it on New Year's Eve, 2003. Teresa Senft (2008) notes that Danni Ashe's less well-known porn-centric site predated Ringley's by two years. However, because Ringley's webcam mixed erotics with less freighted pleasures such as showing her brushing her teeth and working at her desk, her site and online practices are widely accepted as defining the personal webcam as a genre. Such sites do not specifically seek a sexually focused audience; most of the time they show the routine—a flash of activity as the operator heads out the door, an empty room, the eating of cereal at the table, time in front of the T.V., a sheet-draped body asleep in bed. Yet the personal nature of the transmissions and the sense of intimacy they provide can also infuse such activities with a highly charged sense of the erotic. Personal webcams as a genre, to paraphrase Amy Villarejo, defy expectations based on presumed categories: "Twenty-four-hour webcam sites...are neither wholly devoted to what we might call the visibly sexual...nor are they...devoid of connections to that domain or to sexual dissidence" (2004, p. 85).

I focus on webcam culture as it evolved from the late 1990s to early 2000s with specific reference to certain English-speaking, First World gay/queer men with the means of access. I do so in order to reveal the significance of these men's turn-of-the-century techniques and practices for the overall history of websites. These men were among the first to develop the online practices that have come to be known as "lifecasting"—the continual broadcast of a person's life through digital media. Because of advances in webcam technology since its early days, individuals who wish to live online can now easily join commercialized open platforms such as CamStreams.com, UStream.tv, and Justin.tv.[2] The image transmitted by the webcam displays on a viewer's screen, usually within a small box lodged within the larger space of the display. Much of the display contains text-based information punctuated by various images and jpegs. Commercialized sites today provide a customizable template that typically includes the cam operator's profile, schedule of appearances, special events, chat interests, and personal rules for chat that, if broken, result in a visitor's banishment. The number of hours spent online is a key factor in a lifecaster's popularity: 24/7 has the effect of proving one's bona fides. It is now relatively easy and inexpensive to mount a personal webcam. In the late 1990s, however, individuals seeking to do so needed somewhat greater technical knowledge along with the capital to maintain server space to first open and then maintain such a site. Webcam operators' fans would email operators favorite jpeg screen captures of them so that they could then post the images in picture galleries on their sites. Now, operators and registered visitors alike can post and rate video clip highlights of their favorite lifecaster "episodes." The cost of access to server space has decreased, and with access to WiFi, the mobility of webcams also has increased. Some lifecasters, such as Justin Kan, the founder of Justin.tv, have worn a cam attached to a hat or special glasses so that viewers see everything from the wearer's point of view. Most lifecasters position one or more cameras at strategic points in their home so that viewers can observe them within their own personal environments. Some combine both techniques. While technological sophistication has increased since the 1990s, the practices themselves, however, have largely remained the same.

Although I am focused on the use of webcams by gay/queer men during the late 1990s and early years of this century, I am also interested in the ways that these men's online practices worked to complicate the meanings of the term "object" in virtual settings. My analysis is hermeneutical—it seeks to understand the meaningful human actions of webcam operators and fans alike. It is also empirical in its concern to explain this moment of webcam culture as a social phenomenon. And it is critical—I provide readers with insights into the importance of gay/queer men's personal webcams at this historical juncture. These technologically savvy, yet frequently marginalized, men constitute an historical vanguard that turned to the web as a new form of media centrality, visual culture, and as the latest manifestation of the progress myth, and did so in order to perform and therefore *render visible* personal identity claims. As part of a response to cultural homophobia, they made claims to materially exist in the "here and the now" through an ambivalent strategy of virtual visibility: as screen-based moving webcam images, they attained a degree of visibility even as they remained "at a distance" from viewers. In doing so, however, they enacted on the web an ironic and updated version of a much longer history of the dynamics of being "in the closet." Yet at the same time, these men were at the forefront of developing the kinds of now-accepted identity claims more centrally connected to the power of online moving images to fully stand in for or even supplant the individuals or referents behind the images themselves. The practices and techniques developed by these men indicate the historical emergence of a new form of fetish, the "post-representational" online moving image of the webcam operator as digital fetish—what I term the *telefetish*.[3]

The history of these practices is important, for they complicate widely held understandings of the meaning of "object" while also renovating theories of the fetish grounded in colonial-era anthropology that insists the fetish is always a visible material object. The term "object," most often referencing a spectrum of the senses, has a complicated set of meanings. Empiricist understanding holds that an object is "a material thing that can be seen and touched" (*OED*). A divergent, earlier definition, dating from the Renaissance, holds it as "something placed before or presented to the eyes or other senses" (*OED*). Both meanings implicate human sight but the emergence, and increasing naturalization, of the idea of the digital virtual object indicates a return to, or renovation of, this pre-modern meaning in that "something" can mean anything apprehended intellectually but also anything visible or tangible and relatively stable in form. The word "object," then, can reference a material thing *and* a concept or idea. It does not exclude the virtual object from its range of meanings. The rise of the idea that virtual objects exist within digital realms resuscitates the pre-modern meaning: "something" can mean anything apprehended intellectually as well as anything tangible, visible and reasonably stable in its form. The idea of a web-based virtual object indicates the desire to span the modern distance between concepts and objects, ideas and materiality. The discursive value of the idea of virtual objects for academic theorists and information technology sectors alike pushes web-based images toward seemingly greater experiential materiality while also conveying the value of exchangeability between the material and the virtual that lies at the core of what we mean by "simulations."

With respect to the history of early-adopting gay/queer webcam operators, the idea of virtual objects increasingly influences the way viewers experience real-time web-based images of human beings as exceeding the bounds of representation per se. The emergence of the digital telefetish allows viewers and operators alike to experience an operator's digital image not only as a representation—an idea—but also as a seemingly material and animated sign/body, an index or post-representational trace of the operator's embodied reality.

GOING LIVE: WEBCAMS IN THE LATE 1990S

SeanPatrickLive launched in January 1997 and was shut by its operator, Sean Patrick Williams, in the spring of 2001. Williams's fan base was never as wide as Jennifer Ringley's (his site received between 35,000 and 80,000 hits a day), but the Washington, DC-based operator holds the likely distinction of being the first gay/queer man to go live on webcam. His site drew considerable mass media interest. Journalists asked why so many repeat viewers found the largely banal images of the telegenic Williams eating, smoking, sitting in front of his computer, or sleeping of such interest. Williams replied, "People see a reflection of themselves....You watch TV or movies, and you see people wake up wearing makeup, looking perfect. I wake up and, well—it's not pretty. I think people find that refreshing" (Kennedy, 1998). "What I'm trying to show is that what goes on in somebody else's life always seems bigger and different and fabulous and a lot different than what's going on in our lives. But that's not true" (Freiss, 1998). And, "It's sort of to let people know that everybody's not as different as we like to think" (Long, 1998). These 1998 print-based interviews deemphasize the same-sex aspect of Williams's webcam site. Nevertheless, as Donald Snyder records in his ethnographic study, Williams viewed SeanPatrickLive as an opening to write about important issues affecting the lives of gay/queer men:

> I remember living in Wilmington, North Carolina, where I grew up, we didn't hear about like gay things; we just heard about the big gay things. And things like, you know, Matthew Shepard got a lot of press, but like the Roanoke thing [a more recent and less publicized gay killing] didn't. This is probably because the guy that was shot wasn't really young and pretty. He was kind of an older weird looking queer, but I think it's just as important to say that. (Snyder, 2002, p. 192)

Seeming to live his life so openly in front of cameras connected to the web arguably made a difference in the lives of his viewers. In 2006, Will, a self-identified middle-aged Bostonian and fan of the site, entered the following assessment on his own weblog:

> Long before 'The Truman Show' dealt with the subject of having one's life telecast to the world Sean Patrick Williams was visible 24 hours a day via webcam. He lived in D.C. and ate, slept, cleaned, worked, cooked, entertained, watched TV and occasionally had sex on cam (although he was far more restrained in the latter regard than many other

> web cammers). As someone who had always been a very private person, I was fascinated by the openness of it all.
>
> What I admired about his cam site was how powerful and courageous a statement it was for a gay man to take the ultimate 'out' position of living a gay life on line for the whole world to see... . He did it naturally and without grandstanding, and somehow it became compelling. One cared, and missed him when he was gone. (DesignerBlog, 2006)

SeanPatrickLive and the majority of the personal webcamera sites mounted by other early adopters are "history," part of an earlier moment of gay/queer engagement with the web, an experiment in virtual visibility at the onset of the web's massive popularization. Google these sites today and you will find next to nothing. One may think of the web as a "great records machine," but it also accelerates the ephemeral logic of the commodity fetish that weeds out and discards as superfluous those aspects of the past no longer judged as having sufficient display and exchange value. My account of these earlier webcam practices, and the role played by implicit and explicit acceptance of the idea of the virtual object within them, therefore, is also a form of gay/queer history that otherwise would be lost.[4]

The specifics within which many gay/queer identities gestate—the realities of the closet, geographic and social isolation, desire for wider social recognition within a climate of "social cruelty" (Carey, 1998, p. 67) expressed through the enactment of legislation such as the U.S. Defense of Marriage Act (DOMA)—coupled to the considerable numbers of men employed in the Information and Communication Technology (ICT) sector or possessing considerable skill with respect to how these technologies and their applications are deployed—contribute significantly to the emergence of these relatively information machine-savvy men at the forefront in using these technological assemblages in novel ways. Snyder's study notes Sean Patrick Williams's "deep connection to technology": his early interest in computers, his work as a programmer and web designer, interest in video gaming, and ownership of five computers and related electronic technologies (2002, p. 183). As Larry Gross observes, "the Internet has particular importance to sexual minority communities.... Tom Rielly, the founder of PlanetOut, explained, 'Traditional mass media is very cost-intensive. Gays and lesbians don't have a high level ownership of mainstream media properties. The internet is the first medium where we can have equal footing with the big players'" (2004, pp. x–xi). Gross notes the intensity of internet use found among gay men and lesbians, a use related both to geographic and social isolation of individuals and to the ongoing potential in many places to be on the receiving end of physic and physical violence if one is too public about one's same-sex orientation or preference.

These men's turn-of-the-century practices reflect directly on the above-noted points. Though many today would like to assert that the most egregious forms of Western homophobic physical violence and socio-economic forms of discrimination had abated by the late 1990s, mounting a personal webcam allowed these men to perform a degree of virtual public visibility frequently denied them in heteronormatively inflected material public settings. The desire for voice and visibility appears actualized as the online moving image that appears to "say," "look at me, in truth I exist." As already noted, spatial ironies abound. In a sympathetic interview conducted in 2000, Williams argued that the internet "is of particular benefit to gay activism because it empowers closeted gays to 'publicly stand up for themselves in a private way'" (Aravosis, 2000).

Using the web to connect with other people in some ways naturalizes the idea expressed in the phrase "absence makes the heart grow fonder," except that the absence has been produced through screen-mediated forms of virtual encounters that precede (if they occur) any face-to-face meetings between actual individuals. The aesthetic qualities of individuals' online performances, confirmed by receiving emailed jpegs of themselves captured by admiring men, allowed these operators to enact the perilous though comforting illusion that they might conceptually relocate to the pure realm of aesthetics, to somehow "live in art" through fabricating a series of performing images that could, if only for a time, meaningfully substitute for actual individuals and embodied social relations.

A webcam operator such as Sean Patrick Williams can also be understood as having attained a quality of niche celebrity status. While being the star of one's own webcam is often the result of fans having conferred on the operator the cultural capital of inherent authenticity, it also highlights the tantalization of their desires—fans and operators alike—for someone so near yet still so far because always beyond material reach—hence the *tele* in the telefetish. While the possibility of telefetishized interactions certainly is available to any webcam operator and his or her fans, in the case of gay/queer operators it is, ironically, this reiterative dynamic of fetishism that, relocated to online settings, itself mirrors the history of the kinds of tantalization of gay/queer men and lesbians promised by liberal social discourse. In the words of Urvashi Vaid (1995), gay/queer men and lesbians are promised a "virtual equality" that offers a veneer of partial acceptance in exchange for conforming to heteronormative standards of "good taste." Gay/queer men express this dominant standard through their performing (in public and in particular in the workplace) of what heteronormativity prescribes as an acceptable or palatable image of gay/queer identity. The promise of virtual equality has contributed, in ways that suggest the continued saliency of Erving Goffman's arguments (1959) about front stage/back stage interpersonal and symbolic dynamics, to an ironic yet ritualized updating of the historically segregated aspects of gay/queer everyday lived realities into sanitized public fronts and more private modes of interacting. Public fronts, tolerable to heteronormative values, remain separate from supposedly less palatable constructions of online gay/queer personae targeted and transmitting primarily to other men identifying as gay/queer. A hybrid array of askance political analyses, images, and sexualities has been relegated, therefore, not only to back stage material fantasy environments that until recently were largely the purview of commercial same-sex bar culture but, increasingly, to online settings experienced as virtual environments. For gay/queer men online, the neither public nor private yet both at once character of the networked settings they may seek to virtually inhabit permits marginalized aspects of gay/queer cultural realities to attain something like a virtual "return to the center"—to achieve fuller visibility and "presence." The gay/queer sign online denotes the shifting material status of its maker: partially hidden yet therefore also partially in view.

TELEPRESENCE AND THE FETISH ONLINE

The use of digital technologies by these men can be understood as an active engagement with a form of magical thinking by technologically well-connected but socially marginalized men who have understood that screen-based identities are not fully "real" even as they have half-hoped they might be, if only for "the moment." Part of this return to magical thinking, or investiture of fetish "spirit" or "aura" in a web persona implicitly positioned as an extension or trace of the operator's actual being, flows from telepresence. The idea of telepresence supports the increasingly pervasive contemporary belief that the internet is more a space than a set of textual engagements and, therefore, that an individual can be both materially "here" at the same time as seemingly "there" by extending agency through, or components of identity into, the virtual space of digital technologies.

As Sandy Stone puts it, "inside the little box are other people" (1997, p. 16). In Lev Manovich's words, "The body of a teleoperator is transmitted, in real time, to another location where it can act on the subject's behalf" (1995, p. 350). The post-representational sense that a subject can transmit her or his "'presence' in a remote physical location" (p. 349) so as to achieve "action at a distance"—a phrase that also can serve as a rough and ready definition of magic—illustrates the enchanted thinking, the Idealist metaphysics, yet also the creative taking up of dominant spatial metaphors entailed in both the placing and experiencing of virtual objects in the form of fetishized online traces or indices of embodied personal identity. Moreover, the uncritical use of the term telepresence by many new media theorists implicitly promotes the power of web networks as akin to a transcendental signifier because such use occludes consideration that the act of telepresence is really akin to creating a sign in the seemingly magical form of an index or actual trace of the individual. Belief in telepresence as productive of some form of material actuality parallels the rise of the digital virtual object and reveals an inherently metaphysical belief that one might move into a sign, move beyond one's history into a virtual topography that in its indexicality nonetheless necessarily remains a wholly aesthetic "terrain" of signification. Such belief invites the individual to become one with the seductive image.

In many personal webcams and websites dating from the turn of the century, gay/queer men presented themselves as living emblems—as combinations of images and texts. Through various image-text combinations, these men became the simulation of both their sexual fantasies and desire for greater socio-political recognition of their extant existence within wider, heteronormatively inflected, everyday worlds. In so doing they implicitly suggested and explicitly depicted the logics underlying the belief that "the online transmission of my fantasy self is the real me." That is to say, they rendered themselves as telefetishes, as virtual objects. The fantasy images these men transmitted via their personal webcams accrued unto themselves an ironic power: like the archaic fetish they were skillfully contrived, made by art, animated by the owners' spirit, and worshiped in their own character to the extent that spatial metaphors of the web promote belief that individuals and their embodied selves can be fully interpellated within the conceptual apparatus of the technology. Networked digital telepresence works to support users' perception of the telefetish sign/body as experientially real. Website participants actually saw the moving images, the traces of the "little people in the box" to whom Stone refers. In my online research for this project I sometimes encountered the assertion that "my fantasy is real" and suggest that the assertion reflects something of this dynamic; it underscores the complex, even contradictory, distinctions and overlaps between experience and materiality.

Telepresence, in its meaning as the virtual presence in the form of a digital sign of an object or person situated elsewhere, has been a powerful means for making sense of the postmodern commodity-body. Recalling the earlier discussion of virtual objects, the early-adopting webcam owner could be both a material body and his iconic telefetish: material and virtual, spatial and textual, symbolic and indexical, archaic and posthuman, either simultaneously or in a now-here, then-there fashion both instantaneous and fluid. Such hybridity was further made possible by the "actuality" of both materiality *and* virtuality (Lévy, 1998; Shields, 2000) coupled with a webcam site owner's experience of spatial separation or structuring that telepresence seemed to authorize between his body and his web persona even as his website visitors may have participated differently when they experienced a conflation of the two. What telepresence further allowed is for webcam owners to experience their own personification as a visible image to themselves. This telefetishistic dynamic, coupled to the experience of self-visibilization it can support, is available to any webcam operator regardless of identity formation. The implicit reliance, however, by early gay/queer webcam operators and fans on moving images of operators as a way to assert and experience forms of self-visibilization helped foster the sense that these images were capable of transmitting a crucial aspect or existential trace of who these men *already actually* were. The seemingly alive sign, taking the form of a visible, moving image of the operator, could, if only for a time, supersede an odious personal and collective history of being refused or socially excommunicated for whom one is.

What are some of the implications flowing from such a complex juncture? For a webcamera site owner, the images of his body as a cosmopolitan surface, as a commoditized "work of art" on the web, can be equally telepresent to himself as a flowing series of virtual objects. The way they appear to him on a screen can have an equal measure of use value for him as for fans and other viewers. As such the metaphysics of telepresence allows his commodified sign/body something akin to a spirited and haunting return of the trace as an embodied image even as it also has provided him the sense of distance between the iconography and his embodied presence that modern sensibility still demands. This also raises issues of automation and the automaton and was a central reason provided by Sean Patrick Williams for his early engagement with webcam technology and the experience it offers: "It really will run by itself on its own. I think that's why it keeps going" (Snyder, 2002, p. 184). Williams is referring to the automated technology upon which a webcam operation relies. By inference, this includes its generation of the image of the telefetish on the display.

Constructing one's web-based telefetish within an already fetishized technological environment is a doubling—the making of what the fetish theorist William Pietz refers to as a "power object" (1993, p. 145)—and hence a potential negation of any sense of falseness. This may explain the déjà vu quality of fascination on the part of some webcam operators who kept weblogs attached to their sites and who posted comments roughly along these Lacanian lines: "watching myself on the camera is very interesting...to be able to see myself from another's point of view." In such comments that implicitly acknowledge the impersonal power of the gaze as a two-way process, one may see how representation, simulation, or symbolization collapses into a post-representational sense that there really is (at least a trace of) another who watches.

Some of these dynamics are illustrated by the personal narrative of Jason Weidemann (2003), one of the few gay/queer men to have published in print a personal account of his experiences as a webcam operator. Weidemann divided webcams marketed to gay/queer men into three categories: "single cams run by one person [like his own]...portal sites, where subscription to a single commercially run website offers access to several different amateurs; and webcam houses...largely put together by entrepreneurial college students, which started appearing in the late 90's." About his experience he writes,

> I usually tried to be live on camera about two hours a day. Most of the time I must have been boring to watch. I typed papers and wrote e-mails, napped in the nude, ate ramen noodles. The camera was perched above my computer, focusing on my face, but occasionally pointed toward my bed when I read or slept. Often I was sexually explicit, stripping and then masturbating. I usually began the night with the mindset of "putting in my hours," but the thought of men across the world watching me, and some finding me beautiful, usually shifted my attitude to enjoying being sexually explicit on camera. (p. 14)

Weidemann further explained how he came to name his site As You Gaze upon Me.

> The title for my website was undoubtedly influenced by the cultural studies classes I was taking at the time, where the term "the gaze" was making the rounds. As I understood it at the time, the "gaze," always masculine, represented the flow of power between voyeur and subject: the gaze was an act of consumption directed against an object by a more powerful subject. This concept may have merit, but the gaze worked differently for me. It was an act I courted and a part of my own erotics as an exhibitionist. Rather than seeing "the gaze" as a source of insidious power, I came to see it as a major component of my own pleasure. (p. 14)

With its use of the conjunction "as," Weidemann's site name implied the simultaneous viewing practices of many individuals, his own included. Weidemann courted the gaze of his viewers. His desire elicited it, and he experienced it as a welcome message that constituted a major component of his pleasure. The form of the gaze to which he refers is consonant with Laura Mulvey's classic theorization about popular cinema's production and reproduction of the "male gaze." Mulvey appropriated psychoanalytic theory to demonstrate "the way the unconscious of patriarchal society has structured film form" (1975, p. 6). For Mulvey the subject of the gaze is male. The object of his gaze is female. Mulvey tied her theory to the cinematic apparatus. The cinema is a technology of projection and, consonant with analog film technology's mechanisms, "practically all traditional narrative cinema treats woman's body as a projection of male vision" (Huyssen, 2000, p. 208). In her assessment of the potentially positive political aspects of women's webcams, however, Michele White notes that, while privileged forms of looking continue to inhere in "certain kinds of heterosexual white masculinity," other more gender-neutral conceptions of the gaze admit "that compliance and power can be the products, as well as the instigators, of the gaze" (2003, p. 9).

What might it mean for men to gaze on men, not within the cinema of projection but through the web of transmission? Mulvey notes that the conventions of cinema create a hermetically sealed world "indifferent to the presence of the audience" (1975, p. 9), but Weidemann was anything but indifferent. Slavov Žižek argues that the gaze "denotes at the same time power (it enables us to exert control over the situation...) and impotence (as bearers of a gaze, we are reduced to the role of passive witness...)" (1990, p. 2). White proposes that webcam viewers interacting with the screen, "become collapsed with the computer and may fail to distinguish where subject ends and object begins" (2003, p. 20). Her argument suggests the post-representational experiential quality of certain webcam practices, and that the gaze/gazer binary and the histories of oppression related to it are unstable in networked settings. In any case, looking itself indicates the desire that results in the gaze. In webcam settings, the technology of transmission, including its automated features that were so appealing to Sean Patrick Williams ("it really will run by itself on its own"), incorporates the networked gaze. In Weidemann's case, this gaze included Weidemann who could watch himself watching himself as a telefetish—a display that served to experientially fuse image to screen and sign to body. As the operator, too, Weidemann formed part of the anonymous socio-technical assemblage that may surveil the world or focus in on another in a manner that inverts understanding Weidemann and other operators as only exhibitionists. As a man, he possessed the objectifying power of the gaze identified by Mulvey. As a man who desired, and therefore at times sexually objectified, other men, however, Weidemann moved back and forth between Mulvey's gendered positions. In networked settings he was neither subject nor object, sender nor receiver, sign nor body, but all of these at once. He looked into the camera, desiring the male gaze that sought his image on the screen because it visibilized his existence not only as a subject but also as an auratic object—something powerful to be worshiped in its own right, seemingly transhistorical yet, like all forms of the fetish, of value only for a time. A time that for him has now passed.

Transmitting images of bodies in the form of moving graphically inflected signs that are experienced as psychically (but not actually) real introduces the possibility for an individual to be experienced phenomenologically as a telefetish both by others and him or herself. He or she can fetishize the trace of others. The differing forms that individuals adopt to transmit an experience of themselves in these online settings point back indexically—like pointing fingers—to themselves. Why is this recursivity important? All individuals engaging the web remain this side of the interface. But as chat and webcam participants they can experience seeing themselves as a networked sign/body; in cosmopolitan fashion, they experience becoming an image, a virtual object with exchange value courtesy of an assemblage of information machines that maintain separation between people even as they transmit a virtual experience of coming together. As Giorgio Agamben notes, this separation allows for language and communication practices to gain autonomy from actual bodies (2000)—a separation, again, of the practices of consciousness from the thing itself, of conception from the perceptive faculties that first inform it. *Cogito ergo sum.* Beginning in the mid-1990s, web technologies allowed the networked digital image, increasingly positioned and experienced as a virtual object, to acquire a historically significant quality of semi-independent liveliness seemingly worthy of fetishization in its own right. With this status the online image could then start to be imagined as a social relation in itself that somehow also exists beyond both history's purview and its disciplining functions.

POST-MORTEM

The webcam practices discussed above developed when it first became possible for an individual to mount a personal webcam or website. Personal webcams could not have emerged earlier; the technology upon which they relied was not yet in place. But why did these sites all close? Several factors merit mention. A closed website can suggest the end of one's "fifteen minutes" as a standard outcome of celebrity. Celebrity itself is also a commodity. Like all commodities, celebrities and the experiences they entail always stale date. If the gay/queer webcam operators discussed above were early adopters of digital technologies allowing for new modes of self-depiction and formation, then the logic of commodity culture also suggests that the most avant-garde of such operators may have moved on to the "next thing" as part of a broader fetishization of the idea that new equals better.

It should also be noted that the webcam operator who looks too closely into the screen at his own telefetish performance risks mistaking his own appearance-as-experience for being itself. In so doing he risks inundation as a latter-day Narcissus by the very power of the telefetish's animating aura that made it so compelling to transmit in the first place. I raise the Narcissus myth not in the way that contemporary culture mistakenly reduces it to a cautionary against self-love but to call upon its wider instruction that extending oneself ever farther into space increases the risk of becoming numb (McLuhan, 1964, p. 59). This is the self-deception that arises from confusing oneself with an illusion. The prefixes of Narcissus and narcotics are the same. Drinking from a pool, Narcissus became narcotized after gazing into his own mirrored reflection which he mistook for an extension of himself. The continuing salience of the ancient myth reveals the possibility that individuals, webcam operators included, can become aware of the psychic dangers of narcotization about which it warns.

Early adopters are often early adapters. While thousands of gay/queer personal and pay-per-view commercial webcam sites continue to open, operate, and close according to their owners' schedules and the market's logic of exchange, many gay/queer men have turned to weblogs and blogging to enact forms of political hopefulness and their continuing negotiations of telefetishism, digital networks, and the wider world. The origins of contemporary blogs lie in early 1990s Usenet discussion lists, "threads," and postings, but unlike these earlier purely text-based representations, blogs are frequently image rich. Contemporary weblogs exemplify the shift from a late-1990s understanding of the web as a platform for production to more recent promotion of it as a platform for participation. I note this as it is worth considering that the webcam telefetishes discussed above became exhausted not only when operators experienced too great an alienation from the "here" but also when they ran out of steam. (This is easily understood by too-busy readers expected to maintain personal websites who have been asked, "When was the last time you updated your site?") The webcam operators I have discussed were their sites' main attractions. Williams's aforementioned comment that his webcam site "really will run by itself on its own" notwithstanding, while fans did email jpegs for inclusion in image galleries, the ceaseless work of producing content—including labeling, mounting, and maintaining the images—fell to operators alone. Contemporary blogs also attract followers focused on the operators and also require work to maintain. Yet unlike the earlier webcam sites discussed here, weblog operators operate much more intertextually. They need not produce their sites' entire contents. In other words, they need not do all the work required for their sites to remain interesting.

Money is also a factor. Sean Patrick Williams's decision to not charge user fees created a serious financial burden for him. He was charged a fee each time a fan downloaded an image. The more his site gained popularity, the more media attention he received, the greater his monthly costs. Toward the end he was paying out of pocket almost $2,000 a month to keep his site running. These costs were only partially defrayed by early forms of banner advertising and other sales of merchandise (Snyder, 2002, pp. 190–191).

A final factor meriting consideration is that the webcam sites I have discussed were all American and their duration coincided not only with increasing expectations about how the web-based image might bolster identity claims but also fell within the specific spatio-historical moment marked by the somewhat liberal U.S. political climate of the Clinton years. I note the passing of these sites as roughly coterminous with the coming to power of a radically more conservative administration much more hostile to the interests of gay/queer men and women and much more inclined to cooperate with nongovernmental cultural forces adamantly opposed to any extension of visibility for these men. Compared to the way that webcams render gay/queer bodies directly visible, gay/queer weblogs, with their arguably more politically cautious reemphasis of the textual component, suggest the contingent value of more indirect forms of discourse and expression than the image itself during hostile cultural and political moments. Images have hardly "gone away." But the reinvention and reformulation of these virtual spaces temporally accord with the socio-economic, political and cultural circumstances influencing gay/queer lives and experiences in actual places. The telefetish, then, always remains open to change lest its presence become too prescriptive, its specific form and associations too limiting, exhausting. or dangerous, its potential for epiphany a bit too everyday.

The techniques and practices mounted by early gay/queer webcamera operators and their fans and other viewers indicate that the icon of a webcam operator, as a visible and fetishizable sign/body, is a virtual object that can become as desirable to viewers as the embodied operator who can also see himself transmitting as a sign/body to other participants. Seeing oneself seeing oneself can induce a seemingly transcendental experience of awe. The gay/queer webcam operators I have discussed were early adopters interested in transmitting the truth of their existence to virtual publics through virtual environments. Their online practices further indicate how, at the beginning of the web's popularization, webcam operators could witness a self-produced trace of themselves transmitting the affective potential of their "true" multiplicity within a sign system—including all the things they were not yet, or might desire to become. If only for a time...

NOTES

1. Farewell, seminal coffeecam. (2001, March 7). *Wired*. Retrieved January 26, 2009, from http://www.wired.com/culture/lifestyle/news/2001/03/42254
2. For a history of lifecasting, see http://en.wikipedia.org/wiki/Lifecasting_(video_stream). Retrieved January 26, 2009.
3. I have written in greater depth about the intersection of digital media and online fetishism, ritual, and telepresence. See Hillis, 2009.
4. The loss of "parts" of the past scarcely begins with the ephemeral nature of websites that have closed. The very idea of a canon, for example, points to the reality that if "the best" has been conserved, then "the rest" has been discarded as superfluous. In this way memory and history

imbricate one another in unsettling ways often unacknowledged by those with a stake in cementing what will constitute the historical record for future readers and generations.

REFERENCES

Agamben, G. (2000). *Means without end: Notes on politics,* V. Binetti and C. Casarino (Trans.). Minneapolis: University of Minnesota Press.

Aravosis, J. (2000, October 23). StopDrLaura.com. *The New Republic.* Retrieved January 26, 2009, from www.wiredstrategies.com/tnr_sdl.htm

Carey, J. (1998). Political ritual on television: Episodes in the history of shame, degradation and excommunication. In T. Liebes and J. Curran (Ed.), *Media, ritual and identity* (pp. 42–70). New York: Routledge.

DesignerBlog. (2006). Retrieved March 14, 2008, from http://designerblog.blogspot.com/2006_06_01_designerblog_archive.html

Freiss, S. (1998, September 1). Cyber voyeurism, *The Advocate.* Retrieved February 8, 2003, from www.advocate.com/html/stores/0199/0199_patrick.html

Goffman, Erving. (1959). *The presentation of self in everyday life.* New York: Anchor Books.

Gross, L. (2004). Foreword. In Campbell, J., *Getting it on online: Cyberspace, gay male sexuality, and embodied identity.* New York: Harrington Park Press.

Hillis, K. (2009). *Online a lot of the time: Ritual, fetish, sign.* Durham: Duke University Press.

Huyssen, A. (2000). The vamp and the machine: Fritz Lang's *Metropolis.* In M. Minden and H. Bachmann (Ed.), *Fritz Lang's* Metropolis: *Cinematic visions of technology and fear* (pp. 198–216). Rochester: Camden House.

Kennedy, H. (1998, July 16). Thanks to digital video cameras, voyeurism is the latest internet rage. *New York Daily News.* Retrieved February 8, 2003, from www.newstimes.com/archive98/jul1698/cpa.htm

Lévy, P. (1998). *Becoming virtual: Reality in the digital age,* R. Bononno (trans.). New York: Plenum.

Long, T. (1998, August 3). Livecam web sites bring us the excitement, boredom of others' lives. *The Detroit News.* Retrieved February 8, 2003, from http://detnews.com/1998/accent/9809/03/08030008.htm

Manovich, L. (1995). Potemkin's villages, cinema and telepresence. In K. Gerbel and P. Weibel (Ed.), *Mythos information: Welcome to the wired world* (pp. 343–353). New York: Springer-Verlag.

McLuhan, M. (1964). *Understanding media.* New York: McGraw-Hill.

Mulvey, L. (1975). Visual pleasure and narrative cinema. *Screen* 16 (3), 6–18.

Pietz, W. (1993). Fetishism and materialism: The limits of theory in Marx. In E. Apter and W. Pietz (Ed.), *Fetishism as cultural discourse.* Ithaca: Cornell University Press.

Senft, T. (2008). *Celebrity and community in the age of social networks.* New York: Peter Lang.

Shields, R. (2000). Virtual spaces? *Space and Culture* 4/5, 1–12.

Snyder, D. (2002). 'I don't go by Sean Patrick': On-line/off-line/out identity and SeanPatrick-Live.com. *International Journal of Sexuality and Gender Studies* 7 (2/3), 177–195.

Stafford-Fraser, Q. (1995). The Trojan Room coffee pot: A (non-technical) biography. Retrieved January 26, 2009, from www.cl.cam.ac.uk/coffee/qsf/coffee.html

Stone, A. (1997). *The war of desire and technology at the close of the mechanical age*. Cambridge: MIT Press.

Vaid, U. (1995). *Virtual equality: The mainstreaming of gay and lesbian liberation*. New York: Anchor Books.

Villarejo, A. (2004). Defycategory.com, or the place of categories in intermedia. In P. Church Gibson (Ed.), *More dirty looks: Gender, pornography and power*. London: BFI Publishing.

Weidemann, J. (2003). Confessions of a webcam exhibitionist. *The Gay & Lesbian Review Worldwide* 10 (4), 13–17.

White, M. (2003). Too close to see: Men, women, and webcams. *New Media and Society* 5 (1), 7–28.

Žižek, S. (1990). How the non-duped err. *Qui Parle* 4 (1): 2.

CHAPTER 6

Self-portrayal on the Web

Dominika Szope

> With the media, it is similar to the way Roessler, as an endo-physicist, establishes it in regard to consciousness. We swim in it like fish in the ocean, we absolutely need it, and for precisely this reason, it is basically not accessible to us. All we can do is create cuts in it so as to gain operational access to it. (Zielinski, 2002, pp. 48–49)

What at first confronts us with a certain forlornness is fortunately relativized by Zielinski himself. For the "cuts" he speaks of can be defined as "built constructions" and discerned as media worlds. We encounter them as apparatuses, as programs, as media-based forms of expression and realization such as films or web sites. They can be found in various environments and places, at different times and in different degrees of density. This, however, also turns historiography into a tricky affair. Where to start, where to end, and which criteria to select and interpret? It appears as if historiography cannot elude arousing the suspicion of satisfying, above all, the historiographers themselves. The view of history in the sense of a science of history is always subject to changes that make it difficult, if not impossible, to formulate a generally valid definition. So the suspicion arises that even scientific historiography ultimately proceeds, narrates, structures and interprets according to ideological or subjective criteria. In which sense, then, is history written and by whom?

Media history, under the "umbrella" of which a historiographical examination of the web would take place, initially means the entirety of historical descriptions of media processes. The main emphasis is on the history of media inventions and on a history of the changes in sensorial perception effected by media. Due to this focus, modes of description have evolved over time that now appear quite clear. One speaks of chronical, as well as biographical and autobiographical descriptions. The third pillar of this "classical" notion is formed by structural descriptions that seek to avoid the deficits of person-related approaches.

The development of the internet can hardly be described as targeted and linear. The past has shown that the emergence and dissemination of technological innovations is complex and that many groups are involved in the process of development. Hence, innovations are influenced by the situation of origin but also by a large number of coincidences. This course of events becomes visible in the many attempts to write a history of the internet, which time and again are forced to make concessions. For Manuel Castells, it is the accelerating speed of this "revolution" that immediately makes any description outdated (Castells, 1996). John Naughton (1999) and Janet Abbate (1999) also show the impossibility of producing a comprehensive, not to speak of a final, historiography of the internet. While Abbate concentrates on the events prior to 1990 and only briefly mentions subsequent developments, Naughton focuses on personal stories with an authoritative account of where the net actually came from, who invented it and why. Although the World Wide Web is only one possible way of using the internet, the characteristic seems to continue here, where the development of technological innovations is often complex and a large number of persons and groups are involved who partially compete with each other. Moreover, the speed of these processes—and the lack of transparency—appears to have increased due to the impact of social software.

Recent media historiography makes an effort to establish more complex forms of description in which different approaches are combined. In this context, Hickethier speaks of "longitudinal examinations, cross-section examinations, tables, exemplary case descriptions" etc. (Hickethier, 2003, p. 351). Zielinski's approach, on the other hand, proposes a kind of an-archaeology of the media, in which he attempts to keep the concept of media as open as possible. His approach can be understood as a collection of curiosa, of found material "in which something flashes up that constitutes their inherent luminosity and at the same time refers beyond the meaning or function they had in their context of origin" (Zielinski, 2002, p. 49). He also proposes, among other things, to distance oneself from the promise of continuity and no longer seek a confirmation of the old in the new. Instead, the issue is to discover what is new in the old.

The web has experienced decisive changes through social software applications, for example, the increase in options to act has freed the individual from his or her hitherto mostly passive role and led to further developing one's capacities to act. In particular, the possibility of portraying one's own self has strongly increased under the conditions of social software, both in terms of quality and the variety of modes of portrayal. In the past ten years, the presentation of the self has concentrated on the scriptive representation on one's own homepage, in chats and forums. With platforms such as YouTube, myvideo etc. having become established, the possibility of pictorial self-portrayal has reached a new dimension. There is a broad variety of such portrayals, ranging from the trivial articulation of protagonists usually filming themselves, to their accounts of everyday life or certain events, to carefully staged appearances that are not seldom expanded by acting and scenic efforts. What all portrayals seem to have in common is that they reveal the necessity of asserting one's own self.

I would like to take up Zielinski's concept of an-archaeology and—expanded by a juxtaposition—apply it to the phenomenon of self-portrayal on the web. The example of GreenTeaGirlie turns out to be a current "attractive condensation" (Zielinski, 2002, p. 45). Both her appearance and the context of her performances seem to indicate a new trend in the daily life of the web. I would like to juxtapose this relatively new presentation of the self with an example from 1996 which, in my opinion, can be regarded as a predecessor of today's self-portrayals on YouTube. The webcam of Jennifer Ringley, the so-called JenniCam, led to numerous responses and discussions at the time it was broadcast but also years later. It therefore seems to be a further "condensation" worth examining anew. By juxtaposing these two examples, I hope not only to reveal the early webcam recordings such as the JenniCam as predecessors of today's self-portrayals as they are produced by means of social software; fully along the lines of the premise formulated by Zielinski, to discover the new in the old, I also expect to formulate a new aspect by examining the JenniCam.

A CURRENT "ATTRACTIVE CONDENSATION"—GREENTEAGIRLIE

GreenTeaGirlie "entered" the platform YouTube in March 2007. Spiegel-Online editor Konrad Lischka wrote:

> Blue eyes, light brown hair, tanned skin, sparkling white teeth, and 170,000 viewers on her first day on YouTube—she can only be an actress. At least that's what many You-

> Tube users believed, who on March 25 viewed GreenTeaGirlie's ten-second film on the video portal. The twenty-year-old tea vendor from the Midwest greeted her audience with the words: 'Hello, I'm new here. I hope you welcome me.' But only few did so. From the onset, KallieAnnie was suspected of being an actress, a surreptitious advertiser, a liar (Lischka, 2007)

This response comes as no surprise, for GreenTeaGirlie, alias KallieAnnie, had to take on a difficult legacy. Just a year earlier, a first "star" had dominated the platform: Lonelygirl15 mimed the lonely teenager living in a U.S. hinterland, being taught at home and with hardly any friends. She was attractive and divulged things of her life that never revealed too much, but sufficed to rally around her a large number of supporters from the YouTube community. Just a short while later, Lonelygirl15 turned out to be a fake, an actress named Jessica Rose acting according to a concept of two young filmmakers. GreenTeaGirlie's "predecessor" was revealed as a fake, but as opposed to numerous claims, she turned out not to be an "advertising vehicle" (Patalong, 2006a). Yet the reactions of the YouTube users reflected the confusion of the platform and brought the hitherto latent question of authenticity to the fore (Patalong, 2006b).

GreenTeaGirlie did not let herself be disturbed by these first accusations made against her. She understood them largely as constructive criticism and viewed the distrust she was confronted with as an experience one must face as a YouTube newcomer. She produced comparatively short and rather inconspicuous videos in which she gave an account of her day, danced, or put her outfit or hairdo up for discussion. But like Lonelygirl15, she was quite attractive and that, among other things, probably provoked numerous sceptics.

After criticism became increasingly louder, she disappeared from the scene. Connections to a tea shop as well as to the domain vidstars.com, an agency promising to make plugs in the videos of YouTube stars, seemed to reinforce the initial suspicion. But it all turned out quite differently. Research conducted by the *Los Angeles Times* in May of the same year revealed that GreenTeaGirlie, who calls herself KallieAnnie, actually exists, that she really sells tea in a shopping mall in Salt Lake City. The connections made to Dragonwater and vidstars.com, however, were false. The *Los Angeles Times* found out that they originated from a certain Matthew Foremski, who along with two other YouTube users just a few months earlier had uncovered the true existence of Lonelygirl15. He registered the false website of GreenTeaGirlie.com, caused confusion in regard to Dragonwater and alleged plugs. All this occurred within a period of three months; GreenTeaGirlie disappeared and took her videos with her. The only videos of her still on view were those stored, altered and "commented on" by several critics—mostly to her detriment. In early 2008, GreenTeaGirlie again appeared on YouTube, this time after one more experience of life.

GreenTeaGirlie again brought up the discussion on the issue of authenticity that had been raised ever since Lonelygirl15. One could see that the YouTube community was becoming increasingly uncertain. Her appearance made the problem related to the entire community obvious and—after the first scare caused by Lonelygirl15—tangible as well. Comments and discussions on GreenTeaGirlie indicated a sort of self-purification of the community, which now seemed to start reflecting more and more on their platform, a sign of a new attitude towards the reception of self-portrayals. But the platform's creed, "Broadcast Yourself," does not seem to be called into question, because the way in which one "broadcasts" oneself is up to each individual person. Hence, both critical and appeasing voices can be heard.

It is the decision of the individual users whether they actually reveal their innermost thoughts and feelings or put on a mask and fool the others. Yet in any case it seems to be a role play. Already in the 1960s, Goffman established that in everyday life we perform characters and thus stage ideas to convey a certain impression of our selves. Even if this notion was not new to him, he was convinced that under the aspect of performing one can understand, in particular, the structure of our selves (Goffman, 1959). Goffman characterized the performed self "as some kind of image, usually creditable, which the individual on stage and in character effectively attempts to induce others to hold in regard to him" (Goffman, 1959, p. 252). What appears decisive here is his observation that the imputation of a self, namely once one has gained an impression of the individual, does not stem from its owner,

> but from the whole scene of his action. A correctly staged and performed scene leads the audience to impute a self to a performed character, but his imputation—this self—is a *product* of a scene that comes off, and is not a *cause* of it. The self, then, as a performed character, is...a dramatic effect arising diffusely from a scene that is presented, and the characteristic issue, the crucial concern, is whether it will be credited or discredited. (Goffman, 1959, pp. 252–253)

It would be possible, then, that in the case of YouTube, the "honest" sender, when giving an account of his or her everyday life, does not notice that he or she performs a character by making an effort to convey a certain image of the self to the audience. A closer examination of GreenTeaGirlie clearly reveals that Goffman's explication—that the imputation of a self does not stem from its owner but from the whole scene of his or her actions—is still highly topical. In this sense, the question pertaining to the authenticity of the self-portrayals on YouTube always seems to be connected to an analysis of the context. At the same time, GreenTeaGirlie's appearance supports the prevailing assumption that the question of authenticity of the person who is creating a portrait of him/herself, i.e., of self-portrayal as such, is almost impossible. Therefore, it must be considered whether the question should not be directed toward the authenticity of the picture instead. It would then first be analysed as an actual picture, to then arrive at further insights into the person presenting a self-portrayal. In my view, GreenTeaGirlie appears to strengthen this approach. There are signs of a new trend that prompts the YouTube community, in this case, to search for new frameworks of orientation.

The production of GreenTeaGirlie's videos—as one can conclude after viewing them—is subject to a fixed scenario. In most cases she sits in front of a wall flanked by two dark doors. The frame of the shots varies from recording to recording, at times showing her entire upper body, at others only down to her décolleté. In some videos she stands up, but she is always careful not to move outside the focus of the camera, which is set up directly on the table she is sitting at. It is impossible to say whether it is built into the computer or not. It is clear, however, that this is also where the computer, which she uses to control the camera, is located. In only a few cases does the camera angle deviate from this position, for example, the video in which she asks the audience about her hairdo and requests feedback. Here, the camera is set up specifically for this purpose. The design of the individual videos is therefore quite clear. The outer effect of the protagonist is at the fore. Her outfit is pretty inconspicuous but always carefully chosen, as is the case with her hairdo. GreenTeaGirlie never appears without make-up. Her gestures are very limited, she refrains from supporting what she says by referential or discursive motions. It is her voice alone that lends an effect to what she says. GreenTeaGirlie seeks to act in an authentic way. She does without an accompanying background, speaks directly to the camera, and refrains from mannerisms. Her appearance is carefully thought out all the same. In her effort to assert herself, she cannot avoid exposing herself to the suspicion of staging.

It thus becomes clear that her appearance is structured, there is very little or no room for chance happenings, "uncontrolled" actions, or slips. The picture she wants to convey of herself appears to be marked out. But at the same time, and this would be the decisive factor of this approach, the picture gives us an impression of GreenTeaGirlie's state. By attempting to draw a certain picture, she simultaneously embodies a current idea of her self.

In regard to self-portrayal, GreenTeaGirlie is just one of the many examples increasingly appearing on the Web today. The possibilities of social software—the easy handling of the applications—are used to draw attention to one's existence. The "window" that opens up to the private sphere of the "sending" person seems new in its manner, and yet we find it familiar.

JENNICAM—A WINDOW INTO THE PRIVATE LIFE

The beginnings of the webcam are known. After connecting a camera to a computer equipped with a video frame grabber, Quentin Stafford-Fraser and Paul Jardetzky chose as their subject a coffee pot next to the lab. They wrote a simple client-server program that captured images from the camera every few minutes and distributed them on a local network, thus enabling people to check if coffee was there before going to the coffee machine. By the mid-1990s, there were already hundreds of webcams on the Net, especially those offering a view of places. The early webcams were set up mainly in the United States, Europe, and Japan, but soon they also appeared beyond the digital main street—in South Africa, Pakistan, or Mexico (Levin, Frohne & Weibel, 2002). Campanella states in 2004 that "with webcams any modem jock could prove that when the sun went down in Atlanta, Georgia, it was only just coming up over Tokyo Bay. As the electro-optical matrix expanded around the globe, it became possible to simultaneously watch the sun rise and set at the same time—something that, as little as a decade ago, would have been the stuff of science fiction" (Campanella, 2004, p. 58). The private use of webcams was not long in coming either. One of the probably best-known girlcams, which soon became a paysite, however, was JenniCam, who in 1996 appeared on the Net and to whom Victor Burgin dedicated an interesting analysis. As one of the first to do so, Jennifer Ringley installed a camera on her monitor that at regular intervals uploaded stills of her college room without her intervening. From this point on, one could take part in her life in the room, watch her write, read, or sleep—in short: be part of her private sphere. Ringley by no means found this to be a burden: "I don't feel I'm giving up my privacy. Just because people can see me doesn't mean it affects me. I'm still alone in my room, no matter what" (Jennifer Ringley in Burgin, 2002, p. 229). After she graduated from college, the camera was first turned off, but it appeared again a short time later in her new apartment: "I felt alone without the camera" (Jennifer Ringley in Burgin, 2002, p. 231). Burgin describes the situation of the viewer as someone gazing through a window, thus acknowledging his or her own voyeurism. The protagonist in turn, Burgin says, identifies her camera as a mirror, prompting him to make the accusation of exhibitionism. For Jennifer Ringley, JenniCam is "a sort of window into a visual human zoo" (Regener, 2002).

The window as a metaphor appears extremely helpful in examining the presentation, as it indicates an action in which the viewer directs his or her gaze. But in this context, the question immediately arises as to the extent to which one can speak of a viewer or observer. Would it be justified to speak of a voyeur who in this case exploits a "one-sided" gaze?

The symbolic window in which the world presents itself has been represented in panel painting ever since the fifteenth century. Leon Battista Alberti is often cited in regard to this interrelation of window and painting. Hans Belting points out that, for a long time, the everyday gaze out of the window represented the so-called *perspectiva naturalis,* as if it were natural to view the world through a window. In his work *De Pictura,* Alberti in 1435 writes: "First, I draw an orthogonal square of whatever size on the surface to be painted; I assume that this is an open window, through which I contemplate what is to be painted" (Alberti, 1435). Since then, the metaphor of the window continued to be used. In his art-theoretical essay (1465), Filarete remarks that the reader should always first imagine himself standing in front of a window. With Leonardo, the perspective is the way of viewing the world through "*un vetro piano e ben' transparente*" (a smooth and very transparent glass), and Dürer translates the neologism "perspective" from the Middle Ages as "looking through." In Western culture, the window gaze became a paradigm of the picture when painting evolved on the threshold of the modern age. Belting points out that, with the analogy to the window, the painting became the most important picture medium in Western culture. It was introduced in the role of a symbolic window that privileged the gaze to the outside as a mode of perception. I would like to once more cite Belting:

> In such pictures, we look at the world from a position inside the house. What such a painting explicates is the view to the world. Yet what the same painting implies is our invisible point of observation. Hence, the reciprocal reference of inside and outside actually establishes a rule of the Western history of images. ...Here, perspective as a concept has a universal meaning. Only the person standing at a window or door can look through. The window allows the viewer to be on the one side with his body and on the other with his gaze. (Belting, 2000)

Monet painted his views of Paris from a window of the Louvre; Daguerre shot the Boulevard du Temple in 1838 from the window of his house, and there are certainly numerous other examples of pictures made under these conditions up into the twentieth century. The metaphor has survived until this day and marks the option of "connecting" to another world. In cinema the edge of the frame was to be kept as invisible as possible, but TV and video brought their own interpretation of the gaze through a window: "You really have the nice illusion that the screen of your receiver is a kind of window, an additional window in your apartment—a window opening a view to the world for you" (Brüggemann, 1989, p. 300). The integration of the TV set in a spatial environment passed into the situation of the computer. What offered and still offers itself here, very much in contrast to the black box of cinema, is a framed view allowing several windows in one: "From the transparent view of the 15th century, the image-window in the 20th century changed into luminous multi-windows which, however, still bear the traces of Alberti's concept: the image on the monitor is also mathematically structured, broken down to small, rectangularly arranged image dots" (Hüsch, 2001). So, despite the limitation of the projection screen and the angle of vision, we continue to have the impression of being able to penetrate the outside world. The window on our monitor enables us to venture into Jennifer Ringley's private sphere, to look into her private space and participate in her daily life.

Hence, the metaphor of the window is relevant in that it identifies us—the viewers—as subjects, thus allowing us to simultaneously observe ourselves. We look through the window of the monitor at a different world, with the "window frame" maintaining the necessary distance so that we don't become part of that world ourselves. Therefore, our position can indeed be compared to that of a voyeur. In current presentations, as with GreenTeaGirlie, the window can also be discerned due to the largely unchanged technological conditions. But the orientation of the performers themselves also reveals that they firmly align their appearance and scenario with the "format" of this window in order to address the viewers and attract their attention. They take care not to have too many things happen outside of the camera's radius, thus preventing the viewer from possibly discerning a "suspicious circumstance." In this way, their actions are subject to a kind of control, provoking the suspicion that the intention of the self-performer could go beyond pure exhibitionism. However, by calling into question an exhibitionist attitude, doubt is thrown on the accusation of voyeurism, in the case of both Ringley and GreenTeaGirlie.

ASPECTS OF A NEW TREND IN THE OLD ARRANGEMENT

Burgin sees the camera of Jennifer Ringley, as the "sender" of the images, as having the function of a mirror. The mirror metaphor is used here for the control screen depicting the camera image on one's own monitor. This approach is not unusual and can be found, in part quite obviously, in current portrayals as well. When GreenTeaGirlie asks the audience about the quality of her outfit and gets up from the chair to show her entire body, she never leaves the radius of the camera and gazes in a concentrated way in its direction, whereby her eyes are not always directed to the camera but a bit lower. When presenting her hairdo, she also appears to be looking at it while describing it. The screen thus becomes a mirror in which the person performing repeatedly checks her looks and actions, possibly reconsiders broadcasting it and is thus able to at least initially assess its effect on the audience. As a form of control, an inspection of one's own performance, the control screen serves a very practical purpose.

Jennifer Ringley also uses the control screen to examine the shots, but unlike GreenTeaGirlie she has no direct influence on the images. The camera records stills in a certain temporal sequence and directly sends them to the net. There is no possibility of reconsidering or editing the shots. So while GreenTeaGirlie can prepare for her appearance, Ringley cannot. Yet if one follows Goffman's assumption, one can assume that Ringley also unconsciously stages herself. The fact that gives a strong impression of authenticity by allowing a view into her private space could be in line with Ringley's insight that her remaining daily life goes unobserved by the web community.

JenniCam is probably the most famous example of the early encounter with exhibitionism and voyeurism as well as the criticism associated with it. Of course, there are much more harmless examples from that time that partially still exist. As in 2001, Anne-Sofie still presents herself at work in front of her computer. The webcam is located a bit above her, so that while sitting at the computer she never looks at the camera to make eye contact with a user. The camera is running on the side and at regular intervals sends a photo. The same possibility offered itself to Doris, who is now no longer on the net with a webcam, however. In 2001 one could watch "Doris at work," with the webcam showing a "melancholy, concentrated view fixed to the screen" (Regener, 2002). At this point, Regener notes: "As if Doris had to follow a common convention, the webcam doesn't present itself as a stimulating highlight on her site, but describes an annoying restriction through the camera: 'Darn, now I even have to behave decent when I'm online, or did you maybe expect something exciting?'" (Regener, 2002).

The webcams provide not only a new temporal factor by enabling real-time. Regener also attributes to them the illusion of movement in time and changes in the image:

> For example, Holger's webcam. Whenever you switch on, you believe to be in a moving image, although you always only see the same detail of Holger, the same image that changes only minimally: Holger looks at the monitor, in the background a wooden door with a white frame and light-blue, slightly striped wallpaper....The mise-en-scène in front of the computer suggests an activity (the same carried out by the viewer) and in general the symbol: being online. (Regener, 2002)

Doris' statement is a sign of the awareness of being observed and again brings us back to Goffman's assertion that in everyday life we act our selves in front of others. It can therefore be assumed that Ringley, too, in her awareness of being observed, develops a kind of behaviour that would differ from the situation of not being observed. The suspicion that the performance goes beyond pure exhibitionism is reinforced here. Further indications of this were provided ten years later by GreenTeaGirlie, who conceives her appearances, carries them out accordingly and thus seems to pursue other aims than to lure viewers with exhibitionist affectations.

The videos of GreenTeaGirlie reveal that she is strongly concerned with her own looks, with the effect that her appearance and her body have. A certain degree of narcissism suggests itself. And indeed—American psychologists claim to have found out that young people are becoming increasingly narcissistic. An analysis of 16,000 American college students who between 1982 and 2006 answered questions posed in the psychological test Narcissistic Personality Inventory revealed that students have never been more narcissistic than they are today. The narcissism values of two thirds of the students in 2006 are higher than those of the students in 1982. The psychologist Jean Twenge speaks of a "Generation Me" (Twenge, 2006) and contradicts the hopes of her predecessors that those born after 1982 would belong to America's next greatest generation filled with optimism and a sense of duty. She argues that persons born after 1970 are more self-centered, lack respect toward authorities, and are more depressed than previous generations. In her view, the notion of feeling self-esteem is more important than success and that they therefore place their selves above everything else. According to Twenge, this attitude has led to young people opining that any dream can be realized, but it has not prepared them for the experience that this is not necessarily the case.

What Twenge addresses in regard to the sentiment of an entire generation is explicated by Alain Ehrenberg, interestingly enough, in the context of depression.

> The career of depression starts the moment when the disciplinary model of controlling behaviour, which in an authoritarian and forbidding manner assigned roles to the social classes and both sexes, is given up in favour of a norm that demands everyone to take personal initiative: obliging them to become themselves. (Ehrenberg, 2004, p. 4)

While it was a priority in the 1950s to conform and not stick out as different, the situation has now changed enormously. Various studies—predominantly from the United States—support the proposition that now individuality, one's own opinion, and one's own will are what goes to make up present-day coolness and prestige: only in this way is it possible to stand out from others and attract attention. Ehrenberg is of the opinion that the right to choose one's own life and the demand to become one's self set the individual in constant motion. The regulating borders to maintain internal order are thus shifted. According to Ehrenberg, it is now less about obedience than about decisions and personal initiative. One is no longer set in motion by an external order; one must now fall back on one's own initiative and mental capacities. The ideal individual is no longer judged by his or her obedience but by his or her initiative (Ehrenberg, 2004).

Twenge and Ehrenberg lead to an aspect that gains validity not only in regard to the current self-portrayals as can be found on YouTube, it can also be applied to those conveyed via webcams. Both current and earlier self-portrayals benefit from the technologies of their times, and yet they are above all references to one's own existence. It is all about pointing to one's presence, bringing one's person and identity to the fore and attracting attention to them. In this respect, the self-portrayals on YouTube today as well as the earlier webcam broadcasts of simple "desk images" are expressions of "asserting oneself" (*sich-geltend-machen*). The public is characterized as a communication community of "affected ones" who make the effort to assert their presence and with it their concerns as well. This model was already formulated in 1928 by Dewey in *The Public and Its Problems*. Dewey's remarks on the public are based on a guiding notion of pragmatism according to which the lonely doubting ego of a René Descartes is replaced by the idea of coping with real problems of action in a cooperative way (Fetscher/Münkler, 1987). Bruno Latour indirectly supported this theory in 2005 with his concept of thing politics. Latour says that with this neologism he wants to "designate a risky and tentative set of experiments in probing just what it could mean for political thought to turn 'things' around and to become slightly more realistic than has been attempted up to now" (Latour, 2005, p. 14). In the process, he falls back on the metaphor of an object-oriented democracy and assumes, that "we might be more connected to each other by our worries, our matters of concern, the issues we care for, than by any other set of values, opinions, attitudes or principles" (Latour, 2005, p. 14). Among the examples he gives is the discussion on Turkey becoming a member of the EU or the closing down of the factory of the daughter, thus addressing issues in which passions, outrage and opinions appear, which evoke that kind of realism. So where facts fail, Latour resorts to concerns so as to reveal a radical change in our understanding of objectivity.

Latour's thoughts within the frame of the exhibition and political theory are of interest here because social software applications appear to conform with Latour's hypothesis. In the form of self-portrayals and expressing one's sensitivities on YouTube.com or myspace.com, the turn from matters of fact to matters of concern manifests itself (Latour, 2005, p. 23). Like no other medium, social software enables dealing with concerns and therefore with things that shape our present-day society. Social software and its consequences seem to realize what Latour hopes for from thing politics. When assembling is "no longer limited to properly speaking parliaments but extended to the many other assemblages in search of a rightful assembly" and "the assembling is done under the provisional and fragile Phantom Public, which no longer claims to be equivalent to a Body, a Leviathan or a state" (Latour, 2005, p. 41), then realism is achieved. The webcam images mark the beginning of this dealing with concerns by allowing insights into a private situation. The person recorded thus signals his/her willingness to open the "door" and allow a view of his/her sensitivities. At the same time, he/she asserts him/herself on the web, his/her name is now connected to a face and no longer gets lost in the anonymity of a chat or forum. He/She becomes a member of the public that shares its (everyday) life beyond the limits of face-to-face communication and thus begins transferring concerns to a new space.

The early application of webcam technology was largely "overshadowed" by the debate on surveillance, which apparently led to disregarding other aspects of the medium. One of the most popular studies on private webcam recordings to date is most likely Victor Burgin's analysis of the JenniCam. Upon reading it again, however, his focus cannot fend off the suspicion of being overpowered by the theme of surveillance. The characterisation of the JenniCam with concepts of exhibitionism and voyeurism seems to be insufficient. In the juxtaposition with the example of GreenTeaGirlie, it therefore appeared necessary to stress picture-theoretical aspects of the window and mirror metaphors instead of psychoanalytical ones so as to be able to point out the principle of "asserting oneself." For even if Ringley provokes an extremely private self-portrayal, one cannot disregard her intention of self-portrayal and the act of "referring to oneself" that is associated with it. The mode of attracting attention seems relatively simple, and yet it indicates—quite in the sense of the society of control—the existence of the sender and signals the highlighting of one's own individuality. What GreenTeaGirlie today achieves with social software applications in an almost professional manner began with the early webcam recordings.

Under a historiographical aspect, the examination of a present-day example thus leads to projecting current insights on an earlier example and in this way to attaining new insights. In his model of an an-archaeology, Zielinski himself mainly discusses examples of the past and characterises them as "qualitative points of change" (Zielinski, 2002, p. 45) that indicate a paradigm change. It would probably be too much to mark the examples mentioned here as such. Yet it seems that in detaching oneself from chronological accounts, for example, one gains the freedom to reveal further, hitherto perhaps unnoticed or neglected aspects of the material being examined.

REFERENCES

Abbate, J. (1999). *Inventing the internet.* Cambridge. MA: MIT Press.

Alberti, L. B. (1435). *De pictura / Über die Malkunst.* In Bätschmann, O. (Ed.) (2002). (p. 93). Darmstadt: Wissenschaftliche Buchgesellschaft.

Brüggemann, H. (1989). *Das andere Fenster. Einblicke in Häuser und Menschen: Zur Literaturgeschichte einer urbanen Wahrnehmungsform.* Frankfurt a.M.: Fischer-Taschenbuch-Verlag.

Burgin, V. (2002). Jenni's room: Exhibitionism and solitude. In Levin, T. Y., Frohne, U. & Weibel, P. (Eds.), *CTRL_Space: Rhetorics of surveillance from Bentham to Big Brother* (p. 229). Cambridge, MA: MIT Press.

Campanella, T. (2004). Webcameras and the telepresent landscape. In: Graham, S. (Ed.) (2004). (p. 58) *The cybercities reader.* London: Routledge.

Castells, M. (1996). The information age: Economy, society and culture. Vol.1. *The rise of the network society.* Cambridge, MA: Blackwell Publishers.

Ehrenberg, A. (2004). *Das erschöpfte Selbst: Depression und Gesellschaft in der Gegenwart.* Frankfurt a.M.: Campus Verlag.

Fetscher, I. / Münkler, H. (Eds.) (1987). *Pipers Handbuch der politischen Ideen.* Bd.5, Munich/ Zurich: Piper-Verlag.

Goffman, E. (1959). *The presentation of self in everyday life.* New York: New York University Press.

Hickethier, K. (2003). *Einführung in die Medienwissenschaft.* Stuttgart: Metzler-Verlag.

Hughes, T.P. (1998). *Rescuing Prometheus.* New York: Pantheon Books.

Hüsch, A. (2001). *Der gerahmte Blick: Zu einer Geschichte des Bildschirms am Beispiel der Camera Obscura.* Dissertation, Karlsruhe: unpublished

Latour, B. (2005). From Realpolitik to Dingpolitik or how to make things public. In: Latour, B./Weibel, P. (Eds.), *Making things public.* Cambridge, MA: MIT Press.

Levin, T.Y., Frohne, U. & Weibel, P. (Eds.) (2002). *CTRL_Space: Rhetorics of surveillance from Bentham to Big Brother.* Cambridge, MA: MIT Press.

Lischka, K. (5/7/2007). Fast zu schön um wahr zu sein. *Spiegel-online.* Retrieved December 10, 2008, from http://www.spiegel.de/netzwelt/web/0,1518,481447,00.html

Naughton, John (1999). *A brief history of the future: The origins of the internet.* London: Weidenfeld & Nicolson.

Patalong, Frank (9/13/2006a): Der Name der Rose. *Spiegel-online.* Retrieved December 9, 2008, from http://www.spiegel.de/netzwelt/web/0,1518,436783,00.html

Patalong, Frank (9/11/2006b): Nur falsch ist wirklich echt. *Spiegel-Online.* Retrieved December 10, 2008, from http://www.spiegel.de/netzwelt/web/0,1518,436070,00.html

Regener, S. (2002). Upload—Über private Webcams. In: Chi, I / Düchting, S. / Schröter, J. (Eds.), *Ephemer_Temporär_Provisorisch*. Essen: Klartext.

Twenge, J. (2006). *Generation me*. New York: Free Press.

Zielinski, S. (2002). *Archäologie der Medien: Zur Tiefenzeit des technischen Hörens und Sehens*. Hamburg: Rowohlt-Verlag.

Video

GreenTeaGirlie. (2007). *Curly Q*. Retrieved September 12, 2008, from http://de.youtube.com/watch?v=xxGudb4lr-k

GreenTeaGirlie. (2007). What do you think? Retrieved September 12, 2008, from http://www.youtube.com/watch?v=czTL7yvg_bA (12/9/2008)

Homepage Anne-Sofie. Retrieved September 12, 2008, from http://www.anne-sof.com

PART THREE

Web Industries and Media Institutions

CHAPTER 7

Web Industries, Economies, Aesthetics: Mapping the Look of the Web in the Dot-Com Era

Megan Sapnar Ankerson

Just a few years after the dot-com bubble burst, a new way of thinking and talking about the web began to dominate discussions of the internet's post-crash resurrection. Known by the moniker "Web 2.0," the idea was that "participatory media"—blogs, social networking, wikis, and user-generated content—were initiating a radical shift in the way the web was being produced, consumed, designed, organized, and financed. With a mantra to "harness collective intelligence" (O'Reilly, 2005), Web 2.0 ventures embraced a paradigm that drew strength from the activity of users: by uploading, sharing, and tagging their video clips, bookmarks, digital images, and opinions, users replaced experts and professionals as the chiefs of content creation.

While some dismiss the term as pure marketing hype, eagerly adopted by new media companies looking to distance themselves from the failures of earlier dot-coms, others point to the success of YouTube, Flickr, Facebook, and Wikipedia, arguing that these sites represent a different kind of logic than those of the failed initial public offerings (IPOs) in the late 1990s. In any case, the collapse of the internet bubble figures prominently in this distinction between 1.0 and 2.0 eras. When Dale Dougherty, the vice president of online publishing at O'Reilly Media, coined the term in 2003, he was surveying the state of the web industry after the bust and noted that the surviving dot-coms had websites that shared certain characteristics. "Could it be that the dot-com collapse marked some kind of turning point for the web, such that a call to action such as 'Web 2.0' might make sense?" he wondered (O'Reilly, 2005). Agreeing that it did, O'Reilly organized the first Web 2.0 conference in October 2004.

Since then, the notion of Web 2.0 has attracted significant attention, inspiring avid believers in the "wisdom of crowds" (Surowiecki, 2004) and inciting critics who decry "an endless digital forest of mediocrity" (Keen, 2007, p. 3). The term has been used to describe technologies that enhance collaboration, software languages that make these kinds of websites possible, and business plans that propose funding "bring-your-own-content" schemes (Boutin, 2006). Web 2.0 has also been understood as a design aesthetic: neutral palettes accented by vibrant high-contrast colors, significant use of white space, rounded corners, starbursts, lower-case san-serif fonts, and the subtle use of drop shadows, gradients, and reflections have all been cited as defining features of the Web 2.0 site.[1] While arguments ensue concerning the appropriate application of the term, the discourse surrounding Web 2.0 encompasses web industries, user experiences, design trends, new technologies, and types of content.

But too often, these practices are analyzed ahistorically as if they are the next obvious step in the future of web forms. The very notion of "Web 2.0" propagates an understanding of "Web 1.0" as the outdated, buggy past, one that needs an upgrade in order to function smoothly. If the old web was characterized by an era of mass hysteria driven by heartfelt (but, in retrospect, irrational) beliefs in a New Economy and vertiginous stock valuations for otherwise unprofitable companies, the new version represents an eminently more social (some say narcissistic) web designed to let the user take the spotlight, celebrating the amateur's ability to broadcast herself. But how is it that these particular values came to dominate the "turning point" after the crash? In order to better understand the context in which Web 2.0 came to flourish, we should start by revisiting Web 1.0 as a phenomenon of the dot-com boom, investigating websites as cultural artifacts that respond to particular socio-historical, economic, and industrial contexts.

Although recent scholarship by cultural geographers and historians of information labor has critically evaluated the business culture, work practices, and geographic clustering of new media industries during the internet bubble (e.g., Neff, 2005; Ross, 2004; Indergaard, 2004), there is surprisingly little academic work that links the social context of the dot-com boom period to a historical understanding of web style, production practices, and organizational structures. In this chapter, I examine the dot-com era not as a lapse of corporate rationality or a beta version of the better web to come but as a significant moment of cultural production where creative teams, corporate ties, and changing organizational structures interact to produce the "look and feel" of the web, a hegemonic discourse that changes over time.

Television and film studies scholarship has benefited from historical investigations that work to account for the "look" of a text alongside the (re)organization of cultural industries; I use this literature as a model for applying these questions to the study of web design industries. In this cultural and economic history of web design, I rely on analytic frameworks borrowed from cultural studies, political economy, and organizational studies in order to investigate how power struggles and industrial logic contribute to the creation of symbolic forms. Drawing on a wide range of sources that include production manuals, trade press, web design books, interviews with practitioners, judges' comments from web awards shows, archived websites, and threads from online discussion boards, I track the emergence of web design in the United States as a new commercial cultural industry, focusing on the ways in which competing interests struggled to control the medium. Instead of proposing a new periodization system for marking changes in the first decade of web design history, I map shifts in design practices and the organization of web industries against the four stages of a classic speculative bubble in order to relate industrial and aesthetic shifts to the socio-economic context of the dot-com boom. This scheme allows me to consider the ways stock market valuation might translate into particular modes of production, industrial organization, and shifts in the dominant discourse surrounding "quality" web design that circulated between 1994 and 2001, the period that came to be known by the retronym "Web 1.0."

Although Web 2.0 has generated interest in the cultural work of web production, the focus on users, amateurs, and fans has largely come at the expense of constructing a historical account of the web's emergence as a commercial cultural industry. This is certainly not to say that histories of amateur web production are not equally significant undertakings. Artist-practitioner Olia Lialina (2005) offers a useful catalog of popular "vernacular" stylistic elements that dominated amateur web design in the mid-1990s, demonstrating the importance of indexing and archiving the visual culture and personal experiences of early web users. As she eloquently puts it: "To me, what defines the history of web is not just the launch dates of new browsers or services, not just the dot-com bubbles appearing or bursting, but also the appearance of a blinking yellow button that said 'New!' or the sudden mass extinction of starry wallpaper" (Lialina, 2007). Indeed, we need many web histories that chart the practices and experiences of ordinary users in situated contexts. The commercial history I offer here centers initially on the United States, where web advertising began, but it extends to a global community of web designers who worked for studios later acquired by transnational corporations. Focusing on the mainstream, hegemonic practices of commercial American web culture is not meant as an endorsement or an indication of cultural value but rather an attempt to map power struggles that have been inadequately explored.

Many political economy critiques of new media rightfully question how power works in the world of cyberspace (e.g., McChesney, 2000; Raphael, 2001), but these accounts often focus primarily on media consolidation and the encroaching logic of market capitalism without even a nod to the industry structures and creative teams that produced the "look and feel" of the commercial web. After all, websites, interfaces, branding strategies, and advertising plans are not merely produced by media giants but through the creative labor of an incredibly wide range of players: freelance designers, boutique web shops, interactive agencies, in-house new media arms of advertising agencies, and global internet consultancy firms. Sometimes these units work in tandem, sometimes they fiercely disagree, and sometimes they undermine corporate siblings or other departments within the same parent organization as they bicker over budgets, talent, or creative power. For all of the attention political economy-minded scholars once showered on efforts like Disney's "Go" portal, how many have considered the role of Left Field, the web shop that produced the site, or CKS, the agency that created the portal's identity and branding? These organizations, and the new media workers who labored within them, deserve more interrogation.

NEW MEDIA CULTURAL INDUSTRIES

Scholarship that investigates the culture industries of television, film, and music recording is often concerned with understanding how the conditions of production shape the boundaries of a text. Broadcast historians, for example, have explored the various ways that industry structures, marketing strategies, and network economics have helped shape the types of programs that were made, their formal structures, and the formatting and scheduling practices employed (e.g., Boddy, 1990; Gitlin, 1983; Hilmes, 1997; Caldwell, 1995). Meanwhile, film historians have analyzed the relationship between economics and aesthetics in Hollywood filmmaking. One of the first and most influential projects to investigate the interdependence of film style and narrative form with the history of the motion picture industry is David Bordwell, Janet Staiger, and Kristin Thompson's *The Classical Hollywood Cinema* (1985), which systematically demonstrates how aesthetic and narrative conventions were formalized and standardized between 1917 and 1960. Bordwell, Staiger, and Thompson analyze classical Hollywood cinema as "a mode of film practice," which "constitutes an integral system, including persons and groups but also rules, films, machinery, documents, institutions, work processes and theoretical concepts" (xvi). The authors explain that a mode of film practice is a context, a total system that brings economic and artistic aims together in order to situate the textual processes of cinema within the historical conditions of a film's production. Borrowing this theoretical model, one could approach the history of web industries in similar terms. A mode of "web practice" then would refer to the entire system that is put in place to produce the web: the range of new media workers (paid and unpaid), the technologies and software employed, the rules and protocols adhered to, the division of labor, the organization of creative teams, and the methods of securing capital. Like modes of film practice, web practices constitute assumptions about how a website should (or should not) look, sound, function, and behave; how it should be built; and how users should interact with it.

While Bordwell, Staiger and Thompson's analysis emphasizes the enduring nature of the classical style in Hollywood filmmaking—a system characterized by narrative unity, realism, and continuity editing made to appear seamless and invisible—web style is fleeting; it seems to operate on "internet time." The way of conceiving the work of web production, the organization of teams, the stylistic devices employed, the site's architecture, the method of navigation, the perceived role of the user, even the proposed purpose of visiting a commercial website, were constantly being negotiated in the web's first decade. New techniques were exemplary for a time and then rejected. New rules dominated for months before being suddenly revised. Web practices in the 1990s, like web pages themselves, were habitually "under construction." The basic features of the classical Hollywood cinema were established early in the history of the American cinema, but not within the medium's first decade. Likewise, it will be the work of future web historians to assess the transformations and continuities of web design over time. While I do not identify a single enduring web practice of the dot-com era, mapping the ideological variations surrounding the industrial, economic, and aesthetic systems in the earliest years of commercial web culture can provide a useful foundation for thinking about post-bubble practices.

STAGES OF A SPECULATIVE BUBBLE

Because I want to emphasize the intersections of industrial and aesthetic shifts with economic ones, I frame these developments through monetary theorist Hyman Minsky's model of a speculative bubble. Although he died in 1996, just as the technology boom was taking off in the United States, Minsky's "financial instability hypothesis" has helped inform the work of economic historian Charles Kindleberger, who suggests that all speculative bubbles go through four basic stages—displacement, boom, euphoria, and bust (2000, pp. 13–16). As the speculation surrounding the dot-com market moved through each of these stages, struggles over who should make websites were deeply intertwined with shifting organizational structures and the jockeying for power between ad agencies, interactive agencies, global consultancy firms, and web boutiques. In many cases, the role of creative labor in the production process was at the heart of these disputes.

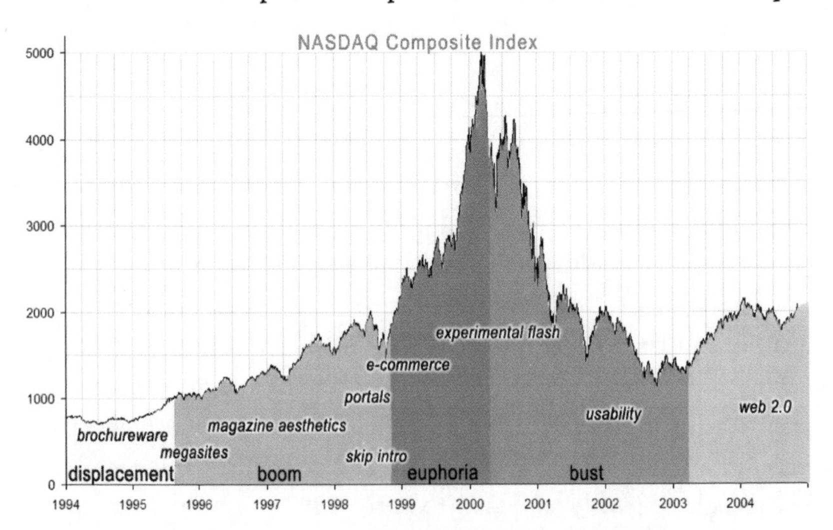

Changes in web style are identified alongside the NASDAQ composite, which is broken down into the four stages of a classic speculative bubble (Source: Stock data is from NASDAQ; the visual mapping of web styles and stages of the bubble are my own).

The chart maps the stages of the speculative bubble alongside the NASDAQ composite, the American stock index dominated by technology start-ups and seen as the main indicator of the dot-com economy. I also indicate changes in web style across this period, but I want to emphasize that formal shifts overlap and persist for quite some time. "Brochureware," "mega-sites," "magazine aesthetics," "e-commerce," and "skip intro" are all examples of once-dominant web practices linked to particular modes of production. These practices rarely disappear, but over time they cease to occupy a dominant position. For example, the discourse of "magazine aesthetics" first emerges around 1996 when more graphic designers became involved in web production. But the notion that "quality" web design bears the characteristics of magazine layout (text organized in columns, cover pages, text-wrap around images, no sound, etc.) has endured and is still widely practiced today. Finally, although not all of the interactive shops I describe were publicly traded, dot-com hype had a very real impact on web design industries in the form of salaries, budgets, workplaces, and perceptions of talent. The organization of these industries, the pitches for client accounts, and the types of creative solutions employed can all be seen as a response to the economic climate.

Displacement

The first stage of a speculative bubble is "displacement," which begins when something changes people's expectations about the future. Borrowing the term from Minsky, Kindleberger describes displacement as an outside exogenous shock to the macroeconomic system that sets off a mania (2000, p. 14). A few well-informed prospectors try to cash in on the vehicle of speculation and make very high returns, attracting the attention of other investors. In the dot-com speculative bubble, the displacement was the 1993 release of the Mosaic browser, which helped the web enter popular consciousness.

To single Mosaic out as the invention that caused the displacement sparking internet fever in the 1990s deserves further interrogation since the notion of any invention that shocks the world into change alludes to a form of technological determinism. As Thomas Streeter (2004) points out, Mosaic was a useful contribution, but it was certainly no more important than some of the other technical contributions and protocols, especially HTTP, that enabled the exchange of web pages in the first place (p. 297). Instead, Streeter suggests that Mosaic created an instant "wow effect" and became the "killer app" of the internet—generating a flurry of interest that shifted the internet from the technical realm to the popular one—because it was, quite simply, pleasurable. And the type of pleasure Mosaic offered, according to Streeter, was "the pleasure of anticipation," in which the anticipation of pleasure becomes part of the pleasure itself (p. 296). Using Mosaic was so pleasurable, Streeter argues, not for what it helped users find online (indeed, there was still relatively little there), but because it inspired them to imagine what else they might see. In other words, Mosaic served as the moment of displacement not for its technical contribution but for providing users with the first prominent site for imagining the future. The graphical space where the user meets the web functions as an ideal arena to display the visual manifestation of internet hype. Perhaps then it is no surprise that as the increasingly graphical web entered popular consciousness, the business of web design became a serious industry. Although the early years of the web are sometimes nostalgically glamorized as the innocent days before the commercialization of internet culture, Mosaic's "wow" effect washed over tech-inclined academics, midlevel managers, journalists, and hopeful entrepreneurs simultaneously. Some of the biggest interactive marketing agencies of the 1990s, Razorfish, Agency.com, and Organic Online all got their start in the year after Mosaic's launch as their founders eagerly anticipated the need for professional site builders who could bring interactive advertising online.

By the summer of 1994, the internet was generating significant attention in the popular press; *Time* magazine even ran a cover story devoted to the "strange new world of the Internet" (Elmer-DeWitt, 1994, p. 50). While advertisers had been interested in interactive media for the past year, they were beginning to realize that the much-heralded "information superhighway" would not come through interactive television trials as originally anticipated but instead would arrive through the computer screen. The launch of ad-supported online magazines HotWired (by Wired) and Pathfinder (by Time Warner) prompted ad agencies to see a new way to reach a generation of young, male, affluent consumers. Just two weeks after the launch of the first commercial browser, HotWired introduced the first model for commercial sponsorship online: the banner ad. Magazine sponsorship was a familiar system that made sense; involvement could be justified to clients and results analyzed and measured. But the work of developing these banners raised new questions: What would happen when users clicked on an ad? Would sponsors have their own "homepages," and if so, what would these pages look like? What would a visitor do there? Even more pressing, who would design these pages?

As the first 14 HotWired sponsors were secured, the work of designing several of the first commercial websites went to Organic Online, a young "web service provider" (as early interactive design shops were called) with very tight ties to HotWired. In fact, two of the four principals of Organic worked at HotWired and helped launch the HotWired site (Cleland, 1994). They split their time between the two companies, a job made easier because Organic, Wired Ventures, and HotWired were all located in the same building. These connections meant that the Organic partners were keenly aware that the first HotWired sponsors would need online solutions beyond the banner ad. Partnering with ad agency Messner Vetere Berger McNamee Schmetterer/Euro RSCG, the Organic team launched five "homepages" on a single day in October 1994 for Volvo, Club Med, AT&T, MCI and Xircom (Schmit, 1996). Limited by browser technology and HTML standards, these sites displayed blue hypertext links on a gray background and used images sparingly due to the slow speed of 14.4 modems. All content was arranged top-to-bottom on the screen with little more than headlines, bullet points, and horizontal rules. Since they were the ones hiring interactive suppliers to follow their creative direction, agencies maintained an upper hand in the early stages of the web media industry. There was the vivid sense that web design was a short-lived career: "We get small companies pitching us left and right, and we use them for formatting computer language. But pretty soon, we'll be doing it all in-house and there won't be a need for them," remarked a partner in the ad agency that hired Organic (Cleland, 1995).

But there were other considerations besides technical limitations. Early sites like those for Volvo and A&T were largely electronic brochures designed to tread a very careful ground. An AT&T executive explains: "We can't call it advertising because of etiquette on the internet that guards against blatant advertising unless you're in a strictly commercial area," such as Prodigy or CompuServe. "We have to play by the rules and be informative," she said (Evenson, 1994). In the spring of 1994, as HotWired was pursuing sponsors for the initial launch, the infamous spam attack by Laurence Canter and Martha Siegel, two immigration lawyers from Arizona, made headlines. After advertising their legal services by flooding thousands of UseNet news groups with messages about their firm's legal services for the "green card lottery," they were bombarded with messages from irate Use-

Net users. In his account of the "virtual lynch mob" that formed in response, Peter Lewis writes in the *New York Times*:

> Angry mobs pursue Laurence Canter and Martha Siegel from network to network. Electronic mail bombs and fax attacks rain down upon them. Their digitized photos are posted on computer bulletin boards around the world, like "Wanted" posters on the electronic frontier. (1994, p. D7)

Given the precedent, advertisers expressed trepidation over commercial spam. Even though legislation in 1992 opened the internet to commercial traffic in the move towards privatization, advertisers were unclear about what kinds of commercial speech were allowed. The "informative" approach, where sponsors offered users product information and corporate relations material, was typical in early 1995, but the social context or industrial logic that informed this practice is rarely explored. Many descriptions of these earliest web pages refer to a phenomenon known as "brochureware," a pejorative term used to denote websites that were mere collections of scanned documents slapped on a web page. But the brochureware phenomenon served a crucial role for commercial organizations testing a new medium and trying to navigate an ill-defined territory between innovation and social norms. Go too far in one direction and businesses risked offending a valuable market; too far in the other and they risked being lackluster. Television scholar Amanda Lotz notes that it has become a truism among media industry pundits that no one in Hollywood wants to be first, but everyone lines up to be second (2004, p. 39). From music genres to video game titles, culture industries across the board balance the fear of failing when trying something new with the desire to be on the cutting edge when a new trend emerges. For many early commercial sites, brochureware was both the safest bet and the clearest acknowledgment that their sponsors were pioneers invested in the newest technologies and cultural forms.

However, although the brochureware approach was "typical," it was not evaluated in business journals, newspapers, and the trade press as "quality" web design. The sites that earned the most positive reviews were those characterized by narrative, community, and interactivity. One site that received substantial attention for these qualities was Van der Bergh Food's site for Ragu spaghetti sauce, "Mama's Cucina," which was covered in over 50 publications from the *New York Times* to the *Scotsman* in the first five months after its March 1995 launch.

The tongue-in-cheek homepage was produced by two small interactive agencies working with Van der Bergh staffers without the help of J. Walter Thompson, the ad agency that held the Ragu account. The homepage featured a cartoon image of the fictional "Mama," who welcomed users to her kitchen with a rotating series of phrases to give viewers the impression that there was always new content. These greetings ranged from the philosophical ("All problems in life get simpler after rigatoni and meatballs") to the intimate ("I think that new butcher, Willie, has a crush on me. He slipped me an extra veal chop last week for free"), a strategy designed to give the character a strong voice while avoiding a hard sell online (Taylor, 1995). With content that included a travelogue of Mama's favorite Italian cities, a section on Italian art and architecture, user-submitted stories from "around the family table," and short audio clips of Italian language, the site had an estimated price tag of $35,000. Even the legal disclaimer was written in a folksy family-style tone: "Mama's niece Ana, the lawyer, wrote this next part: Copyright 1995 Van de Bergh Foods Inc. All rights reserved." Most prominently, the site featured recipe swapping, a staple of early internet culture. One reviewer notes that the food-centered focus has a long tradition dating back to UseNet groups of the late 1970s, a time when recipes were one of the few things that could be posted online with impunity, since they escaped cross-border copyright laws (Standage, 1995, p. 17). This put Mama's Cucina "very much in the spirit of the internet—it does not just advertise; it tries to enhance the community feeling of the Net, rather than merely the fortunes of its creators" (Standage, 1995).

Boom

After the internet was completely privatized and the NSF handed control of the backbone to private telecom corporations in May 1995, all pretenses of limitations on commercial use were lifted. As displacement transitioned to boom that summer, big investments in corporate websites made headlines as marketers scrambled to stake out a place online. Many analysts mark the Netscape IPO in August of 1995 as the official start of the dot-com boom. The company had issued 5 million shares, but brokers were backed up with orders of 100 million; prices jumped from the initial offering of $28 a share to $72 within minutes of opening. The phenomenon was covered extensively in the mainstream news, effectively introducing the World Wide Web portion of the internet to the general public (Cassidy, 2003, pp. 84–86). As businesses rushed to establish a web presence, the demand for site designers outstripped supply, triggering a flood of new start-ups that set up shop in metropolitan centers like New York, San Francisco, and London (Rifkin, 1995). This surge in commercial interest in the internet following privatization also translated into much bigger productions. Ambitious new sites were conceived as "megasites," massive arenas designed with individual sections, usually bestowed with trendy nicknames (i.e., "The Fly Zone," "The Loop," "Soundbox") for users to congregate.

After the Netscape IPO, the megasite was used as an indicator that media companies or major brands were serious about their commitment to the web. Working with ad agency Foote, Cone & Belding and interactive agency R/Greenberg Interactive, Organic developed a site for Levi's in the fall of 1995 that won huge acclaim. Compared to the brochureware sites of only a year before, the Levi's 1995 megasite was an enormous international style magazine that existed in four languages and offered specialized content for European and American visitors. With reports on street fashion from around the world, a behind-the-scenes look at the history of Levi's ad campaigns in different countries, and features on youth culture topics, the site was designed to exploit "the transcontinental nature of the internet" and "help reinforce the brand's position as an international presence" ("Case Study," December 18, 1995). In order to tap into their 17–24-year-old target, the website featured all the latest technologies: "real-time audio, full-motion video, and virtual reality (VR) navigable spaces" ("Case Study," 1995). "We want consumers to feel as though they've literally walked into the Levi's brand and have immediate, interactive access to us," explained the VP for Corporate Marketing at Levi's ("Levi's new web," Oct 12, 1995). The price tag was well over a million dollars—cheap, compared to traditional advertising.

But little more than six months after the Levi's site launched, the project was scrapped and held up as an example by Levi's and others of "an outdated, expensive, and unwieldy way to market on the Net" (Voight, 1996, p. IQ30). The huge images and multimedia features far outstripped the average home user with modems ill-equipped to handle the high bandwidth multimedia content. The discourse surrounding commercial web production shifted from an emphasis on technology and multimedia to magazine aesthetics and slimmed-down graphics. It was not coincidental that this new discourse of quality design accompanied a rush of graphic designers with print experience to the web. In 1996, a flood of web design books (as opposed to general internet and HTML authoring titles) appeared on the market by a new breed of experts who could address the principles of design.

In these first two years of commercial web design work, ad agencies were stewards of the brand, and hence, keepers of the power. They regarded interactive start-ups as filling the same role as the TV production companies that made their clients' commercials; agencies would maintain all client relationships but farm out much of the responsibility to third-party companies (Cook, 1997). In turn, web service providers worked hard to cultivate synergistic relationships and repeatedly expressed outright admiration in the trade press for the consumer insights that traditional agencies could deliver (Taylor, 1996). Even after major marketers like Reebok, AT&T, and Coca-Cola began bypassing their long-time Madison Avenue agencies and heading straight to the interactive shops to navigate their entry into cyberspace, traditional agencies were unruffled. Ad agencies and new media developers coexisted rather peacefully from 1994 to late 1996, in part because client spending in interactive media was miniscule compared to the huge budgets for traditional media. Furthermore, because web production was considered software development instead of creative labor, agencies felt that it didn't fit into the traditional agency structure (Taylor, 1996).

By 1997, however, these once-tight partnerships were becoming strained. Animosity between agencies and interactive shops grew as agencies realized online media billings were growing faster at independent upstarts than at traditional media departments (Hodges, 1997). As the dot-com boom intensified, interactive work was seen as essential to future growth. Ad agencies wanted to avoid being dwarfed by dot-com start-ups, which were now portraying agencies less as collaborators and more as "bloated and inflexible" "slow-moving dinosaurs" ("Keep Ad Agencies," 1997). Feeling vulnerable, many agencies scrambled to establish in-house interactive units or to align with digital production companies.[2] As traditional and interactive agencies wrestled to protect their claim to the web, each group began to offer the services of the other. No longer wanting to be seen merely as web designers, large and mid-size web shops re-positioned themselves as strategic marketing and brand consultants. So began the massive restructurings of ad agencies and web shops, both industries wanting to be seen as full service agencies capable of delivering internet strategy. By the end of 1997, a number of mergers and acquisitions were well under way; the trend would only intensify in 1998.

Euphoria

As boom transitioned to euphoria, the stage where technology stocks were gobbled up purely for resale and prices began to soar, websites also got bigger, more sophisticated, and more expensive. Organic's founder explained:

> In 1994, we could get away with sketching simple brochure sites on paper and being awarded the business on the spot for $20,000. Today, that site would involve a blend of brand marketing, direct marketing, customer-service databases, content databases, competitive analysis, quantitative and qualitative modeling, supply-chain integration, banking interfaces, encryption, media plans and application software development....Twenty thousand dollars would barely cover the travel and expenses of the pitch team. (Nelson, 1998)

In order to serve as a "one-stop shop" that could handle the demands of e-commerce, the industry experienced a massive wave of mergers and acquisitions that resulted in the formation of mega-agencies.[3] Restructuring brought a range of companies from web hosting experts to credit card security firms together with ad agencies and interactive shops under the same parent company—a sometimes turbulent project that made old arch-rivals suddenly siblings embroiled in contests over budgets and expertise. These re-structurings aimed to control the market by achieving economies of scale, but acquiring talent was another important rationale for acquisitions. By buying other interactive agencies in strategic locations, companies aimed to quickly assemble a productive team of experts for less money than it would take to hire talent piecemeal and pay head-hunters for the service (Nacham, 1999). But keeping creatives—who were now expected to punch in and work regular business hours—on board after a merger was another story (Rafter, 1999). Many left for smaller independent boutiques that offered more freedom and less bureaucracy.

So while industry consolidation meant more consulting expertise and backend support, there was a pressing concern that the mergers took a toll on creativity. E-commerce was all the buzz in the late 1990s, but shopping carts and secure servers did little to convey the sexiness and excitement that new media evoked. Small, specialized design boutiques were greatly sought after for individual projects, due to the edgy, avant-garde, creative images they conveyed (Williamson, 1999). As dot-com saturation peaked, commercial sites found it increasingly difficult to stand out. In order to "cut through the clutter," websites with sophisticated motion graphics, sound, and advanced interactivity were in extremely high demand as these techniques were believed to augment the interface to the brand. Therefore, at the very top of the pay scale were designers who demonstrated a strong mastery of the industry-standard web animation software Flash, distributed by the software company Macromedia. Because Flash websites looked so different from the web that most users were accustomed to, clients began demanding Flash integration into commercial websites as a way to achieve "stickiness," another buzzword of the late 1990s that emphasized web content that compelled users to stick around.

In this climate of heavy consolidation, freelance designers and small boutiques used their creative technical expertise in Flash to win back some of the power that had been ceded to the mega-agencies. But for many designers, Flash solved a number of important technical problems as well as economic ones. Working in the heat of the Browser Wars, it was not unusual for web designers to create up to four different versions of the same website to accommodate users with different browsers. Flash offered "cross browser compatibility" by allowing developers to create a single website that displayed uniformly across multiple browser versions. Not only did this substantially reduce the cost of development, it also provided designers with a new skill-set they could use to negotiate higher salaries. Furthermore, Flash's vector format made full-screen, high-resolution motion graphics possible for a fraction of the file size as a bitmapped image—which meant the technology could be used to efficiently deliver a rich media experience to modem users. And unlike HTML, Flash offered a designer total control over typography and screen space. But perhaps more than anything, Flash appealed to a whole new vision of the web, one that was vastly different from the static, silent, textual form that imitated the aesthetics of print magazines. One of the first all-Flash websites to appeal to this new logic of web design was Gabo Mendoza's 1997 website, Gabocorp.com. Following a dramatic countdown while the site loads, the opening text cinematically fades in: "You are about to enter a new era in website design. This is the new standard for all things to come. Welcome to the new Gabocorp." The dramatic sound effects, full-screen graphics, and animated buttons inspired an outpouring of imitators and parodies that proliferated for the next two years. The "new era" that many sites heralded was one that emphasized creativity, expression, and impact; websites were no longer to be used but experienced.

By 2000, Flash had become a multimillion-dollar industry of not just software sales but numerous Flash-related products: training videos, workshops, magazines, books for inspiration, books for learning code, and even books by design studios that analyze, in step-by-step fashion, the process of conceiving, building, and coding their websites. Popular events such as the semiannual "Flash Forward" conference and Film Festival generated a massive interest in the software and in the practice of web design. Meanwhile, the image of the web designer was being reinvented as a web auteur, whose work was seen as the artistic expression of a director with a unique vision. When creative director Gene Na, a judge of the 2000 *Communications Arts Interactive Design Annual* awards, was asked, "How do you see the skill sets of designers evolving as more bandwidth becomes available?," he responded:

> As the web moves more towards broadband it will require less of a designer and more of a director who understands [the language of] 'cut' and editing and flow of information....The experience can reach emotion; there's anticipation while you...watch a special loading screen and then there's a build up. Things like that...will pull more of a viewer connection to the information. (Na, 2000)

At the height of dot-com euphoria, the "hottest" designers and agencies became sought-after brand names commanding extraordinary sums for their involvement in a project. Like a film producer's determination to secure the right director, web project managers eagerly hunted for the best interactive talent who could give a website the right "look and feel."[4]

Bust

It was not long after the stock market reached its tipping point in March 2000 that the discourse surrounding Flash registered a noticeable chill. Critics like usability expert Jakob Nielson lambasted Flash as "99% Bad" in October 2000 for "encouraging design abuse" and "breaking with web fundamentals," charges that expressed a biting exasperation with the bells, whistles, and lengthy animated website intro sequences that were seen as part of the "look what I can do" bravado of the Flash aesthetic. Because the same new media production tools were being used by both professionals and amateurs, the distinctions between hobbyist and creative professional were significantly blurred. Peter Lunenfeld describes the "myth of web design" as the enchanted discourse that was attached to the notion of the creative individual in the dot-com era, inspiring countless non-professionals to think of themselves as "designers" who were forging ahead on the cutting edge of technology and internet culture (2005). But after the crash, this mythos and its visual expression in Flash also came to signify everything that was wrong with irrational exuberance: the greed, hubris, self-importance, renovated lofts, foosball tables, hedge funds, and 20-something millionaires that would soon come to be regarded as extravagant relics of the dot-com bubble. As commentators reflected on the state of the web industry after the bust, some pointed to the links between web design and the indulgences and excesses of dot-com IPOs, a "boys club" atmosphere of "tech fetishism, aggression, and adolescent image-consciousness that quickly became the dominant culture of web media agencies" (Mahoney, 2003).

Once-fashionable Flash introduction sequences became denigrated as "skip intro" style websites, named after the obligatory button added to sites should users want to skip the animation and enter the website. To combat the software's reputation for gratuitous animation, Macromedia went on damage control by heavily promoting the notion of "Flash usability," which involved educating the Flash community about the importance of "user-friendly" experiences. Macromedia hired Flash nemesis Jakob Nielsen to develop "best practice guidelines for creating usable rich internet applications with Flash." The software was revamped further in 2002, adding features designed to help Flash developers "be more productive while ensuring their work is both usable and accessible" (Macromedia, 2002). Macromedia also unveiled a broad usability awareness strategy that included a "Design a Site for Usability contest" and a usability resource center featuring tips, a showcase of best practices, test case analyses, whitepapers on developing user-friendly Flash content, and inspirational quotes from prominent Flash designers preaching usability. New Flash development books published in 2002 forcefully articulated "right" and "wrong" ways to use the technology. In *Flash: 99% Good*, for example, the authors advise:

> When applied correctly, Flash can be an effective tool for engaging and retaining users....When used incorrectly for things like lengthy intros, mystifying navigation, and

> annoying, unstoppable audio, then Flash becomes a nuisance, leaving users with a bad taste in their mouths. (Airgid & Reindel, 2002, p. 5)

The message was not to abandon Flash but to reimagine the craft of web design in a post-crash climate that emphasized the user and downplayed the expert designer.

CONCLUSION

By tracking shifts in web practice—the connected system of creative workers, new media industries, rules and assumptions, work processes and methodologies, software and "best practice" guidelines, aesthetics and stylistic forms—I tried to show how the "look of the web" is not merely the result of technologies that inaugurate new creative practices or business decisions (foolish or wise) that structure the work of web production. Nor is the post-crash pursuit of "user-centered" design a simple case of web designers "coming to their senses" at last. Instead, it is a register of a new dominant discourse that gained momentum in response to shifts in the larger socio-economic context. Without historicizing the role of the professional designer/auteur in the late 1990s, Web 2.0 simply becomes the latest buzzword, an obvious evolution towards a more advanced web.

Remembering "Web 1.0" is not an exercise in nostalgia but a way to illuminate a discursive web connecting aesthetics, ideologies, economies, and industries. As we move forward to analyze the new media cultural industries of the Web 2.0 era and the shift to user-generated content, we would do well to see these new logics as not unconnected to the power struggles of the recent past. By historically grounding these practices, we can help remind future generations of internet researchers that it is by no means "natural" or "obvious" for any media forms to work or look the ways that they do. Understanding the social, technical, and industrial climate of new technologies and new cultural forms can give us a better picture of the complex relationships between texts and their contexts.

NOTES

1. Web design blogs have been very vocal about this. See "Design Patterns: Badges, Tag Clouds, Huge Fonts." (2006, September 3). *Smashing Magazine*. Retrieved June 1, 2009 from http://www.smashingmagazine.com/2006/09/03/Webdesign-trends-badges-tag-clouds-enormous-fonts/

2. In 1997 Saatchi & Saatchi spun off its new media capabilities into a new unit called Darwin Digital and Young & Rubicam consolidated their interactive operations to form a stand-alone interactive unit, Brand Dialogue, in a partnership with WCJ.
3. Most prominently, interactive group Modem Media merged with the ad agency Poppe Tyson, creating a powerhouse agency with offices in four continents. Likewise, internet technology firm USWeb merged with interactive marketing agency CKS in a $300 million deal that united more than 1,500 employees combined.
4. For example, in *Hi-reS!* (2007) designers Florian Schmidt and Alexandra Jugovic describe how director Darren Aronofsky hired them for the *Requiem for a Dream* website after he was shown their experimental project, Soulbath.com, and wanted a similar style for the film's website.

REFERENCES

Airgid, K. & Reindel, S. (2002). *Flash 99% good: A guide to Macromedia Flash usability*. New York: McGraw-Hill/Osbourne.

Boddy, W. (1990). *Fifties television: The industry and its critics*. Urbana: University of Illinois Press.

Bordwell, D., Staiger, J., & Thompson, K. (1985). *The classical Hollywood cinema: Film style and mode of production to 1960*. New York: Columbia University Press.

Boutin, P. (2006, March 29). Web 2.0: The new Internet 'boom' doesn't live up to its name. *Slate*. Retrieved June 1, 2009, from http://www.slate.com/id/2138951/

Caldwell, J. (1995). *Televisuality: Style, crisis, and authority in American television*. New Brunswick, NJ: Rutgers University Press.

Case Study: Reaching the world with a web site. (1995, December 18) *Internet Week, 1*, 1.

Cassidy, J. (2003). *Dot.con: How America lost its mind and money in the Internet era*. New York: Perennial.

Cleland, K. (1994, December 12). Upstart company snares big Internet marketing gigs. *Advertising Age, 65*, 4.

Cleland, K. (1995, March 13). Fear creates strange bedfellows. *Advertising Age, 65*, S16.

Cook, R. (1997, August 22). Campaign interactive: Behind the hype. *Campaign*.

Elmer-DeWitt, P. (1994, July 25). Battle for the soul of the Internet. *Time, 144*, 50.

Evenson, L. (1994, October 20). Wired to go online. *San Francisco Chronicle*. p. E1.

Gitlin, T. (1983). *Inside prime time*. New York: Pantheon.

Hilmes, M. (1997). *Radio voices: American broadcasting, 1922–1952*. Minneapolis, MN: University of Minnesota Press.

Hi-Res. (2007). *Hi-ReS! Amantes sunt amentes*. London: Gestalten Books.

Hodges, J. (1997, August 4). Media buying & planning: Online work leaves media department, goes to specialists. *Advertising Age, 68*, S20.

Indergaard, M. (2004). *Silicon Alley: The rise and fall of a new media district*. New York: Routledge.

Keen, A. (2007). *The cult of the amateur*. New York: Doubleday.

Keep ad agencies in the loop when marketing goes interactive. (1997, January 10). *Interactive Marketing News*, p. 1.

Kindleberger, C. (2000). *Manias, panics, and crashes: A history of financial crises*. New York: John Wiley & Sons.

Levi's new Web site explores, reflects and shapes global youth culture. (1995, October 12). *Business Wire*.

Lewis, P. H. (1994, May 11). Sneering at a virtual lynch mob. *New York Times*. p. D7.

Lialina, O. (2005, January). *A vernacular web*. Paper presented at the Decade of Web Design Conference, Amsterdam. Retrieved June 1, 2009, from http://art.teleportacia.org/observation/vernacular/

Lialina, O. (2007). *Vernacular web 2*. Retrieved June 1, 2009, from http://contemporary-home-computing.org/vernacular-Web-2/

Lotz, A. (2004). Textual (im)possibilities in the U.S. post-network era: negotiating production and promotion processes on Lifetime's Any Day Now. *Critical Studies in Media Communication, 21*, 22–43.

Lunenfeld, P. (2005, January 21). *19?? To 20?? The long first decade of Web design*. Address given at the Decade of Web Design Conference, Amsterdam. Retrieved June 1, 2009, from http://www.decadeofWebdesign.org/prints/lunen.html

Macromedia. (2002, June 3). *Macromedia and usability guru Jakob Nielsen work together to improve Web usability*. Press Release. Retrieved June 1, 2009 from http://www.adobe.com/macromedia/proom/pr/2002/macromedia_nielsen.html

Mahoney, K. (2003, June 7). Bubble or extinction burst? *LOOP: AIGA Journal of Interactive Design Education*. Retrieved June 1, 2009, from http://loop1.aiga.org/content.cfm?Alias=bubbleessay

McChesney, R. (2000). So much for the magic of technology and the free market. In A. Herman & T. Swiss (Eds.), *The World Wide Web and contemporary cultural theory* (pp. 5–35). New York: Routledge.

Na, G. (2000). Judge's Remarks. *Communication Arts Interactive Design Annual, 6*. Retrieved February 1, 2007, from http://www.commarts.com/ca/interactive/cai00/jur_gn.html

Nacham, S. (1999, February 5). Melding cultures: No easy task when companies marry. *Business Week*.

Neff, G. (2005). The changing place of cultural production: The location of social networks in a digital media industry. *The Annals of the American Academy of Political and Social Science, 597*, 134–152.

Nelson, J. (1998, November 9). From bedroom to boardroom, the net grows up. *Adweek (Eastern Edition)*. p. 160.

Nielsen, J. (2000, October 29). *Flash: 99% bad*. Retrieved June 1, 2009, from http://www.useit.com/alertbox/20001029.html

O'Reilly, T. (2005, Sep 30). *What is Web 2.0?* Retrieved June 1, 2009, from http://www.oreillynet.com/pub/a/oreilly/tim/news/2005/09/30/what-is-Web-20.html

Rafter, M.V. (1999, January 25). Keeping the peace. *The Industry Standard*.

Raphael, C. (2001). The web. In R. Maxwell (Ed.), *Culture works: The political economy of culture* (pp. 225–249). Minneapolis, MN: University of Minnesota Press.

Rifkin, G. (1995, November 27). Increasingly, top designers are drawn to the Web. *New York Times*, p. D7.

Ross, A. (2004). *No-Collar: The humane workplace and its hidden costs*. Philadelphia: Temple University Press.

Schmit, J. (1996, July 29). Organic Online's reputation grows. *USA Today*, p. 7B.

Standage, T. (1995, July 21). Menu masters of the cyberian beanfeast. *The Scotsman*, p. 17.

Streeter, T. (2004). Romanticism in business culture: The Internet, the 1990s, and the origins of irrational exuberance. In A. Calabrese & C. Sparks (Eds.), *Toward a political economy of culture: Capitalism and communication in the twenty-first century* (pp. 206–306). Lanham, MD: Rowman & Littlefield.

Surowiecki, J. (2004). *The wisdom of crowds*. New York: Doubleday.

Taylor, C. (1995, September 25). Mama says: Anatomy of an interactive ad. *Mediaweek, 5*, IQ26.

Taylor, C. (1996, April 1). Ordering interactive a la carte. *Adweek, 37*, 29.

Voight, J. (1996, November 18). Organic online. *Brandweek, 37*, IQ30.

Williamson, D.A. (1999, July 26). Agencies left in cold as marketers expand online. *Advertising Age, 70*, S26.

CHAPTER 8

Analysing and Comparing the Histories of Web Strategies of Major Media Companies—Case Finland

Tomi Lindblom

Strategies are essential objects of any historical research, irrespective of whether they relate to wars, economic development, or organisational changes. Strategies are—or at least should be—an essential part of research into media history as well. However, they have received far less attention in media studies than in other fields of science.

Strategies have a relevant role in the functions of any company. As Michael P. Porter (1980) emphasises, every company that competes with other companies in some sector or another has a competitive strategy that includes active choices, which have been made intentionally, and passive choices, which usually occur by accident in the course of many years.

It is also important to note that a strategy is not only a plan, but it may also be a pattern, i.e., consistency in behaviour over time. Organisations develop plans for the future, and they also evolve patterns out of their past (Mintzberg, 2000).

The internet and other forms of new media have forced media companies to change their strategies and to include the web in their strategic planning. Since the mid-1990s the web has had a remarkable role in the strategic work of media companies—irrespective of whether the company operates in the print or broadcasting sector.

Research into new media histories has mainly concentrated on analysing what has happened and why. The interest in the "why" question applies also to research concerning the web strategies, which has mostly focused on examining why media companies have spread their operations to the web.

Chan-Olmsted and Jung (2001) have, for example, found that U.S. television's goal for new media is to penetrate the existing market with better customer service via the internet and more online features that would enhance the network's off-line products. Sääksjärvi and Santonen (2002) outline five reasons for newspapers to spread their operation to the internet: developing an online newspaper, earning extra income, improving cost-efficiency, gaining skills in the new technology, and strengthening the brand. Consterdine (2007) found four factors that stood out as major strategic objectives for the websites of various magazines: to create new revenue streams, to expand the audience, to attract new readers for the print products, and to build a community around the brand.

Apart from studies with the "why" question as their main theme, research into web development or business strategy research related to it has been quite rare. However, the questions of "who," "when," and "how" have an essential part in studying the media strategies—not only to research single media companies but also to compare the strategies of various companies.

In Finland three major media companies will serve as good examples in comparing the web strategies: Alma Media (including the commercial TV company MTV3), Yleisradio (YLE, the Finnish Broadcasting Company) and Sanoma (including *Helsingin Sanomat*, the largest daily in the Nordic countries).

In this chapter I suggest that analysing and comparing the history of web strategies in media companies consist of six elements:

1. Writing the history by analysing various published and unpublished materials and by interviewing persons involved with the web functions in the companies.
2. Classifying the companies' web strategies by their intensity into four categories: active strategy, careful strategy, permissive strategy, and passive strategy.
3. Analysing the problems which the companies may have faced when adopting web strategies into their functions by the seven-element-classification made by Allan Afuah and Christopher L. Tucci (2002).
4. Analysing the companies' web strategies by the classifications of convergence made by Everette E. Dennis (2003) and Graham Murdock (2000).
5. Analysing the success of the companies' web strategies by the classification of Sylvia M. Chan-Olmsted and Louisa S. Ha (2003).

6. Making the comparison by combining the results of the study of the financial development of the companies' web functions, the intensity of the companies' web strategies in the periods of convergence, and the problems of adopting the web at the companies.

WRITING THE HISTORY OF THE WEB STRATEGIES

Writing the history of the web strategies of the companies gives answers to the questions of when, by whom, and with what large-scale financial and human resources did the companies start their web functions, and how the strategies have changed over the years.

We should pay special attention to how widely the companies had mentioned and emphasised their web functions in their strategic statements over the years. The researcher should analyse, e.g., all available published material (including annual reports, press releases, printed interviews, and articles). It's also important to read all earlier research related to the companies and even to interview various people who are working closely, or have at one time done so, with the companies' web functions.

In Finland Alma Media was born of the merger of the Finnish newspaper company Aamulehti and the largest commercial TV Company MTV3 in 1997. Aamulehti and MTV3 as well as YLE were the first major actors to start planning their web functions in 1994, and they all launched their first public sites a year later. For Sanoma it took a bit longer: the company had done some web experiments in 1995, but actually started their web sites in 1996.

Alma Media was very active regarding the web and invested a remarkable amount of money in this sector in 1994–2002. The company set up many web functions, bought new media companies, and also invested a lot in the web business until the end of the new media hype in late 2001. However, because it was haemorrhaging money, the board of the company decided in 2002 to replace its CEO and abandoned most of the functions that had been built up under earlier management.

In 2003 Alma Media made many large-scale changes in its organisational structure, e.g., by abolishing its Interactive division. At the same time, all MTV3-branded web functions, which in 1999 had been transferred to Alma Media's Interactive division, were returned to MTV3. The company was so stunned at getting back the web functions that it announced its intention to concentrate on strong growth in the internet service provider business. As a result, MTV3 was as unsuccessful as Alma Media had been a couple of years earlier, and the termination of many of MTV3's web functions cost the company about 2.4 million euros in 2004 because of its own internal new media hype.

Sanoma was very careful with its web strategy in the mid-1990s. It was only in late 1999 that the company added force to its web strategy by buying other companies and building up various online and mobile services and portals. By the end of 2001 Sanoma had started to see signs that everything was not going to go as well as had been expected. The number of new media employees was halved, and there were no more investments in the web. There was no need for the board of the company to order cutbacks, because the executives of Sanoma's new media department began scaling back operations as soon as they saw that the hype would soon be over. This spontaneous action meant that, unlike in Alma Media, no Sanoma executive was forced to depart.

The board of The Finnish Broadcasting Company YLE emphasised the importance of the new distribution channels by declaring, as early as 1997, that the company give every Finn an equal opportunity to use the new communication services. YLE emphatically announced that its strategy was oriented toward a digital future and that the internet service had an important part in it.

YLE did not participate in the new media hype by building new units or functions but remained calm and moderate throughout. YLE is a public service broadcaster and was not able to buy new media companies in the same way that its commercial rivals did. However, it was able to test various web ideas with co-operative partners like Telecom Finland and Microsoft.

When researching the web strategies of the media companies note the claims that only a few, if indeed any, of them have even had a strategy for the web. They may have had some kind of strategy for their traditional functions, but in many cases they did not have one for the web until its growing popularity made them realize that a more proactive approach was needed.

This was the situation of most companies in the early years of the web, as many of them had just drifted into web operations as a result of the eagerness of some of their employees more than through companywide strategy work and planning. Especially in the mid-1990s many media companies seem to have gone into the web more by accident than by any real design. In Finland, for example, MTV3 and the Finnish Broadcasting Company (YLE) launched their first web pages only because of individual employees' eagerness, not as a result of major strategic planning by the board of the company. Hedman (2002) has noted that the same situation in some media companies in Sweden.

An essential part of the strategy is valuation of the company brand. Some media companies value their brand so highly that they use it in every possible product (in Finland, e.g., MTV3 Internet, MTV3 Teletext). However, some companies think their brand is not strong enough and for that reason are building entirely new ones, like Time Warner's Pathfinder. A new brand may become a hit but can also be a disaster (in Finland, e.g., Alma Media's Port Alma and Sanoma Corporation's 2ndhead mobile portals in 2001–02, which never caught on with the public and were soon closed after millions of euros were lost).

CLASSIFYING THE WEB STRATEGIES

There are many methods available for researching the strategies of media companies, for example, Everett M Rogers' (1995, p. 10) theory of diffusion. According to Rogers, "Diffusion is the process by which an innovation is communicated through certain channels over time among the members of a social System."

Rogers theorises that innovations spread through society in an S curve, as the early adopters select the technology first, followed by the majority, until a technology or innovation becomes common. Rogers (1995) divides consumers into five different categories: innovators (2.5%), early adopters (13.5%), early majority (34%), late majority (34%), and laggards (16 %).

The above-mentioned categorisation has also been used to analyse the media industry by reducing it to three categories (Alström, 2005), the first of which includes extremely aggressive companies that want to be part of anything new. These companies have a lot of risk capital and are not afraid to fail. The second category contains slow followers, which are curious but calm and want to maintain their market position and income. The companies in the third category are the "living dead," as they are very careful not to make any rapid movements.

Rogers' categorisation (1995) and the three-stage version of it are, however, not the best ways to analyse the media strategies. These models do not take account of, for example, the employer's role in the development of the strategies nor do they allow for the fact that a strategy may vary a lot over the years.

The business strategies may be classified into four different categories: active, careful, permissive, and passive.

An active strategy describes a company's stated aim of achieving strong growth in the new media business and even wanting to be the market leader locally or globally. A company with an active web strategy has also made large-scale investments in the web, created new functions, and strongly developed old ones. In Finland an active strategy has been vigorously pursued by Aamulehti Corporation since the mid-1990s, and by Alma Media until the beginning of the twenty-first century. As a good example of this, Aamulehti announced in their press releases and interim reports as early as 1995 that it had a great interest in following the development of the web.

A careful strategy exists when the company has been speculative or even hesitant about the web business. The company may have carried out some experiments and may have even launched some web pages, but the dominant strategy has been cautious about the possibilities of the web. This kind of company has not made huge investments nor created large-scale web functions. In Finland a careful strategy was pursued by Sanoma during the early years of its web functions. The company, and especially its largest newspaper *Helsingin Sanomat,* was still wondering, as late as the mid-1990s, whether it should go online at all.

A permissive strategy sees the web as no more than a by-product, but rather than forbid it, the management allows eager employees to start web projects. The main business of the company is TV broadcasting and newspaper publishing, and this focus has been the key point in all its decision making. Even if the company has had some web products, they have not had a place in its strategy. The top management has merely allowed new functions to be launched if employees pushed for it, but has not earmarked new funding or hired new employees for web strategies. In Finland, MTV3 in particular pursued a permissive strategy from 1994–96. The web was mentioned only in two short sentences in the company's annual report in those years. The web was not mentioned as part of the company's strategy nor even in its future prospects.

A passive strategy means that the company has done nothing in the web business, and the web has not featured in its strategic statements. Even if there have been some web functions under the company's trademark, these may have been totally outsourced. In 1999–2003, Finland's MTV3 belonged to this group, as its web functions were completely outsourced to Alma Media Interactive, a company controlled by MTV3's parent Alma Media.

ANALYSING THE PROBLEMS IN ADOPTING WEB STRATEGIES

Adopting web strategies into any company's functions is not by any means an easy thing to do. As Rolland (2003) says in "Convergence as Strategy for Value Creation," it is not only the technologies and business ideas that are novel: the whole revenue-generation model is totally new. And it is not possible to gain more revenue if the company does not understand how to operate in the convergenced branch.

The management of a company, and especially the management's strategic vision, have a huge effect on how the web functions in those traditional businesses are succeeding. In the worst case scenarios, however, the new technology changes the traditional assets of the company from benefits to disadvantages where the web is concerned.

Allan Afuah and Christopher L. Tucci (2002, pp. 204–208) list seven main problems that companies may face when adopting new media into their traditional functions: dominant managerial logic, the competency trap, fear of cannibalisation and loss of revenue, channel conflict, political power, co-opetitors (e.g. important advertisers') power, and emotional attachment. This categorisation may also be utilised to analyse the web strategies.

Afuah and Tucci describe strategies related to all industries and businesses and discuss how new technology makes its way into businesses in general. However, the classification may also be broadened to analyse especially the functions and strategies of media companies. The seven-element classification should even be developed further if it is to become a better tool to analyse web strategies. The competency trap may be divided into static and over-active sub-classes and the co-opetitor power into slow and rapid variants. Furthermore the emotional attachment should be divided into defending and downgrading versions.

Dominant Managerial Logic

A dominant managerial logic—a common understanding of how best to do business as a manager in the firm—usually comes to the fore. The longer a management team has been in the industry, and the more successful the firm has been, the more dominant and pervasive the managerial logic is. This logic can be a competitive weapon, but it may also prevent managers from understanding new things like the rationale behind new technology and the web.

In Finland, Alma Media gradually became more and more a prisoner of the dominant managerial logic in the beginning of the twenty-first century. The company had invested strongly in the web and year after year the management promised huge future profits. The promises formed the dominant managerial logic, which believed all new product launches to be successes, every investment to be profitable, etc. However, that proved not to be the case and the CEO was forced to leave.

Competency Trap

A company may face the competency trap even if its managers do grasp the importance of new technology. A company in a static competency trap understands the benefits of new technology but cannot change its functions in order to execute new models in the best possible way. The firm's managers may, for example, not want to invest enough money to build a new logistics system for their online store as they believe that the old one still works very well for all possible purposes.

A company in an over-positive competency trap presumes to know everything about the new technology and rushes to invest in anything related to it. This happened in Finland in the late 1990s when the managements of both Alma Media and Sanoma Corporation thought they had a terrific web strategy and declined to heed any doubts. A competency trap prevented those companies from anticipating the imminent burst of the new media bubble, and caused them huge losses.

Fear of Cannibalisation and Loss of Revenue

The internet often renders many of a firm's existing products or services noncompetitive. The new products/services often offer better customer value than the old one, which leads to the cannibalisation of existing ones because fewer customers would buy the older product. The fear of cannibalisation often makes firms reluctant to adopt technologies such as the internet.

In Finland this was the case inside Sanoma Corporation, whose *Helsingin Sanomat* went online in 1996, one or two years after its strongest rival newspapers did. One of the main reasons for this delay was the fear of cannibalisation.

Channel Conflict

New media render some existing distribution channels and some sales skills obsolete. Channel conflict often occurs because existing sales forces and distributors fight hard against the new channels rather than see their revenues go to them.

This often takes place when, for example, the sales office of the traditional medium is unwilling to start selling advertisements for the online publication—and reluctant to support a new special online sales department for fear that their own income would be reduced.

Political Power

Political power means the ability to have one's preferences or inclinations reflected in any actions taken in the firm or organisation. On the practical level, this would mean a power struggle within management, the departure of some directors, and changes in organisational structures. In my view, however, political power also means a struggle between the managing director and the board of the company. If the board is not satisfied with the CEO, it may change the leadership. For example, the CEO of Alma Media had made huge investments in 2002 in the web, but because the board was no longer willing to wait for the massive profits that he had promised, he was forced to leave.

Co-opetitor Power

The customers, suppliers, complementors and, e.g., huge advertisers with whom a firm has to compete and cooperate also play a role in how successful it can be in adopting the internet. If a firm's customers do not want the new technology, it risks adopting too late. If its customers are strong and powerful enough, the firm will tend to listen to them in an effort to satisfy their needs and may face the braking effect of co-opetitor power.

Listening to co-opetitors too much can be detrimental to a firm's adoption of new technology and the company faces being rushed into things. A media company may, for example, look too much abroad (and think, for example, that if this is a success in Sweden, it will absolutely be a success in Finland), or it may rely too heavily on a consultant's ideas instead of its own capacity.

Emotional Attachment

Some general managers may feel that their position is stronger with the company's present technology and products. They may have such a strong emotional attachment to the existing technology that they delay innovations like adoption of the web. If, e.g., a media company has been a print house for decades, it may be too hard for it to change course towards the web. That is what happened in the case of Sanoma, which was the last huge media company in Finland to start an online publication.

On the other hand, the new leadership may downgrade emotional attachment just to show that everything done by the old leadership was wrong. It rejects everything that is old and traditional even if the experience derived from it might still have been very useful for the company. The management of MTV3 tried to downgrade emotional attachment, as the web functions under its brand moved back to MTV3's administration in 2003. The management of MTV3 seemed to downgrade all that had been done previously in Alma Media, and it changed the leadership and the strategy of its web functions in order to imprint its own idiom on them.

ANALYSING THE WEB STRATEGIES THROUGH CLASSIFICATIONS OF CONVERGENCE

The third way to analyse the history of the web strategies is the classifications of convergence made by Everette E. Dennis (2003) and Graham Murdock (2000). Dennis mentions the four stages of convergence, which include an incremental awakening in the early 1980s, the period of early adoption in the 1990s, followed by nearly uncritical acceptance by the end of that decade, and finally in the early 2000s what he calls "presumptions of failure."

The four stages mentioned by Dennis describe quite well the attitude that Finnish media companies adopted towards convergence in the years in question. The Finnish Broadcasting Company was very interested in the web during part of the period of early adoption in the 1990s. Nearly uncritical acceptance by the end of the 1990s, the web strategy of Alma Media during those years, did very well, and the presumptions of failure in the early 2000s was characteristic of Sanoma.

Dennis made his classification system in the beginning of the twenty-first century, and, therefore, it doesn't cover the period of stabilisation and moderation that followed the years of new media hype in 2002–04. During these years of a more careful approach, many media companies scaled back their web functions and stopped their investments, especially in the cases of Alma Media and Sanoma. Therefore, the period of stabilisation should be added as the fifth stage to Dennis' classification.

Graham Murdock (2000, pp. 35–39) identifies three trends of convergence in today's media: the convergence of cultural forms, the convergence of communications systems, and the convergence of corporate ownership.

The best examples of the convergence of cultural forms are the websites on the internet as well as media art. The audiences are no longer passive but are active navigators, choosing their own routes. Convergence of communication systems means that different solutions will be offered to audiences for their computers, mobile phones, and other portable devices. The convergence of corporate ownership is linked to the two previous convergences. As the technological developments make convergence of the content possible, the media companies see a very good opportunity to combine.

The classifications of convergence made by Dennis and Murdock and the problems in adopting web strategies by Afuah and Tucci may be combined regarding the intensity of the web strategies of Alma Media, Sanoma and YLE, as table 1 shows.

The table shows that none of the companies noticed the various forms of convergence in their strategies in the late 1980s. This is only to be expected as convergence was not that well known as a phenomenon then, and as Murdock (2000) points out, e.g., the convergence of cultural forms was, even in the early years of the twenty-first century, only just emerging. The convergence of cultural forms is absent also from the strategies in the time of early adoption as well as the convergence of corporate ownership for the same reason.

The stage of uncritical acceptance in the late 1990s was essential for both Alma Media and Sanoma, both of which strongly believed in the convergence of communications systems and corporate ownership. Alma Media was facing the problems of a dominant managerial logic and a competency trap. Sanoma was even fearing cannibalisation and facing an emotional attachment. Also YLE believed in all three forms of convergence, but it has not been added into the table, as the company was not among those that uncritically accepted convergence.

	FORMS OF CONVERGENCE		
STAGES OF CONVERGENCE	*Communications systems*	*Cultural forms*	*Corporate ownership*
1980s *An incremental awakening*	———	———	———
1990s (first years) *Early adoption*	———	**ACTIVE STRATEGY**: ALMA MEDIA (AND YLE). *ALMA MEDIA: Dominant managerial logic, Competency trap.*	———
1990s (last years) *Uncritical acceptance*	**ACTIVE STRATEGY**: ALMA MEDIA AND SANOMA. *ALMA MEDIA AND SANOMA: Dominant managerial logic, Competency trap.* *SANOMA: Fear of cannibalisation, Defending emotional attachment.*	**ACTIVE STRATEGY:** ALMA MEDIA AND SANOMA. *ALMA MEDIA AND SANOMA: Dominant managerial logic, Competency trap.* *SANOMA: Fear of cannibalisation, Defending emotional attachment.*	**ACTIVE STRATEGY**: ALMA MEDIA AND SANOMA. *ALMA MEDIA AND SANOMA: Dominant managerial logic, Competency trap.* *SANOMA: Fear of cannibalisation, Defending emotional attachment.*
2000s (early years) *Presumptions of failure*	**CAREFUL STRATEGY:** ALMA MEDIA AND SANOMA. *ALMA MEDIA AND SANOMA: Dominant managerial logic, Competency trap.* *ALMA MEDIA: Political power.*	**CAREFUL STRATEGY**: ALMA MEDIA AND SANOMA. *ALMA MEDIA AND SANOMA: Dominant managerial logic, Competency trap.*	**CAREFUL STRATEGY:** ALMA MEDIA AND SANOMA. *ALMA MEDIA AND SANOMA: Dominant managerial logic, Competency trap.*
2002-2004 *Period of stabilisation*	**ACTIVE STRATEGY:** YLEISRADIO AND MTV3. **CAREFUL STRATEGY:** ALMA MEDIA AND SANOMA *ALMA MEDIA: Political power.* *MTV3: Competency trap, Downgrading emotional attachment*	**ACTIVE STRATEGY:** YLEISRADIO AND MTV3. **CAREFUL STRATEGY** ALMA MEDIA AND SANOMA. *ALMA MEDIA: Political power.* *MTV3: Competency trap, Downgrading emotional attachment*	**CAREFUL STRATEGY:** ALMA MEDIA AND SANOMA. *ALMA MEDIA Political power.*

Table 1: The classifications of convergence made by Dennis and Murdock combined with the intensity of the web strategies and the problems of adopting the web at Alma Media, Sanoma, and YLE.

At the stage of presumption of failure in the early years of the twenty-first century Alma Media and Sanoma moved to follow a careful web strategy. Both companies faced failure with the convergences of cultural forms and communications systems and, in the case of Alma Media, even with the convergences of corporate ownership. Problems in the way of adopting the web were a dominant managerial logic and a competency trap in both companies.

In the period of stabilisation political power took centre stage in Alma Media. MTV3 believed especially in the convergences of communications systems and cultural forms and the management faced both a competency trap and a downgrading of emotional attachment. YLE was complying with an active web strategy at the same time as Sanoma was acting carefully.

ANALYSING THE SUCCESS OF THE WEB STRATEGIES

Chan-Olmsted and Ha (2003) say that companies will succeed in integrating the web into their traditional business if they have developed a multi-channel business model that produces income from various sources. This kind of model must utilise the web to broaden the basic business (e.g., by e-commerce and/or online advertising), to increase the profit (e.g., by effectiveness), and increasing the number of customers (or getting them to buy more). To be successful with the web strategy requires all of the above-mentioned elements to become reality (table 2).

	ALMA MEDIA	SANOMA	YLE
Broadening the basic business	E.g., Electronic commerce and online ads.	E.g., Electronic commerce and online ads.	No E-commerce nor online ads.
Raising profit	Some efficiency but huge losses.	Losses but smaller than in Alma Media.	No losses.
Increasing the number of customers (visitors)	Number of visitors increased in 1994–2004.	Number of visitors increased in 1994–2004.	Number of visitors increased in 1994–2004.

Table 2: Integrating the web into the basic business.

As the table shows, the integration of the web into the basic business has been quite successful in all the companies in the study. Alma Media and Sanoma got into e-commerce in 1994–2004 and started publishing online advertisements. Even if they both suffered some financial losses due to unsuccessful investments, both of them also developed their organisation and business procedures and achieved some economies as a result. Both companies have attracted more and more visitors to their web sites.

It would, however, be very difficult to analyse the success of the web strategies merely by noting the elements mentioned by Chan-Olmsted & Ha. Concentrating on, e.g., broadening the basic business would leave YLE out of the equation, as it does not sell advertising on its web site.

Comparing the increase in profits is also quite problematic. Does being profitable mean that the web is yielding more money than the company is investing in it? If so, how should all the investments be measured, i.e., is the company profitable only after it has earned more money from the web than it has invested in it during all the years of its history? Or is it enough only to look at how the company's share price has risen or just to count the increase in the number of customers? (Pavlik, 2001).

Increasing the number of customers would be basically the easiest way to compare companies, but it is still quite problematic, because the companies have used different ways of counting online visits and visitors, and there is more than one specific that could be easily compared.

MAKING THE COMPARISON BY COMBINING THE RESULTS

As there are no clear instruments to measure success in the web strategies, the success should be analysed with respect to each of the following elements: (1) the financial history of the web business in the companies, (2) the intensity of the web strategies by stages of convergence (Dennis and Murdock), and (3) the problems that the companies have faced in adopting the web into their businesses (Afuah & Tucci).

Financial History of the Web Business in the Companies

If the web strategies of Alma Media, Sanoma and YLE were measured by the success of the financial outcome, the biggest success stories over the years 1994–2004 would seem to be both YLE and Sanoma. In 1995–2001 Alma Media had to bear the loss of a total of 230 million Finnish marks (Finland's currency before the euro), which the company spent on its web and other new media functions; that sum comes to around €38 million. MTV3 (a division of Alma Media at that time) was even forced to declare losses of 2.4 million euros in 2004 because of the web. At Sanoma, the losses were a lot smaller and YLE did not suffer losses at all.

The Intensity of the Web Strategies by the Stages of Convergence

The success of the web strategies can be examined more thoroughly (at least in Finland) only after the decisions made at the stage of uncritical acceptance in the final years of the 1990s. At that time both Alma Media and Sanoma started to invest heavily in various web functions. The crucial year for the intensity of the web strategies seems to be 2000, when both Alma Media and Sanoma continued their active strategy, but YLE decided to move from an active to a careful one.

Now, when analysing the strategies of the companies in retrospect, the biggest mistakes Alma Media made in 1999–2001 seem to be its massive investment programme and the Port Alma mobile portal launched by the company. Most of the investments proved to be worthless, and Port Alma had a lifespan of only about 18 months. Sanoma was wasting its resources on its 2ndhead mobile portal and a massive online portal called Lumeveräjä, which never gained any popularity with the audience.

The biggest losers at the stage of uncritical acceptance in the late 1990s were Alma Media and Sanoma. The biggest winner seems to be YLE, which did not make useless investments.

Alma Media became, however, one of the biggest success stories in the early years of the millennium, during the period of presumption of failure, as the company changed its strategy and was able to make its new media division profitable after swingeing cutbacks. Even Sanoma enjoyed success in its new media strategy then, as it switched to a careful strategy rather than the active one that it had been pursuing.

The biggest loser in the period of stabilisation was MTV3, which got the web functions bearing its brand back from Alma Media and incurred huge losses with its active strategy, competency trap and efforts to downgrade emotional attachment.

The Problems the Companies Have Faced in Adopting the Web into Their Businesses

A dominant managerial logic and a competency trap have proved most harmful for the companies. Fear of cannibalisation and a preserving emotional attachment may have delayed the adoption of the web, but the delays have ultimately caused little—if any—damage to the companies. Political power had a positive effect on Alma Media, but a downgrading emotional attachment had the opposite influence in MTV3's case.

To summarise, it can be noted that none of the companies has been absolutely better than any of the others in their web strategy during the period 1994–2004 as a whole. However, the biggest success seems to have been enjoyed by YLE, which avoided the worst case combination, i.e., an active strategy compounded by a dominant managerial logic and a competency trap.

YLE avoided unnecessary investments mainly because, as a public service broadcaster, it was not able to implement them as effectively as its commercial rivals were able to do. The company did, however, follow the development of the web functions very carefully and participated in many co-operative projects with companies like Microsoft as early as the 1990s. YLE has also been very active and innovative in its web functions. It was, for example, the first company in Finland to launch journalistic mobile services. In 2009 the company had a collection of in total more than 13,000 audio and video files on its web site (www.yle.fi/elavaarkisto) making it one the largest on-line archives in Europe.

CONCLUSION

When researching media histories one should not forget media strategies. New media, like radio in the 1920s, television in the 1960s, and the web in the 1990s, have all had a remarkable role in the strategic work of media companies. It's also important to understand that all media companies consist of people—managers and employees—and that both groups may have a remarkable effect on the strategy.

Some individual employees' eagerness may be vital if a company is to adopt new ideas and functions. At the same time, managers following a dominant managerial logic and falling into a competency trap may massively harm the company.

It is also important to understand that a strategy is not a stable element but evolves over the years. The intensity of a strategy also changes, as do its effects on the company's functions and success. Changing the strategy may cause many problems for the company, and it is very important to understand the possible negative effects that the new functions may have on traditional ones.

Strategies have been an area to which far less research has been devoted in media studies than in other fields of science. Now, however, it is time to change this approach in media research. Let us not face a competency trap, a fear of cannibalisation, nor an emotional attachment in our approach to media strategy research, and let us switch from a passive or careful strategy in research to an active one.

REFERENCES

Afuah, A. & Tucci, C. L. (2002). *Internet business models and strategies. Text and cases.* Second Edition. Boston: McGraw-Hill.

Alström, B. (2005). Från tidningsföretag till innehållsföretag. In: Hvitfelt, H. & Nygren, G. (eds.): *På väg mot mediavärlden 2020. Journalistik, teknik, marknad.* Tredje upplagan. (pp. 139–157). Lund: Studentlitteratur.

Chan-Olmsted, S. M. & Ha, L. S. (2003). Internet business models for broadcasters: How television stations perceive and integrate the internet. *Journal of Broadcasting & Electronic Media,* Vol. 47, 597–617.

Chan-Olmsted, S. M. & Jung, J. (2001). Strategizing the net business: How the U.S. television networks diversify, brand, and compete in the age of the internet. *The International Journal on Media Management.* Vol. 3 (IV), 213–225.

Consterdine, G. (2007). *Routes to success for consumer magazine websites.* FIPP.

Dennis, E. E. (2003). Prospects for a Big Idea—Is there a future for convergence? *The International Journal on Media Management,* Vol. 5, (1), 7–11.

Hedman, L. (2002). Medieföretag utan strategier. In: Alström, B. & Hedman, L.: *Nyheter 2020—Rapport 2. Dagens medieföretag—morgondagens affärsidé* (pp. 56–70). Sundsvall: Mittuniversitetet.

Mintzberg, H. (2000). *The rise and fall of strategic planning.* London: Prentice Hall.

Murdock, G. (2000). Digital futures: European television in the age of convergence. In: Wieten, J., Murdock, G. & Dahlgren, P. (eds.): *Television across Europe: A comparative introduction* (pp. 35–57). London: Sage.

Pavlik, J. V. (2001). *Journalism and new media*. New York: Columbia University Press.

Porter, M. P. (1980). *Competitive strategy: Techniques for analyzing industries and competitors*. New York: Free Press.

Rogers, E. M. (1995). *Diffusion of innovations*. Fourth Edition. New York: Free Press.

Rolland, A. (2003). Convergence as strategy for value creation. *The International Journal on Media Management*, 5, (1), 14–24.

Sääksjärvi, M. & Santonen, T. (2002). Evolution and success of online newspapers: An empirical investigation of goals, business models and success. In: Monteiro, J.L., Swatman, P.M.C. & Tavares, L. V. (eds.): *Towards the Knowledge Society, eCommerce, eBusiness and eGovernment*. Proceedings of the Second IFIP Conference on E-Commerce, E-Business, E-Government (I3E 2002), October 7–9, 2002 (pp. 649–665). Lisbon: Portugal.

CHAPTER 9

BBC News Online: A Brief History of Past and Present

Einar Thorsen

The BBC News website has gone from being a late (official) entry on the internet in 1997 to one of the most popular online news sites in the world and an integral part of the BBC's digital strategy—widely referred to as its 'third broadcast medium.' The evolution of the BBC website is an incredibly rich and diverse history—some even argue, that more than an evolution, the site represents a revolution in web history (Barrett, 2007).

BBC Online acts as an umbrella for a wide range of websites and subsections, supporting the Corporation's various broadcast channels (television and radio) and programmes, in addition to other specialist websites (e.g., education, archives, traffic, and travel). This chapter is concerned with the news website, BBC News Online, which is one of the success stories of the Corporation's online strategy. In taking a historical approach to the study of this website, the chapter concentrates on technological innovation and 'user-generated content' (the BBC's preferred term for audience material), which have played a central role in the evolution of BBC News Online (for a more focussed policy discussion, see Thorsen, Allan & Carter, 2009).

This chapter explores four key historical periods of the BBC News Online website—the Corporation's online activities leading up to the official launch in 1997, the early years of the website that led to it being firmly established as one of the world's most popular news sites, the eight months from December 2004 to July 2005 when user-generated content was propelled from a culinary add-on to centre stage, and finally the present-day experimentation with technology that is starting to fulfill early visions of integrated multimedia storytelling. With rapidly evolving online news forms and practices, it is equally important for a web history to document the present. Technological experimentation and projects that might never see the light, could conceivably be forgotten or overshadowed if documented retrospectively. The chapter concludes by highlighting areas of concern in relation to preserving the BBC News Online website as a historical web artefact and the human processes associated with its evolution.

AUNTIE GOES ONLINE

The Corporation's focus in terms of new technology adaptation was firmly fixed on the traditional broadcast media, the digitisation of these, and the role of cable and satellite broadcasting (cf. Goodwin, 1997). The early development of the BBC website was not guided by policy per se but rather by the foresight and dedication of BBC technical staff. Brandon Butterworth (member of the BBC design and development team at the time) in particular was a central driving force in the early years and the person who registered the bbc.co.uk domain name in October 1991 (Butterworth [1999], n.p.).

Butterworth (Butterworth [1999], n.p.) had registered with the Defense Data Network Network Information Center (DDN NIC) in January 1989 and received a Class B address to cover the entire BBC network. He set up internet access in mid-1989 as bbc.uucp (Unix-to-Unix Copy, a legacy system used for internet connectivity) with dial-up access via Brunel University—a service only made available to the BBC development group. Butterworth also describes how he was originally not allowed by the UK academic naming body, NRS, to register anything other than a UK domain (.co.uk) and was required to have a director sign the domain application form to prove that it was legitimately coming from the BBC (it was signed by C. Dennay, director of Engineering at the time).

The domain was originally used for internal communication, although Butterworth solicited content from around the BBC to create proof-of-concept websites.

> As new technology, such as streaming, became viable I enticed more to join....It was symbiotic—I needed content to test the technology, producers needed technology to deliver new services, the public was hungry for content and their use justified our efforts. (Butterworth, cited in Barrett, 2007, n.p.)

BBC Education was the first to capitalise on the opportunity, "recognising that it could enhance learning beyond the broadcast in the same way as leaflets, books and events" (Barrett, 2007, n.p.). George Auckland, education producer at the time, recalls having to teach himself HTML programming in order for the Education team to produce a companion website for their television programme *The Net* in 1993—without anyone's permission announcing the URL at the end of the programme (ibid., 2007, n.p.). The *BBC Networking Club*, another BBC Education project, launched in June 1994 and started to formalise the arrangement—acting as a means to get members of the public connected to the internet and more importantly the early BBC content.[1] Starting in 1995, several of these early projects also sought to use the internet as a means of interacting with members of the public during live television and radio programmes.

> Email feedback seems trivial now, but being able to respond to a programme and have the presenter respond to you on air was far simpler to do than a phone-in. IRC [Internet Relay Chat] questions into live political chat shows hooked News and Radio 3's Facing the Radio programme produced live from user-generated content and streamed the programme. (Butterworth, 2007, n.p.)

The BBC News and Current Affairs team published a dedicated site for the 1995 budget speech, entitled Budget '95, in collaboration with the Press Association.[2] The news and audio links were, however, all directed to the Press Association site, and the promise of live coverage never materialised (see Belam, 2005). Experimentation continued in August 1996 when the BBC published a party conference website, including a live uninterrupted audio feed (unlike the programme breaks on radio and television) and 'wall-to-wall coverage' (Butterworth [1999], n.p.). The event that really propelled the development of the BBC News Online project, however, was the surprising popularity of the dedicated Budget 96 site, which was launched in November 1996.[3] The site contained background information on the budget (analysis, history, and procedures—with an associated quiz), RealAudio streams and some 28 news reports (published in the period November 11–27, 1996), details of the main measures and reaction from key political parties. There was also a section dedicated to answering emails from members of the public (eleven were published with associated responses from experts on the *Money Box Live* panel) as well as transcripts of the Radio 4 *Budget Call* programme, where listeners had called in to ask questions about the budget.

At this stage the BBC website was still destined to become a commercial operation. The impetus for this came in part from a White Paper entitled *The Future of the BBC*, published by the Conservative government in 1994, "which urged the BBC to expand into new media and to become more commercial, in order to both make up its financial shortfalls and to forge a bridgehead for British media into global markets" (Born, 2003, p. 66). When exploratory talks with Microsoft about a potential partnership stalled "after the software giant suggested it might like some editorial input" (Smartt, 2007, n.p.), BBC management instead opted to have a commercial presence (using the domain beeb.com) through an existing deal between BBC Worldwide and computer company ICL. However, following the successful renewal of the BBC's Royal Charter in May 1996, John Birt (director general at the time) pulled out of the deal with ICL at the last minute in December 1996, deciding instead to make news and sport public service offerings (Barrett, 2007, n.p.). The decision was to have an incredible impact on all of the BBC online activities and was described by Jem Stone (BBC Future Media and Technology executive producer) as "the most important thing he ever did" (cited in Barrett, 2007, n.p.).

The BBC's Election 97 site went live on March 17 when then Prime Minister John Major announced May 1 as the election date.[4] Birt's decision to pull out of the ICL deal and the popularity of the Budget 96 website helped the BBC News team justify the creation of a dedicated election website. However, as Butterworth recalls, the approval was only issued some six weeks before the election, leaving the people working on the project little time to prepare (Butterworth, 2007, n.p.).

Upon launch the BBC published a news report, together with an audio clip of Major's announcement (just shy of 17 minutes long). Subsequently, about 5–10 news reports were published most days leading up to the election. Beyond news reports, the BBC also provided lists of the various constituencies, details of all candidates, and party profiles. These profiles formed the vast majority of the approximately 8,000 pages published on the site. They were created automatically using a proprietary Content Production System (CPS, originally built in three days, it gradually evolved and still forms the basis of the BBC News website), which "turned live Ceefax and Election system feeds into html for each constituency and candidate" (Butterworth [1999], n.p.).

Background issues were also explored, including an archive of past elections, analysis of campaign issues, including a tool allowing comparison of party manifestos, and finally detailed information on the election procedures. Throughout the site were links to audio content published in Real Audio format. On polling day, results were published on a special 'live' page that was updated continuously.

Despite politicians and the political parties not making much of an attempt at engaging with voters on their sites, the BBC requested feedback both on the quality of its website and on specific election issues. The BBC published a handful of this feedback in a section entitled "You say!," which would in 2001 become "Talking Point" and in 2005 "Have Your Say." The BBC also invited users to submit questions which were then put to politicians and published in a 'forum' section. However, only five politicians and Bill Bush, the head of the BBC Political Research Unit at the time, actually answered questions. Other interactive features included an early attempt at recreating Peter Snow's Swingometer and some more basic calculation forms to predict outcomes based on percentage of overall vote, as well as a quiz-based game entitled "Have You Got What It Takes to Be an MP?" Many of the features were not fully developed or were indicative of innovative forms of use being held back by technological limitations.

The Election 97 site was considered a great success internally and BBC News quickly established Politics 97 as a follow-up site,[5] which included the first public screening of the Hong Kong handover (Butterworth, 2007). The site was essentially a response to the positive performance of other news sites (including CNN) and was only intended as a stop-gap while another team worked on the full news site (Butterworth, [1999]). It was, however, the death of Diana Spencer (Princess of Wales) and Dodi Al-Fayed in a car crash on August 31, 1997, which finally demonstrated the need for and justified the investment in BBC News Online from a governance point of view. The tribute site, which was hastily put together overnight, received an estimated 7,500 emails on the topic and all were published.[6] Bob Eggington, project director of BBC News Online at the time, recalled how this response made him realise the importance of incorporating citizens' voices.

> It was a huge revelation to me that people wanted to participate and what they wanted to read was what they, not the BBC, had written. (Bob Eggington cited in Barrett, 2007, n.p.)

Butterworth was still leading the technical development and described the impulsive reaction by management to finally commit to a BBC News Online site as follows:

> By a week later—September 10th—the response to the Diana coverage had convinced everyone that the Internet would be big and that the BBC would be there—properly. With an October deadline, there was no point continuing with meetings. A committee wasn't going to make it. A ninja squad was needed.
>
> I got a small bucket of cash and got told to do whatever was needed. (Butterworth, 2007, n.p.)

The site ended up being less ambitious than 'the great ideas' the design team had originally intended. Mike Smartt, BBC News Interactive's editor-in-chief for the first eight years, recalls how the original design for the BBC News Online site was rejected three weeks prior to launch on the basis that it would "take several hours to render on people's screens down ponderous dial-up connections" (Smartt, 2007, n.p.). BBC News Online eventually went live in an official capacity on November 4, 1997, with the main BBC Online website going live on December 15, 1997.[7] Originally the BBC was granted a one-year trial by the Department for Culture, Media and Sport (DCMS), which was then ratified a year later (Barrett, 2007, n.p.).

FORMALISING BBC NEWS ONLINE

Despite its late official arrival on the scene, the BBC quickly established itself as the leading British content site on the internet—mitigating some of the early criticism the BBC received in relation to adaptation of new technology (Goodwin, 1997). By March 1998 the BBC News website recorded 8.17 million page impressions, and by June of that year BBC Online offered 140,000 pages of content, of which about 61,000 consisted of news (Allan, 2006, pp. 37–38). The BBC News website became known internally as the "third broadcast medium" (Allan, 2006, p. 37), though Smartt described the site more pragmatically as a dynamic newspaper or a hybrid of formats:

> When I was asked in the early days what BBC News Online would become I used to say: a national and international newspaper, updated every minute of every day, with the best of TV and radio mixed in. (Smartt, 2007, n.p.)

While the analogy of a hybrid newspaper is useful in relating to the predominantly text-based format of the web at the time, the BBC's commitment to the internet was very much based on extending its public service values to the online domain. These public service values are often surmised as "inform, educate and entertain," based on the BBC mission statement that has remained largely unchanged for the past 80 years (BBC, 2007, n.p.). The 'historical' functions of the BBC more specifically also incorporate "informing democracy and citizenship," to ensure people "have the necessary knowledge to make informed decisions," and the BBC acting "as a technological pioneer" (Graf, 2004, pp. 68–69). While the latter explains how the BBC's online presence was able to develop before its official launch in 1997, the former point demonstrates the reason why such technological innovation often concerned forms of internet use that would increase the interaction between the Corporation and British citizens.

The Corporation's submission to the license fee review panel in March 1999 articulated for the first time in a formal capacity what the Corporation perceived to be the core elements of BBC Online.

- The provision of news and information
- The role of trusted guide to the internet, helping users to enjoy the full potential of the internet
- The development of communities of interest, based around BBC content
- The opportunity for viewers and listeners to provide feedback on programmes and services
- The provision of a range of educational sites and services
- Local and regional content

(cited in Graf, 2004, p. 69)

News and information were at the forefront, while the third and fourth points demonstrate the importance of interactivity and civic engagement, which are positioned in the report as a core objective to delivering on the BBC's public service obligations. Interestingly the interactivity is stated as being between the BBC and members of the public ("feedback"), as well as between members of the public themselves ("communities of interest"). These social elements have a stated purpose of "re-enforcing democratic values, processes and institutions" (cited in Graf, 2004, p. 70). The strategy of developing BBC Online as a public service offering was also a long-term commitment to future generations because, in the words of Bob Eggington, "that's where young people are going" (Bob Eggington cited in Allan, 2006, p. 35). This view was shared by the independent panel on the future funding of the BBC in 1999, which rejected external pressures to turn BBC

Online (including news and sport) into a commercial operation as they expected it:

> . . .to become a core part of the BBC's public service in the next few years. We also expect that closer convergence will take place between websites and broadcast services, so that the BBC's domestic audience will increasingly access BBC output via the website. (Davies et al., 1999, p. 65)

National elections represent perhaps the most explicit opportunity to demonstrate commitment to civic engagement and technological innovation, not least because waiting for the next election allows the BBC considerable time to prepare its online strategy. By the 2001 UK General Election the internet was starting to have a significant impact on campaigning and even the BBC announced (in its *Guidance for All BBC Programme Makers during the General Election Campaign*) that "[t]his will be the first full Online election." The BBC's Vote 2001 site provided several animated interactive features and two key sections for civic engagement. The first of these, "Talking Point," allowed citizens to post comments on a range of pre-defined issues and questions. This section can essentially be seen as an attempt at facilitating debate between ordinary members of the electorate and is the precursor to the "Have Your Say" section during the 2005 election. The second feature, entitled "Forum," was vastly improved from the Election 97 equivalent. Essentially an extension of the "Talking Point" feature, Forum allowed citizens to submit questions to the BBC, a selection of which would then be put to politicians by one of its correspondents. The BBC this time also commissioned ICM Research to conduct regular online surveys of a 2,000-strong voters' panel, aimed to be representative of the UK adult population and not just internet users. The feature was dubbed "Online 1,000" and contained a new issue every month, and every week in the three weeks leading up to the election. Again there was also an opportunity for pre-voters to get involved—this time they were invited to suggest political policies on the Newsround feature "If U Were Prime Minister," which, according to Coleman (2001, p. 683), received several thousand posts.

The Vote 2001 site registered around 500,000 page views every day throughout the campaign, with a massive surge to 10.76 million on polling day, June 7, and results day, June 8 (Coleman, 2001). The latter figures exceeded the BBC's previous record, interestingly achieved by the 2000 U.S. presidential election. Coleman concluded, "[p]eople go on the web for breaking news (such as election results) and personalised information (such as their constituency results)" (Coleman, 2001, p. 683). Overall, however, the internet had little decisive impact on the 2001 election. According to a survey carried out by Mori only 7% of respondents claimed to have used it to look for election information, compared with 74% for newspapers and 89% for TV. Only 4% of respondents said it had a 'great deal' or 'fair amount' of influence on their voting decision (cited in Chadwick, 2006, p. 161).

If the 2001 election represented a carefully planned execution of an online news strategy, events later that year epitomised the other extreme of this spectrum. In the hours following the two planes crashing into the Twin Towers on September 11, online news websites in the United States (including those of CNN, MSNBC, ABC, CBS, and Fox News) buckled due to the unprecedented number of visitors and remained largely inaccessible throughout the day (for a detailed discussion, see Allan, 2002). Unable to access domestic news websites or seeking alternative perspectives, many people in the United States turned to international news providers. The BBC News website received a large proportion of this traffic as one of the most popular news websites in the world at the time. Together with a dramatic increase in number of hits from the UK (many of whom were accessing the internet from work), the BBC began to struggle to maintain its online presence as well. Brendan Butterworth described how the surge in traffic originally seemed like a malicious hacker attack:

> I was sat in an operations meeting when the pager went off and didn't stop: something big was happening. There was a massive influx of traffic to the site—a DoS [denial of service] attack, it seemed. Damion called us back: "*there was this plane...,*" We turned on a TV and saw a burning World Trade Center Tower. Then another plane. (Butterworth, 2007)

The template for the BBC homepage was not designed to cope with a breaking news story of this magnitude, and "all that could be done was to edit the three promotional slots on the page to carry news of the unfolding events" (Belam, 2007). Eventually, the technical team bypassed the content management system altogether and uploaded small HTML updates via FTP—some of which were still available years later if you knew the exact URL. The BBC website was reshaped and focused only on the single story—Mike Smartt, BBC's new media editor-in-chief at the time, recalled:

> We decided to clear everything off the front page, which we've never done before and concentrate all our journalists on the story....Most important to us were the audio and video elements. It was among the most dramatic news footage anyone has ever seen. The ability to put all that on the web for people to watch over again set us apart. (Smartt, cited in Allan, 2002, pp. 130–131)

The BBC's servers experienced hits in the millions, far surpassing the record set during the election earlier that year, and staff struggled to maintain its presence online despite having streamlined the content.

> Our New York server farm was two blocks from the WTC site; it survived but suffered as power failed. The dust eventually clogged the generators and there were problems getting in fuel. The only outage was in the days after; we covered that by moving all traffic to London. The sites were designed to operate as hot spares for each other. We had planned around London suffering at some point, but it was the opposite. (Butterworth, 2007)

Many people ended up seeking alternative sources, not least various forms of citizen journalism that flourished with eyewitness accounts, personal photographs and the occasional video footage (see Allan, 2009). Whereas citizen journalism played an important part in reporting the crisis as it unfolded, the BBC was *not* able to capitalize on 'user-generated content' (the Corporation's preferred term for any audience material) in the same way. That is, although the BBC claimed to have received thousands of emails from people witnessing the events, only two of these reports led to live news interviews with people in New York (Wardle & Williams, 2008, p. 2). Another nine 'moving eyewitness accounts' were published as part of the in-depth online feature "America's Day of Terror." It was not until three years later, prompted by another crisis, that citizen journalism really came to the fore of the BBC's website.

AUDIENCE MATERIAL GOES MAINSTREAM

In the space of eight months three key events propelled citizen journalism and audience material (or 'user-generated content') into the mainstream and transformed the way BBC News Online handled public participation online: the Indian Ocean tsunami in December 2004, the UK general election in May 2005, and the London bombings in July 2005.

The Sumatra-Andaman earthquake that led to the tsunami on December 26, 2004, is generally considered to be one of the most powerful ever recorded and killed at least 283,000 people. It was one of the deadliest earthquakes of all time—second only to the Chinese Shaanxi earthquake in 1556, which killed some 830,000 people. Very few Western news organizations had reporters nearby who could respond quickly to the disaster. Early coverage, therefore, tended to focus on the administrative response to the crisis as a stop-gap (Allan, 2006). Even when Western reporters did make it to the scene, their access was restricted by the same logistical problems facing aid workers, with additional criticism levied at them for a top-down, or 'helicopter journalism,' approach to the crisis (Schechter, 2005). Instead ordinary citizens recorded the events as they witnessed them unfolding and provided the vast majority of the extraordinary imagery used in the mainstream media (see Allan, 2009; Gordon, 2007; Liu, Palen, Sutton, Hughes & Vieweg, 2009).

While the BBC cleared its broadcasting schedules to make room for extended bulletins and special programmes, the website provided extensive contextual information, including graphics explaining why earthquakes occur and a seven-page animated guide to the tsunami. The BBC received thousands of emails containing eyewitness accounts, some including digital photographs and even video shot using mobile phones. Audio-visual material was used to illustrate news packages, while emails sent to the news website were read out on BBC News 24. The BBC News website also used its "Talking Point" section, now retitled "Have Your Say," to help people establish contact with missing friends or relatives. The message board was incredibly popular, receiving more than 250,000 hits on the first day alone. Using the website in this way was new territory for the BBC. Matthew Eltringham, an assistant editor on the site, explained:

> This has grown out of nothing—but we've managed to reunite six sets of people so far. One Dutch man found his brother via a Vietnamese woman living in Stockholm. (cited in Price, 2005)

The Indian Ocean tsunami provided an unexpected source of audience material through the citizen-journalism efforts of people caught up in the tragic events, providing important lessons about audience interactivity in preparation for the

2005 UK general election. Following its early attempt in 1997 and more successful execution in 2001, the BBC again created a dedicated site for its election coverage—this time entitled Election 2005. As in previous years, BBC News Online featured several sections to complement its traditional news coverage, designed to give citizens a more in-depth knowledge of election issues. Compared to the election sites of 1997 and 2001, the BBC in 2005 significantly improved the opportunities for ordinary citizens to post their comments and influence debates. The "Have Your Say" section contained 53 topics across 68 pages, with a small minority of news and feature articles also containing comments posted by citizens. These pages combined attracted 7,684 comments. The "UK Voters' Panel," created in collaboration with breakfast television, was seemingly an evolution of the "Online 1,000" feature from the 2001 election. However, this panel consisted of only 20 voters who had been asked in advance to contribute their views 'in text and in video, using 3G mobile phones,' throughout the election. There were nine different debate topics with an average of six panelists publishing a response on each occasion. Citizens could discuss each of these entries, and the section attracted some 524 comments.

New for this year was the BBC's election blog, entitled "Election Monitor," which announced on the main page that it aimed at 'bringing you first-hand reports from around the country from our team of correspondents, as well as the best of the newspapers, choice morsels from the web, and your e-mails.' The blog finished on 276 posts (in addition to the main holding page), of which 189 received one or more comments from members of the public, totaling 783 comments across all blog posts.

The level of interactivity and opportunities for citizens to engage in debate on the Election 2005 site demonstrates that the BBC clearly recognises its role in facilitating spaces for public dialogue. In relation to new media services, the BBC *Statements of Programme Policy for 2005/2006* devotes an entire section to 'democratic value':

> In line with the new remit, our news and information service will be aimed primarily at creating democratic value and civic engagement, complementing the BBC's broadcast news coverage across all subject areas. (BBC, 2005, p. 40)

Justifying the interactive features in terms of public service, the editor of BBC Interactivity, Vicky Taylor, argued that it is "much better if you're getting your audience telling you what they think than just the officials or people in power…it's a form of democracy—more people get their chance to have their say about something" (Taylor, 2007).

While the internet might not be perceived as having had a significant impact on the election outcome, the BBC has certainly had a considerable impact on citizens' online activities. On average, 550,000 people visited the Election 2005 site each day of the campaign, though this only represented 10% of all BBC News Online users (Ward, 2006, p. 17). On election day, May 5, the number of unique visitors to the election site tripled to 1.5 million, with the figure doubling on May 6, when the results were published (Ward, 2006, p. 17). During the 2005 campaign, BBC News Online accounted for 78% of all UK internet news traffic, about one in five of the total election news audience (Ward, 2006, p. 10). Blogging, which had featured noticeably in the U.S. presidential election the year before, attracted only 0.5% of the online audience during the election (Ward, 2006, p. 11).

Echoing the sequence of events in 2001, the UK election was to be followed a few months later by another devastating terror attack—this time in London. At approximately 8:50 AM on July 7, 2005, three bombs exploded within a minute of one another on the London Underground. Initially it was not clear what was happening, with early reports from Reuters suggesting it could have been a power surge. At 9:47 the fourth bomb detonated on a double-decker bus in Tavistock Square, and just over an hour later the police formally announced there had been a coordinated terror attack. The principal source of news for many people already at work was the internet, with iconic images and eyewitness reports again provided by ordinary citizens caught up in the events (for a more in-depth discussion, see Allan, 2006).

The BBC News website was among the first to break the story online. In contrast to September 11, 2001, the Corporation now had "an established process of handing control of the main picture promotional area of the homepage directly over to BBC News in the event of a major story breaking" (Belam, 2007). With the website receiving on average 40,000 page requests per second, it soon became clear that the technical team would have to reduce the content on the page "in order to minimise the download footprint for each page view" (Belam, 2007). The solution was to deploy an experimental 'proof of concept'

XHTML/CSS table-free version, which eased the bandwidth usage, thus allowing a greater number of connections.

Having learnt from the Indian Ocean tsunami some eight months previous, the BBC quickly began soliciting eyewitness accounts and photographs from ordinary citizens. Richard Sambrook, director of Global News, recalls the incredible response:

> Within six hours we received more than 1,000 photographs, 20 pieces of amateur video, 4,000 text messages, and 20,000 e-mails. People were participating in our coverage in a way we had never seen before. By the next day, our main evening TV newscast began with a package edited entirely from video sent in by viewers. (Sambrook, 2005)

The four people responsible for managing 'user-generated content,' whose team had only been set up as a temporary measure for the 2005 election and then made permanent in the aftermath, were clearly unable to cope with the wealth of contributions from members of the public. However, as Sambrook explained, "audiences had become involved in telling this story as they never had before" (Sambrook, 2005).

> The quantity and quality of the public's contributions moved them beyond novelty, tokenism or the exceptional....Our reporting on this story was a genuine collaboration, enabled by consumer technology—the camera phone in particular—and supported by trust between broadcaster and audience. (Sambrook, 2005)

This remarkable admission demonstrates not just citizen journalism coming of age but also an acceptance by traditional news organisations that audience material is integral to online news reporting—not least in times of crisis.

CURRENT DEVELOPMENTS

As this chapter has shown, the BBC News website has to a certain degree been a product of technological experimentation and innovation—sometimes as part of an online strategy, other times in response to crisis events. In order to keep up with the rapidly changing nature of the web, the BBC News website is in a constant state of flux—always evolving and adding new features. Changes to the site include refreshing the visual design and consolidating technical approaches (such as the type of software used to power discussion boards), but also appropriating ideas and forms of practice established elsewhere on the web (blogging, social bookmarking, and Twitter to name a few).

The BBC has recently experimented with inline hypertext powered by Apture. Until now the BBC has only provided links to related websites outside the main story, typically on the right hand side of the page, in order not "to interrupt a news story by sending the reader off the page in the middle of a sentence" (Herrmann, 2008). The new system, however, "shows the related content in a smaller window within the same page, whilst also being quick and simple for the journalists to add" (Herrmann, 2008). For the trial the external sources include Wikipedia articles, YouTube and Flickr content, and the BBC's own pages. Tristan Harris, co-founder and CEO of Apture, states that his idea is for the BBC to "facilitate the discovery of meaningful information...perhaps even make it interesting to a reader who wouldn't otherwise care" (Harris, 2008).

The BBC Innovation Labs initiative has also led to a series of collaborative semantic web projects—such as the Muddy Boots prototype API developed with Rattle Research (see Austin, 2008). The idea is for such a system to scan the news article and automatically provide links to other websites (typically Wikipedia, YouTube, Flickr or IMDB), and extract snippets of contextual information that can be shown without leaving the BBC's site. While the Apture system relies on the journalist highlighting the words or phrases that should be linked to, this latter system would be wholly automated—the challenge being therefore to develop a system that can differentiate between ambiguous names or phrases depending on the context in which they appear. Both systems are intended to enrich digital storytelling and semantic hypertext narratives but also represent an emerging trend in computer-aided reporting that largely automates the task of enriching the news copy—an important element of online news reporting that might traditionally have been perceived as too time consuming.

However, the most significant new online initiative for the Corporation, probably since the launch of the BBC website itself, has been the development of the BBC iPlayer. The service was widely anticipated, in large parts due to continued promises by former director of the Future Media and Technology group at the BBC, Ashley Highfield, that it was 'coming soon.' It was first made available as an 'open beta' download Peer-to-Peer player in July 2007 amidst a cloud of controversy surrounding choice of proprietary technology (Windows Media Player) and associated Digital Rights Management (DRM) issues. The official version finally launched on December 25, 2007. Within a fortnight some 3.5 million programmes had been streamed or downloaded—the most popular being the *Dr Who Christmas Special*, but nearly half of the programmes were from outside the traditional top 50 most popular shows.

The iPlayer has become a remarkable success story in a relatively short period, but it is the integration of this technology with streaming video content (both recorded and live) from within the BBC News website that has really transformed the site. Embedding the live news channel feed creates an entirely new audience experience and redefines the notion of multimedia journalism. By way of example, for the election of Barack Obama in November 2008, the BBC was able to provide online users with an embedded video feed from the BBC News channel, maps showing the results dynamically updated as states were called, and a running text commentary by BBC journalists integrated with selected quotes from the BBC's "Have Your Say" debates, external blogs, and even Twitter updates.

The volume of audience material received by the BBC also continues to increase and set new records—most recently, in February 2009 when the UK experienced its heaviest snowfall in 18 years, leading to widespread disruption across the country. According to Peter Horrocks, head of BBC Newsroom, more than 35,000 people submitted pictures and videos of the heavy snow.

> This was a record both for the sheer number of pictures and almost certainly for the size of the audience response to a news event in the UK. (Horrocks, 2009)

This popularity was also reflected in visitor statistics, with the BBC News website attracting some 8.2 million unique visitors (5.1 million from the UK) on Monday, February 2—which was also a new record. Meanwhile, the BBC News channel had a peak audience of 557,000 viewers—"no doubt boosted by huge numbers of people taking an enforced day off work," as Horrocks points out. However, in a significant demonstration of the convergence between the online and broadcast platforms as mentioned above, there were also 195,000 plays of the BBC News channel live on the website. Without doubt this highlights the dramatic journey of the BBC website, and fulfillment of its original vision in late 1997, as mentioned at the outset of this chapter, when the original designs for the Corporation's news website were rejected as being ahead of their time.

CONCLUSION

The history of BBC News Online is incredibly rich and diverse, and this chapter has only been able to touch upon selected key events with an emphasis on technology and audience material. It should by no means be considered a conclusive account, and further research is required to ensure parts of this early history are not lost. Very little has actually been written about the early years of the BBC website, and what material is there often contains vague, sometimes contradictory, references to what actually took place. Examples of sites prior to 1997 are difficult to access—despite many of these being available on the BBC servers, they are not readily publicised as such and finding the correct URLs involves a healthy proportion of guesswork. Once the original index file has been identified, many of the links contained within are broken and content is missing.

In the early years of BBC Online computer storage costs were high, thus the larger (in size) and more interactive material has in some instances been deleted (e.g., the Swingometer from the 1997 election site or images associated with the 1998 FIFA World Cup site)—perhaps in order to make way for new pages or as a consequence of human error. It was only in August 1999 that John Birt, director general at the time, issued a request to the head of Heritage to "work out what we need to do to preserve the BBC's early work on the Internet" (cited in Smith, 2005, p. 22). The Legal and Historical Internet Archive system that was subsequently put in place interestingly did *not* capture BBC News Online (since this remained online), audio and video content (most of which was stored in its original broadcast format) and dynamic database-driven content (Smith, 2005, p. 23). Although such functionality is said to have been planned for the future, it highlights the danger of an incomplete historical archive.

The Internet Archive's Wayback Machine[8] goes some way to mitigate this, but perhaps even more impressive for the BBC are two independent initiatives. First, Matthew Somerville's BBC News Archive[9] indexes the BBC homepage and news front page every minute and provides a range of visualization tools and a version comparison based on these data. Second, NewsSniffer[10] contains two services—Revisionista, which archives and displays revisions of news items, and Watch Your Mouth, which monitors "Have Your Say" debates (with the intention of spotting comments that have been removed or censored). What none of these systems are currently capturing, of course, are all those submissions from ordinary citizens that were discarded prior to publication—particularly a problem with material submitted to the "Talking Point" and early iterations of "Have Your Say" sections when content was received via email and published manually. There would have been no process for systematically archiving material that was never published.

Finally, any form of web archiving is unable to capture the human processes involved in producing these websites. As demonstrated by this chapter, what illuminates and brings to life any web history are the first-person accounts of the people working on or with the site at the time. The detailed mapping of their experiences is urgently needed to ensure their valuable insight and experiences do not suffer a similar fate to much of the early web material.

NOTES

1. The site was originally published on http://www.bbcnc.org.uk/ (no longer available) to support existing radio and television programmes, and later merged back into the main BBC website (http://www.bbc.co.uk).
2. http://www.bbc.co.uk/budget95/index2.html
3. http://www.bbc.co.uk/budget96/index.htm
4. http://www.bbc.co.uk/election97/index.htm
5. http://www.bbc.co.uk/politics97/
6. http://www.bbc.co.uk/politics97/diana/
7. During this period the BBC News team had also managed to produce another site, dedicated to the 1997 budget, entitled *Budget 97* (URL: http://www.bbc.co.uk/politics97/budget97/).
8. http://www.archive.org/web/web.php
9. http://www.dracos.co.uk/work/bbc-news-archive/tardis/
10. http://www.newssniffer.co.uk/

REFERENCES

Allan, S. (2002). Reweaving the internet: online news of September 11. In B. Zelizer & S. Allan (Eds.), *Journalism after September 11* (pp. xviii, 268 p.). London: Routledge.

Allan, S. (2006). *Online news: Journalism and the internet.* Maidenhead: Open University Press.

Allan, S. (2009). Histories of citizen journalism. In S. Allan & E. Thorsen (Eds.), *Citizen journalism: Global perspectives.* New York: Peter Lang.

Austin, J. (2008). Muddy boots. Retrieved 08.02.2009, from http://www.bbc.co.uk/blogs/journalismlabs/2008/12/muddy_boots.html

Barrett, C. (2007). Revolution not evolution: The birth of bbc.co.uk, Retrieved from http://www.bbc.co.uk/blogs/bbcinternet/2007/12/revolution_not_evolution.html

BBC (2005). *BBC statements of programme policy for 2005/2006*: BBC.

BBC (2007). About the BBC: Mission and values, from http://www.bbc.co.uk/info/purpose/

Belam, M. (2005). Budget 95 on BBC.co.uk. Retrieved from http://www.currybet.net/cbet_blog/2005/04/budget_95.php

Belam, M. (2007). The BBC's homepage on July 7th 2005. Retrieved from http://www.bbc.co.uk/blogs/bbcinternet/2007/12/running_the_bbcs_homepage_on_j_1.html

Born, G. (2003). From Reithian ethic to managerial discourse: Accountability and audit at the BBC. *Javnost—The Public, 10*(2), 63–80.

Butterworth, B. (2007). Brandon's history of online BBC. Retrieved from http://www.bbc.co.uk/blogs/bbcinternet/2007/12/brandons_history_of_bbc_on_the_2.html

Butterworth, B. ([1999]). BBC internet services—history. Retrieved from http://support.bbc.co.uk/support/history.html

Chadwick, A. (2006). *Internet politics: States, citizens, and new communication technologies*. New York and Oxford: Oxford University Press.

Coleman, S. (2001). Online campaigning. *Parliamentary affairs*(54), 679–688.

Davies, G., Black, H., Budd, A., Evans, R., Gordon, J., Lipsey, D., et al. (1999). *The future funding of the BBC: Report of the Independent Review Panel*: Department for Culture, Media and Sport.

Goodwin, P. (1997). Public Service Broadcasting and new media technology: What the BBC has done and what it should have done. *Javnost—The Public, 4*(4), 59–74.

Gordon, J. (2007). The mobile phone and the public sphere. *Convergence: The International Journal of Research into New Media Technologies, 13*(3), 307–319.

Graf, P. (2004). *Report of the Independent Review of BBC Online*. London: Department for Culture, Media and Sport.

Harris, T. (2008). Buzz about apture and the BBC. Retrieved from http://blog.apture.com/2008/08/buzz-about-apture-and-the-bbc/

Herrmann, S. (2008). New ways of linking. Retrieved from http://www.bbc.co.uk/blogs/theeditors/2008/08/new_ways_of_linking.html

Horrocks, P. (2009). Thanks from BBC News. Retrieved from http://www.bbc.co.uk/blogs/theeditors/2009/02/thanks_from_bbc_news.html

Liu, S. B., Palen, L., Sutton, J., Hughes, A. L., & Vieweg, S. (2009). Citizen photojournalism during crisis events. In S. Allan & E. Thorsen (Eds.), *Citizen journalism: Global perspectives*. New York: Peter Lang.

Price, J. (2005). Tsunami test for news teams. Retrieved from http://news.bbc.co.uk/newswatch/ifs/hi/newsid_4140000/newsid_4145000/4145099.stm

Sambrook, R. (2005). Citizen journalism and the BBC. *Nieman Reports, Winter, 59*(4), 13–16.

Schechter, D. (2005). Helicopter journalism: What's missing in the Tsunami coverage. Retrieved 05.02.2009, from http://www.mediachannel.org/views/dissector/affalert308.shtml

Smartt, M. (2007). The days before launch. Retrieved from http://www.bbc.co.uk/blogs/bbcinternet/2007/12/the_days_before_launch.html

Smith, C. (2005). Building an internet archive system for the British Broadcasting Corporation. *Library Trends*, 54(1), 16–32.

Taylor, V. (2007). Interview conducted by Thorsen, E. 27.03.07.

Thorsen, E., Allan, S., & Carter, C. (2009). Citizenship and public service: The case of BBC News Online. In G. Monaghan & S. Tunney (Eds.), *Web journalism: A new form of citizenship?* Eastbourne: Sussex Academic Press.

Ward, S. (2006). What's the story…? Online news consumption in the 2005 UK election [Unpublished]. Oxford Internet Institute.

Wardle, C., & Williams, A. (2008). *ugc@thebbc: Understanding its impact upon contributors, non-contributors and BBC News*. Cardiff: Cardiff School of Journalism, Media and Cultural Studies, University of Cardiff.

CHAPTER 10

(R)evolution under Construction: The Dual History of Online Newspapers and Newspapers Online

Vidar Falkenberg

INTRODUCTION

This chapter presents a historical perspective of the development of online newspapers. A parallel examination of two intermingled processes is necessary: The history of print newspapers' online presence and the creation of an online newspaper with functions similar to those of traditional newspapers. Here, newspaper functionality is not limited to a classical notion of delivering news of current events but includes the newspapers' role as a medium for entertainment, enlightenment, community announcements, and advertising, among other functions. The central argument also presents the notion that online newspapers are not just the websites of printed newspapers but are conceptually independent of their publishers' background, and this simple premise demands a double-pronged historical approach.

The history and development of online newspapers are thus in need of a preliminary clarification. What is the point of departure? Is it the online presence of printed newspapers, or is it the creation of an online newspaper? The first case is easily traceable and directs the search for elements of the past as well as the diachronic examples with their resultant multitude of online strategies employed today. The latter case makes the hunt for origins less clear, as many influences are applicable, in terms of explaining the birth of a particular phenomenon, web sphere, or genre on the web: the online newspaper. The online presence of newspapers on the web starts at a clear-cut point in time—when websites became available—although the prehistory leading up to the publications is of utmost relevance. The birth of an online newspaper is much harder to pinpoint. Many parallel developments take place to form the constituents of what can be defined as an online newspaper, and the long pregnancy is an indication that the birth metaphor may be misleading. There is no clear conception, the pregnancy is partial—an unnatural state for a pregnancy—and it is hard to settle on one undisputable date of birth. What is more, most pregnancies require the joint effort of two, and only two, beings, with one of them carrying the fetus. The websites of newspapers may be seen as results of joining the printed newspaper with the web, the "pregnancy" being the web carrying the responsibilities of nurturing the offspring, and the birth and the "baby" being the publication made available on the web. To explain the making of the online newspaper in more general terms, not as the online products of print newspapers, but as a way of structuring, presenting, and interacting with content on the web, a different metaphor is needed. Instead of conception, pregnancy, and birth, the phenomenon can be explained in terms such as coevolution, metamorphosis, and diffusion. The construct is not to be taken as an exact replacement for the three facets of the birth metaphor but rather as another way of addressing the many developments taking place at the same time: technologically, institutionally, journalistically, economically, and socially, to name the most prominent.

This chapter is divided into three parts: a presentation of the underlying theoretical concepts, a general introduction to the most relevant analytical elements, and finally, a descriptive example, describing the Danish online newspaper sphere.

EMERGENCE AND MEDIAMORPHOSIS

A reasonable introduction to the theoretical framework would be an explanation of the title of this chapter. The much-awaited and anticipated digital revolution of the newspaper (Negroponte, 1995) has turned out to be a process that is progressing much more slowly than expected. The predictions of Bill Gates and Jakob Nielsen, among others, concerning the future of printed newspapers, were too optimistic or too pessimistic, depending on the point of view. In fact, with regard to the newspaper business and their web presence only, the story is not much of a revolution at this point. This aspect of newspaper development can to a larger extent be seen as a transformation from the traditional print newspaper to a newspaper delivered online, a continuation of the ongoing development of newspaper production and the newspaper concept (Falkenberg, 2007b). However, this process is subordinated and related to another process with a much more far-reaching outcome: the transformation of the media matrix from one epoch to another. A media matrix consists of all co-existing media in a given society, and from a historical perspective, five epochs can be identified: oral, writing, print, electronic analogue, and digital (Finnemann, 2008). The transformation under discussion is the change from a system composed of electronic analogue media, speech, writing, and the printing press, to a new epoch characterized by the presence of digital media, along with all previously existing media, and a subsequent adjustment in the functions of old media. In short, this chapter revolves around the evolution of the newspaper within the revolution of the media matrix.

The (r)evolution *under construction* is also a homage to the widespread but short-lived custom of adding an icon depicting a construction worker to a website, presumably to create the impression of being dynamic and undergoing constant improvement. This placeholder for real innovation may, somewhat tauntingly, be seen as a symbol for the newspapers' online presence during the years before and after the turn of the millennium.

Evolution and Emergence

The history of newspapers and the web, both when investigated from an institutional point of view and from the standpoint of the realization of the newspaper concept on the web, must be told with a dual focus on newspapers' online presence and on online newspapers as emerging entities. The combination of evolution and emergence explains both the ongoing evolution of newspaper production and the newspaper product as well as the leap from electronic to digital media, of which the newspaper evolution is a part. This relationship is also present theoretically, where evolution is a central element in explaining emergence. Evolution, understood as the synergetic effects which arise when combining different elements, is necessary for creating emergence. On the other hand, not all synergetic effects cause emergence (Corning, 2002). Thus, the current media matrix has emerged because of the evolution of computer technology and the internet, but some (or most) of the evolutions happening in this same period do not explain this emergence by themselves.

The Emergence of a Fifth Epoch in Media History

The transformation of the media matrix following the introduction of digital media and the internet is the overarching frame of reference for investigating how the existing media technology of printed newspapers and the new media technology of the internet create and enable a novel media form, the online newspaper. It is within this emergence that the online newspaper should be interpreted, not as a result of the printed newspapers' online presence alone. This point is crucial, because the history of online newspapers tends to be told with the newspaper industry as its sole point of departure (Ihlström, 2004; Li, 2005). Undoubtedly a significant story, but to properly explain the characteristics of online newspapers, the other side of the equation demands similar attention. The media-specific traits of the internet, and especially of the World Wide Web, need attention, alongside the traits of the newspaper concept. At the risk of sounding pompous, we should ask not only what the newspaper can do for the internet, but also what the internet can do for the newspaper.

The issue of greatest relevance to this chapter is the identification of the new characteristics of the media that constitute a new epoch: what properties make them different from traditional media, in what directions the new media widen the use of media and mediatization, how the new media add to the complexity of the matrix, and how the new media affect the roles and functions of the older media (Finnemann, 2008, p. 7). These four criteria distinguish one epoch from another and pinpoint the most crucial changes when reconfiguring the newspaper from paper to web. Quoting Finnemann, this includes the ability to "integrate the storage capacities of print media with the transmission speed of electronic media," allowing "synchronous and asynchronous communication [to become] an option within the very same medium"; allowing "new kinds of social interaction"; transforming mediatization "in all spheres of daily professional, private, social and public life...because synchronized interactions can now be performed across distances;" and finally, expanding "the mediated public as well as private space beyond the reach of traditional mass media" (Finnemann, 2008, p. 8). Later in this chapter, the analytical elements of both newspaper web history and online newspaper development will be discussed.

The Mediamorphosis of the Newspaper Online

Metamorphosis, the biological process in which certain animals change physical form, is one of six principles of mediamorphosis (Fidler, 1997). It shares the element of dramatic change with the concept of emergence, but metamorphosis is a transformation which takes place within a defined lifespan and, most commonly, to all members of a species. The inaccuracy of the term is addressed below. Mediamorphosis cannot explain the emergence of the internet itself, but it can elucidate the connections between the internet and other media forms such as newspapers. The six fundamental principles of mediamorphosis encapsulate the essence of these different theories: coevolution and coexistence, metamorphosis, propagation, survival, opportunity and need, and finally, delayed adoption (Fidler, 1997, p. 29).

The first principle states that all media coexist with other media and that these media coevolve and influence each other. This fundamental premise is central to many theories within the Medium Theory tradition, and a short genealogy of the concept is presented in Brügger (2002). The coevolution discussed in this chapter is primarily concerned with the coexistence of the printed newspaper, the websites of printed newspapers, other websites, and the emergence of the online newspaper as a result of this interplay.

The second principle, borrowing the term "metamorphosis" from biology, is used to explain how new media arise gradually from changes in older media, and how the older media continue to adapt in order to survive. A critique of this interpretation is presented below, but the main issue is the process of change.

Propagation, the third principle, is also a biological term, but it is probably used more figuratively here. Fidler states that "emerging forms of communication media propagate dominant traits from earlier forms" and that these traits "are passed on and spread" (Fidler, 1997, p. 29). In the case of online newspapers, this is of special interest when considering the transition of the newspaper paradigm from paper to web. Understanding the newspaper concept as something more than just news is vital for constructing and identifying the online newspaper, and following the traits that move from the print newspaper product to the online newspaper product is central to that process.

The fourth principle, survival, is closely connected to both coevolution and metamorphosis. Evolution is made necessary by changes in the environment and is the only alternative to becoming extinct. But even though the will to survive is strong and the technology may match that of the competitor in quality, the outcome remains uncertain. A well-known example is the battle between the formats for home-video recording, where the first-mover advantage and subsequent development of Sony's Betamax were outmatched by the strategic maneuvering of JVCs VHS (Cusumano, Mylonadis & Rosenbloom, 1992). The survival principle explains one of the reasons that change is initiated, but it does not mean that any given technology will survive.

The fifth principle goes beyond evolution itself and addresses the opportunity and need that must be present for new media to be widely adopted. Fidler states that there must always be "an opportunity, as well as a motivating social, political, and/or economic reason for a new media technology to be developed" (Fidler, 1997, p. 29). This principle is deeply inspired by Winston's supervening social necessities, working as accelerators and brakes for the diffusion of a technology (Winston, 1998). Previously, the question of a future mediamorphosis versus an imminent mediacide seemed to favor a more positive outlook for the future of media companies, even immediately after the internet stock market bubble burst in 2000 (Alves, 2001).

The sixth and final principle, delayed adoption, reflects elements from Saffo's 30-year rule (Saffo, 1992) and Rogers' Diffusion of Innovations (Rogers, 2003). The essence of these concepts is that it generally takes a surprisingly long time for new media technologies to be widely adopted. The suggested time-frame of 20–30 years should not be taken too literally. Even though many of the examples given by Saffo fit the approximate length of a human generation, it should be noted that it can be difficult to define the exact moment of either the point in time when an innovation is presented or the achievement of its widespread adoption. More important is the basic rule that adoption of a new idea takes longer than initially expected.

To sum up, the principles and their sources of inspiration provide a solid basis for analyzing the development of both newspapers online and online newspapers.

The Ambiguous Metamorphosis

The ambiguity of Fidler's interpretation of "metamorphosis," when this concept is compared to its biological origins, deserves additional attention, because it may illustrate the differences between the two different but parallel and similar processes in question. Fidler uses metamorphosis as a sub-element of mediamorphosis, the essence of which is how communication forms and established media enterprises, just as species, evolve to better survive in changing environments (Fidler, 1997, p. 29). Fidler's notion of mediamorphosis contrasts somewhat with the biological understanding of emergence, evolution, and metamorphosis. In mediamorphosis, the process described is both the development of a particular communication technology or media enterprise (metamorphosis) and the evolution from one configuration to the next (emergence). The discrepancy can be explained by Fidler's use of the concept of metamorphosis itself:

> New media do not arise spontaneously and independently—they emerge gradually from the metamorphosis of older media. When newer forms emerge, the older forms tend to adapt and continue to evolve rather than die. (Fidler, 1997, p. 29)

In biology, metamorphosis is used to describe the change in form during normal development, while Fidler's metamorphosis is the explanation for the change in communication media brought about by the introduction of new technology. More to the point: biological metamorphism takes place within the normal development of almost all members of a species; metamorphism as interpreted by Fidler is a part of the evolution of the species (media) itself. The subtle differences between the approaches in this chapter mean that the development of newspapers' online presence is the metamorphosis of the newspaper product, while the development of the online newspaper is the emergence of a new media form as the result of a new epoch. The distinction is not clear-cut, but the dual approach helps to broaden the understanding of newspapers meeting the web. Mediamorphism is a useful concept for understanding media changes, despite the ambiguity surrounding the term, and the combined explanatory power of the fundamental principles and their underlying theories are necessary to fully understand the development of online newspapers.

Tales and Lessons: The Historical Perspective

There is another duality, from the standpoint of chronological flow. The history of the newspapers on the web may be best understood in terms of a chronological flow from the past to the present: it is more or less known what the newspapers were like in the early 1990s, and their web presence can be traced, although the remains are scarce and scattered. From a more or less common point of departure, at least for comparable publications, their development suggests that different paths are possible, but they remain close to their origins and one another for a long time (Falkenberg, 2007a; Jankowski & Van Selm, 2001). The history of the development of the newspaper as a concept on the web is best understood as flowing in the opposite direction. With the present understanding of an online newspaper, the origins of its appearance, its content, and its functionality have to be traced backward to discover how online newspapers have been constructed. These prehistories do not have a single, unequivocal point of departure but instead have a set of independent or co-dependent practices, technologies, and needs, with multiple points of departure and individual, although interrelated, paths of development. In both cases, their futures have a multitude of possible directions. Figure 1 visualizes the dynamics of the two approaches.

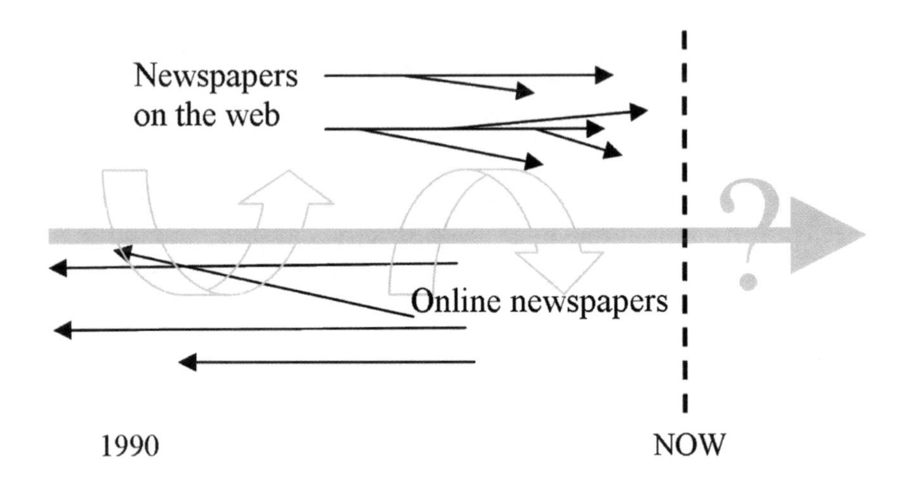

Figure 1: Directions of storytelling

The middle arrow separating the two approaches is not to be taken literally as a division of two unrelated media forms, and the arrows crossing this horizontal line symbolize the mutual influence of the newspaper and the web on the idea of an online newspaper. The dual approach is an attempt to understand how the development of a newspaper format on the web is unfolding. The history of newspapers' online presence lends itself to a chronological narrative in which the events unfold with some degree of coherence, and the connection between the analytical elements explained below is traceable in the physical or digital remainders of the past. The nature of the participants as traditional media companies makes the delimitation easier. Investigating the origins of the online newspaper, on the other hand, requires another course of action. First, because the field of interest is much more loosely defined, actually knowing what to look for and where to look is a bigger challenge. Second, the connections and sources of inspiration are not necessarily preserved and available for investigation. The history from the past to the present is, therefore, a story of where we were and where we went, while the history from the present to the past starts by looking at where we are, and tries to unravel how we got here. However different the metaphors and approaches, the histories are intertwined and co-dependent. The telling of the two sides of the history of newspapers online and explaining the making of online newspapers have one common goal: seeing the whole picture by understanding how the mechanisms of mediamorphism apply to the interplay between old and new media, and between competing and supplementary media.

A DUAL APPROACH

The fundamental questions, when applying the mediamorphic process to the two approaches, are where and how they are similar, and where, how, and why they differ. The interplay between some properties is so heavily linked together, that they must be treated equally. In other cases, each approach is best analyzed with a specific view, reflecting either the forward-looking flow of newspaper web history or the backward-looking flow of online newspaper development. The following section sheds light on which principles and which elements apply equally to both approaches and which elements are more specific to just one of the two.

Some of the mediamorphic principles provide more explanatory power. The most prominent of these are the principle of propagation, the principles of opportunity and need, and the principles of co-evolution and coexistence. Within each of these principles, the analysis must focus on diverse aspects and dimensions. In the case of propagation, it is relevant to investigate the functional and aesthetic features of the printed versus the online newspaper, the main interest being the identification of how key elements of the printed newspaper have persisted, been altered, or been realized in new ways. Ihlström (2004) and Li (2005) present many examples of this propagation, as does most of the empirical research on online newspapers conducted before 2000. The principle of co-evolution and coexistence comprises the very backbone of the argument for a dual focus. The fact that newspapers are publishing online at the same time as the web and its form and functionality are being developed is the *conditio sine qua non* of mediamorphosis and of this chapter. A more detailed analysis exceeds the limits of this text, but inspiration is found, for instance, in Finnemann (2005), Boczkowski (2005), Fagerjord (2003), Bolter and Grusin (2000), and Fidler (1997). The main focus of the present section is the question of opportunity and need, which will be discussed from the standpoint of their respective approaches.

The Newspaper's Web History

The answer, or rather the set of answers resulting from the question of investigating opportunity and need, are to be found by using different methods in various settings. This includes an interest in the institutions, the regulatory and legal actions, the overall media structure, and most of all, the internal and external competition in the newspaper market.

The institutional aspect is naturally extremely relevant, as the changing environment within which an institution operates quickly creates new opportunities which the companies may pursue, or new needs they must attend to. Many leading newspapers are strong institutions with a deep cultural and political impact and influence, and their histories and traditions are central to the study of the development of their online presence. Alongside the institutions themselves, regulatory and legal actions supplement or add infringements or possibilities, affecting both the development and diffusion of the media technology itself and the institution or individual using it. These actions are highly relevant, as both direct and indirect driving forces. Both the institutional aspect and the regulatory conditions are dependent on the local, regional, national, or transnational context and are hard to discuss in general terms, because of the multitude of histories and regulations. The case presented below illustrates some of these aspects in the Danish context.

To a certain extent, the competitive situation is more readily generalized. The internal competition for the online presence of print newspapers has been rival print newspapers' online presence, and the inclination to compete as to speed of delivery has been in constant conflict with the fear of revealing top news stories to rival publications if the news were published before the deadline of the print version (Krumsvik, 2006, p. 286). The external competition takes both anticipated and unexpected forms. The print newspapers were already competing with radio, television, and other print media before the emergence of internet, and new forms, such as portals, aggregators, blogs, and social network sites add to the already challenging environment in which they operate. The competition is additionally challenging, because it consists of more than just another competitor seeking money and time from the same pool of readers and advertisers.

Increasing competition has been haunting commercial media for a long time and does not represent a new threat, although it obviously puts even more pressure on the already challenged media revenue. What is new is the fact that the competitors are now present on the same platform, where the preconditions are more evenly matched than they are on different platforms. Potentially, the amounts of time and space involved are the same, regardless of the size of the publisher. In the newspaper world, the cost of printing and distributing physical copies quickly becomes an economic constraint, as does as the need to adapt the nature of the content to a global audience. Online, the content is globally accessible by default, and the cost of servers and bandwidth is much less although not an entirely negligible restriction. For a typical local newspaper, the benefit of being globally accessible is limited. For a competitor whose primary object is not to deliver a complete bundle of content, as a traditional newspaper does, but a specialized type of content, such as classified advertisements or news of special interest topics, eliminating geographical limitations is the key to entering a new market.

The classified advertisements are an obvious example for two reasons. First, their economic significance for the traditional newspapers has been vital to total revenue and for funding the less profitable, but still more important and prestigious journalism. Second, the technology behind online classified sites is easily transferable across geographical boundaries and even adds to the value of the content. A local newspaper online will not benefit from including local news from every other town or region, but an online classified site can add value by offering access to advertisements from outside the local area, while still maintaining the value of a geographical delimitation as an option in the interface. The essence of this platform-specific property is that newspapers can be challenged on areas where an almost monopolistic dominance previously secured high profitability and made the need for renewal and innovation less explicit.

In addition to the new situation, where the participants compete for readers' and advertisers' attention on the same platform, the transformation into a new media matrix means an additional complexity with the coexistence of the old structure alongside the new structure. For the media companies competing for readers and advertisers on the web, any presence on other media platforms may be affected. In the best-case scenario, the media companies simply move their businesses from one platform to another; in the worst case, they lose their businesses on some or all platforms. Along this continuum it can also be the case that they manage to gain a significant position on the web in terms of number of readers and sale of advertising space but with far lower revenue on both accounts, compared to the print product. Although the total online advertising expenditures and time spent online increase at the expense of other platforms, primarily print newspapers, the biggest growth is seen on the websites of non-traditional media, either among the giants such as Google, Yahoo, MSN, and Facebook or the many small sites of the long tail. To understand the changed competitive situation for the newspapers and for other traditional media companies, it is necessary to gain more detailed knowledge about the traffic of each individual website, and about the total use of the web, compared to that of other media platforms.

The Online Newspaper's Development

The opportunity and need are manifested in the technological development of both print newspapers' online presence and of online newspapers as special entities. Earlier developments in newsprint technologies were influenced by other print media, which, in turn, had a reciprocal influence on them. This collaboration and mutual influence are even stronger online. Web technology in general has had, and still has a large community of developers serving a wide range of industries, and new possibilities spread quickly across domains. The collaborative efforts of formally unconnected users have characterized the development of the internet (Hafner & Lyon, 1996) as well as the web (Berners-Lee & Fischetti, 2000), and the open-source ideology has been a catalyst for the development of, and a supplement to the traditional commercial industry. Naturally, the paths of open-source software and traditional commercial software intersect on various occasions, just like the two approaches of newspapers online and online newspapers. Open-source software does not necessarily imply lack of revenue. Common strategies include revenue coming from services related to the software, from support, training, or documentation and in certain cases from licensing the software itself. A parallel in the online newspaper world is to offer general news content for free, while services such as searching, monitoring, accessing archives, or the sale of related products ensure revenue. Only in cases where the news is of special value can this content be monetized. Another possible interpretation of the open-source paradigm is to consider the journalism itself as a collaborative process similar to open-source development (Bruns, 2005).

In addition to the formalized and clearly visible results, in terms of commercial open-source software such as Red Hat Linux and MySQL Enterprise, or in terms of online news sites, such as Indymedia and Slashdot, there is a less visible consequence of the two practices. It is more difficult to determine how and where coevolution and coexistence actually result in the coexisting forms influencing each other. For instance, the support for tabbed browsing in most present-day browsers has been implemented in and popularized through both open- and closed source software, interchangeably. In parallel, the typical multi-column layout of newspapers has been replicated in newspapers online (Ihlström, 2004), and can also be found in websites from other companies. The ancestry of online newspapers is inevitably responsible for their physical appearance, but there are two related influences worth mentioning: first, influences from other media forms, and second, the influence of the online newspaper on printed newspapers. The menu structure on newspaper websites and online newspapers is influenced by the advances in usability and changes in user preferences, in combination with a need for conserving screen real-estate for content or advertisements. A current trend, therefore, is to replace the left-hand column navigation with a horizontal navigation menu, near the top of the page. Elements borrowed from other forms, such as multimedia content, are controlled with traditional buttons for play, pause, stop, and so on, and the tabbed interface of browser windows is also finding its way into individual parts of the web page. The influences thus come from both other media forms and from other types of websites. Finally, the impact of online newspaper design on printed newspapers is insufficiently documented, but general recommendations for web design, such as shorter segments of text with scanable headlines, may find their way into print newspapers in the same way that other media forms, such as TV news, have adopted a web-like division of the screen resembling frames, grids, or multi-column layouts (Cooke, 2003).

Apart from the way technology is developed, and its cross-influence on layout, a more theoretical issue is illustrated in the model in figure 1, page 240. The obvious conceptual difference lies in the multiple origins of the online newspaper, compared with the more unified point of departure for the printed newspaper's initial online efforts. Although relatively insignificant, the attempt to determine a date for, or even to simply identify, the first online newspaper becomes difficult as opposed to determining when a printed newspaper made the first website. Including the web as a way to delimit the definition of an online newspaper helps narrow the object of study, but it excludes the many predecessors found in other media forms. Carey and Elton describe how videotex and other online services offer an alternative path for explaining the web phenomenon (Carey & Elton, 2009), and Nguyen expands the notion of "online" from the current narrow understanding of an interconnected network of computers to a wider conception of point-to-point communication networks, like the telegraph (Nguyen, 2007). Thus, the basic argument is that the online newspaper has influences predating the web, much as the print newspaper has influences predating the printing press.

A Problematic Division and Demarcation

The complexity of the two concurrent processes, the evolution of the newspaper within the revolution of the media matrix, and the particulars of the dual approach to newspaper web history and online newspaper development cannot be covered in full detail in this short chapter. The main concern, therefore, is to zoom in on the most relevant mediamorphic principles and to clarify how these principles apply in practice to each approach. In reality, it can be hard to maintain a clear-cut distinction between evolution and revolution, or newspapers online and online newspapers, when trying to explain the theoretical background with real-world examples. The significance of their coexistence and reciprocal influence cannot be overstated, but the main intention is to illustrate the benefits of looking at the same part of the world through different lenses. For instance, increased competition is central to both approaches. In fact, the element of rivalry is evidence of how and where the two approaches meet. Before the web became mainstream and acquired an economic significance, internal competition online and external competition between the web and other platforms were present but still inconsiderable. At the same time, it is imperative to note that neither the newspapers online nor the online-only newspapers have been the only, or even the most successful, players in this game. The competition online is heavy, both among online businesses and among the various media platforms.

Just as some of the principles of mediamorphosis are more relevant than others in this case, some of the elements regarding opportunity and need receive more attention than others. The focus on institutions, regulatory and legal actions, media structure, competition, and technology is justified by the strong web presence of traditional media companies and, additionally, their concern for the future. For this reason, vital themes such as the development of online journalism or how readers/users consume and interact with the product are absent from the present discussion.

THE SLOW TRANSFORMATION OF NEWSPAPERS ONLINE AND ONLINE NEWSPAPERS IN DENMARK

The remainder of this chapter presents an example of the theoretical approach explained above, using the Danish development of online news as a case study. Space limitations prohibit a detailed or complete account, and the chronology can only present a rough overview. The main points are that the newspapers' online presence was without competition for a long time and that the emergence of new alternatives primarily challenged the newspaper companies on aspects other than the primary journalistic content. The contenders, apart from the few online-only newspapers mimicking printed newspapers' websites, were public service providers, aggregation services, or niche content websites. The products that come closest to fulfilling the functions of a newspaper on the web are currently polarized by one specific factor: the attitude towards, or actual receipt of, state subsidies. Online newspapers in Denmark are thus divided into three groups on this basis: public service providers, print newspaper websites, and independent online newspapers. These three voices are characterized, respectively, by a public service contract allowing and demanding the web as a distribution platform, state support for the distribution of VAT-exempt printed newspapers allowing their publishers the advantage of reusing content and facilities in the production of online newspapers, and, finally, independent publishers not receiving any support, who are, in their own view, subject to unfair competition, because they compete against state-supported media. The interrelationships among them are complex, because each has favorable assets not held by the others. On the level of functionality and content they are more similar, but uneven economic conditions may skew the balance of power among them. Their biggest problem, however, may be external competition.

A Concise Chronology

The development from the early 1990s and the following 15 years progressed, as suggested in the previous section title, rather slowly. A thorough presentation would start with the preconditions for both newspaper and online production at that time, such as the above-mentioned institutional, journalistic, regulatory, legal, competitive, structural, and technological situations. The much more limited story given here must skip these lengthy explanatory accounts, and focus on the main events. Fifteen years is too short a span to be broken into identifiable periods, so the purpose of the following subdivision is more of a presentational tool than an empirically based classification. According to Saffo, a mapping of his 30-year rule with Rogers' diffusion curve shows that in the first decade there is much excitement and puzzlement but not a lot of penetration; the second decade has a lot of flux and the beginnings of penetration into society; and in the third decade, the technology has become a standard nobody notices (Saffo, 1992, p. 23). These words, written in 1992 as a commentary on the development of personal computers, fit reasonably well with the first decade and a half of online newspapers. However, it is not a given that these 10-year periods will proceed as suggested, nor is the outcome certain. The following four stages of development in Denmark, therefore, represent nothing more than a contemporary delineation of the story thus far:

1993–1996: The web is growing—slowly

1997–2002: Optimism and decline

2003–2006: Slow and steady growth

After 2006: Entering a new stage?

The first period is best described as a playful interaction with a new media technology. The pioneers, either the publications with a professional interest in technology or the tech-minded individuals at more traditional newspapers, experimented with publishing and interacting with their audience online. The management was rarely deeply involved but found the experiments useful as a way of projecting innovation and investment in their future. However, the journalistic and editorial resources and ambitions showed few signs of fulfilling these aspirations.

The second period was a turbulent time, with many global and national events having significant impact on the combination of news and web. Internationally, incidents such as the death of Princess Diana, the Clinton-Lewinsky affair, the dot-com boom and crash, and the September 11, 2001, bombing of the World Trade Center demonstrated the power of the web in terms of speed, depth, and community building (Allan, 2006). On a national level, less dramatic but still decisive events caught the attention of both publishers and the public in Denmark. A far-reaching strike in 1998 affected most print newspapers, by having journalists at work, but not the workers responsible for printing, packaging, and distributing the newspaper. They could, however, publish online. A sudden boost in the number of newspapers with websites and available content on the pre-existing newspaper websites was a wake-up call for publishers and readers alike, and the changed media structure suddenly became more than just an experimental playground. The consequences seemed insignificant, when the printed newspapers were back on the streets after a few weeks, but there were other simultaneous events of greater importance. The leading national newspapers, *Berlingske Tidende* and *Jyllands-Posten* were openly at war over circulation and classified advertising dominance, and this paper-based combat was suddenly exposed to a new threat, when a small internet start-up company cooperated with public service provider TV2 on a job advertisement website, threatening one of the newspaper industry's main financial resources. The newspapers were terrified of the possible consequences of an outside competitor running away with the revenue and traffic generated by online classified advertisements, and they sought to legally prohibit the competition by appealing for stricter regulation of the public service broadcasting institutions' online efforts. Failing in this, the newspapers were forced to cooperate in order to avoid losing control over the growing market.

The newspapers were still quite alone on the market for online news. Online-only newspapers like *Nettavisen* in Norway or rumor-mongers like *Drudge Report* did not exist, so the only other competitors were the public service providers, *Danmarks Radio* and *TV2*. The decline during this period reflected the economic effects following the dot-com crash and September 11, resulting in a loss of advertising revenue and a reduction in newspaper staff and a general lack of enthusiasm about the web as the future of newspapers.

The third period started when optimism slowly returned, and broadband access became the standard internet connection for companies and in private homes. The global events of the period were the terrorist bombings in London and Madrid, and the natural disasters of the tsunami in Asia and hurricane Katrina in the United States. The increased mobility and use of civilians as contributors to mainstream news providers were important to each of these disasters (Allan, 2006). The technological development of mobile media allowed news providers or others to harvest user-generated content, improving the coverage of breaking news and attracting more readers and contributors without hiring many expensive journalists or photographers. In Denmark, micro-payments were once again temporarily discarded as a viable option for user payment, after years of expectations and experiments, but the main issue was the conflict between the Danish Newspaper Publishers' Association and companies aggregating the content of the newspapers' online content without permission. The infamous Newsbooster case (Delio, 2002) has been imprecisely labeled as a ruling forbidding the use of deep links, the links that point directly to an article and not to the front page of a website. The actual legal decision and its consequences are in fact more limited, but the debate is often used as an argument for the newspapers' reluctant attitude towards the internet. This complicated matter has no simple answer or truth, but it does illustrate a more problematic copyright situation, and the struggle to maintain a viable business model on the web. Once again, legal and regulatory actions dominate the discussion.

The fourth and, to date, final period is only now beginning to demonstrate a noticeable change with respect to a situation in which the competition online intensified, and the polarization of online newspaper types, as mentioned above, became more apparent. Most print newspaper companies reallocated resources from print to web, and the public service providers increased their efforts online as well. Some independent online newspapers were showing healthy economic signs and criticized the current system of state support for the media. The global financial crisis dampened the expectations for online advertising growth but did not alter the move towards a situation in which advertisers and readers prefer the internet over newspapers.

The Similarities between Online and Print Newspaper Markets

In general, the online newspaper market in Denmark in 2007 closely resembled the configuration of the print newspaper market. The geographical division of national, regional, and local publications was extended with a niche content publication as suggested by Salwen et al. (2004). In Denmark, the small local online newspapers could be further divided into two groups: independent publishers displaying a wide variety of strategies and the template-based publications of the local newspapers owned by the two dominant publishing houses, Berlingske Media and JP/Politikens Hus. Despite the global nature of the web, geographical position was still important for Danish online newspapers, because their content was aimed at a local audience. The niche-oriented sites tended to be situated in the larger cities, suggesting that even though the content was not geographically bound, the physical infrastructure of technology and the available workforce required the publishers to establish themselves in favorable geographical locations. In short, the balance of power on the web was equal to that of the print newspaper market, with only a few new contenders, mainly the public service broadcasters on the mass market and some independent publishers in the niche markets.

The Principles and Elements Revisited

As indicated previously, the mediamorphic principles are applicable to online newspaper development to varying degrees. Coevolution and coexistence are fundamental to the understanding of how the reciprocal influences occur. Metamorphosis is more relevant to the newspapers online than to online newspapers, but this concept is complicated by the surrounding emergence of the internet and a new media matrix. Propagation is evident but treated more explicitly elsewhere. The survival principle is true of all media challenged by other media, but it does not result in the survival of a media technology on its own. Even if it is compelled to adapt and evolve, it might be made extinct by other media technologies without being technically inferior. The principles of opportunity and need are the most explanatory principles, allowing a wide range of circumstances to be considered. The final principle of delayed adoption is not treated extensively but is implicit as the standard diffusion rate seems to apply.

Within the focus on opportunity and need, and because of the point of departure in printed newspapers' web presence, the institutional aspect, the regulatory and legal actions, the technological development, the internal and external competition, and the overall changes in the media structure constitute the majority of the examples. These are, however, only a subset of possible fields of interest, with the development of online journalism and the user experience as the most notable omissions.

CONCLUSIONS—SHARED PAST, UNKNOWN FUTURE

The purpose of this chapter has been to address the differences between the role of newspapers' online presence and the development of online newspapers. Newspapers online were initially a part of the constant development and evolution of existing technology, improving what was already there. Online newspapers emerged as computers and the internet initiated the transformation from one media matrix to another. The development of the online newspaper has to be explained in the light of both approaches simultaneously, because it is taking place as an improvement on an existing technology, within a newly emerging technology. This process bears a close resemblance to Fidler's mediamorphosis:

> The transformation of communication media, usually brought about by the complex interplay of perceived needs, competitive and political pressures, and social and technological innovations (Fidler, 1997, p. xv).

With the material presented here, it is difficult to say whether the effects and mechanisms of mediamorphism, as applied to online newspapers, are generalizable to all kinds of media development. A prerequisite for mediamorphism is the coexistence of a new technology and strong institutions, either themselves producing content for both kinds of technologies, or at least being in competition with content producers of the new technology. Without this coexistence, coevolution can not take place. Similarly, opportunity and need are necessary for evolution to happen. Mediamorphism lends itself to explanations of situations in which one technology or practice, dominant at a given point, is under pressure from an upcoming contender, as is the case with printed newspapers versus online newspapers. It might be less applicable to other uses of the web. The popularity and widespread use of social network sites might be compared to earlier online and offline forms facilitating social networking, but the principles of mediamorphism seem less relevant, as the tension between the different media forms is less obvious.

The dual focus of investigating first the newspapers online and then the online newspapers as separate but interwoven phenomena is necessary, to broaden this field of study. To some extent they do overlap, but misunderstanding them as identical disregards the difference in origins, as well as the possible future paths of the printed newspapers' web presence and of the online newspapers. Both futures are disputed. The print newspaper is not dead, but its financial situation is continually threatened on several fronts, and at present it is unlikely that it is possible to successfully transfer the blueprint for making and financing newspapers from paper to web. On the other hand, it is still uncertain whether an online newspaper will turn out to be a product desired by users and profitable for publishers. In Denmark, after 15 years of online presence for printed newspapers, their economic status is best described as the transition from one way to make much money to many ways to make no money.

REFERENCES

Allan, S. (2006). *Online news: Journalism and the internet*. New York: Open University Press.

Alves, R. C. (2001). The future of online journalism: Mediamorphosis or mediacide? *Info*, 3(1), 63–72.

Berners-Lee, T., & Fischetti, M. (2000). *Weaving the web: The past, present and future of the World Wide Web by its inventor*. London: Texere.

Boczkowski, P. J. (2005). *Digitizing the news: Innovation in online newspapers*. Cambridge, MA: MIT Press.

Bolter, J. D., & Grusin, R. (2000). *Remediation: Understanding new media*. Cambridge, MA: MIT Press.

Brügger, N. (2002). Theoretical reflections on media and media history. In N. Brügger, & S. Kolstrup (Eds.), *Media history: Theories, methods, analysis* (pp. 33–66). Aarhus: Aarhus University Press.

Bruns, A. (2005). *Gatewatching: Collaborative online news production*. New York: Peter Lang.

Carey, J., & Elton, M. C. J. (2009). The other path to the web: The forgotten role of videotex and other early online services. *New Media & Society*, *11*(1–2), 241–260.

Cooke, L. (2003). Information acceleration and visual trends in print, television, and web news sources. *Technical Communication Quarterly*, *12*(2), 155–182.

Corning, P. A. (2002). The re-emergence of 'emergence': A venerable concept in search of a theory. *Complexity*, *7*(6), 18–30.

Cusumano, M. A., Mylonadis, Y., & Rosenbloom, R. S. (1992). Strategic maneuvering and mass-market dynamics: The triumph of VHS over beta. *The Business History Review*, *66*(1, High-Technology Industries), 51–94.

Delio, M. (2002). *Deep link foes get another win*. Retrieved 12/18, 2008, from http://www.wired.com/politics/law/news/2002/07/53697

Fagerjord, A. (2003). Rhetorical convergence: Earlier media influence on web media form. (Dr.art., Faculty of Arts, University of Oslo). *Acta Humaniora*, no. 178.

Falkenberg, V. (2007a). Decomposing the newspaper: A typology of online newspapers. Paper presented at the Internet Research 8.0, Let's Play, Association of Internet Researchers, Vancouver.

Falkenberg, V. (2007b). Online newspapers as newspapers online. Extending the concept of 'newspaper'. In N. Leandros (Ed.), *The impact of internet on the mass media in Europe* (pp. 105–117). Bury St Edmunds: Abramis.

Fidler, R. (1997). *Mediamorphosis: Understanding new media.* Thousand Oaks, CA: Pine Forge Press.

Finnemann, N. O. (2005). *Internettet i mediehistorisk perspektiv* [The internet in the perspective of media history]. Frederiksberg: Samfundslitteratur.

Finnemann, N. O. (2008). The internet and the emergence of a new matrix of media. Paper presented at the Internet Research 9.0, Rethinking Communities, Rethinking Place, Copenhagen.

Hafner, K., & Lyon, M. (1996). *Where wizards stay up late: The origins of the internet.* New York: Simon & Schuster.

Ihlström, C. (2004). *The evolution of a new(s) genre.* Göteborg: Gothenburg Studies in Informatics, Report 29, September.

Jankowski, N. W., & Van Selm, M. (2001). Traditional news media online: An examination of added values. In K. Renckstorf, D. McQuail & N. Jankowski (Eds.), *Television news research: Recent European approaches and findings* (pp. 375–392). Berlin: Quintessence Publishing Co.

Krumsvik, A. H. (2006). What is the strategic role of online newspapers? *Nordicom Review, 27*(2), 283–295.

Li, X. (2005). *Internet newspapers. the making of a mainstream medium.* Mahwah, NJ: Lawrence Erlbaum.

Negroponte, N. (1995). *Being digital.* London: Hodder & Stoughton.

Nguyen, A. (2007). The interaction between technologies and society: Lessons learned from 160 evolutionary years of online news services. *First Monday, 12*(3).

Rogers, E. M. (2003). *Diffusion of innovations* (5th ed.). New York: Free Press.

Saffo, P. (1992). Paul Saffo and the 30-year rule. *Design World, 24,* 18–23.

Salwen, M. B., Garrison, B., & Driscoll, P. D. (2004). *Online news and the public.* Mahwah, NJ: Lawrence Erlbaum.

Winston, B. (1998). *Media technology and society. A history: From the telegraph to the internet.* London: Routledge.

PART FOUR

Preserving and Presenting

CHAPTER 11

The Aesthetics of Web Advertising: Methodological Implications for the Study of Genre Development

Iben Bredahl Jessen

Advertising messages are part of commercial web culture. As an object of study in a historical and media aesthetic perspective, web ads can be regarded as particular ways of organizing and using existing media and genre conventions in order to attract the user's attention. In this context, this chapter will examine the actual appearance of web ads. The question is how to study the aesthetics of web advertising and its genre development from an empirical basis.

INTRODUCTION

Typologies of web advertising are often proposed as lists of web ad formats according to their graphic form and appearance, including banners, pop-ups, interstitials, hyperlinks, sponsorships, and corporate websites (e.g., Rodgers & Thorson, 2000).[1] However, from a historical perspective, such lists often do not seem to be very helpful in keeping up with the development of media technology, and the appearance of new types of ads requires frequently updated lists with additional formats (cf. Faber, Lee & Nan, 2004). Comparing lists of web ad formats over time can give an impression of how web advertising has evolved on the general level of graphic form but will not reveal much about the representational strategies of web advertising and the ways of addressing the user.

Not only the formats and media technology evolve, it is an inherent logic of the genre of advertising to explore new ways of attracting the attention of the consumer by means of new and surprising constellations. As described by Cook (2001), it is a common characteristic of advertisements to play with and recombine elements of existing genres such as conversation, film, sitcoms, cartoons, or jokes, to create distinctive modes of expressions associated with specific brands or products. Furthermore, following the logic of remediation (Bolter & Grusin, 1999) and, with reference to, e.g., web advertising, Leckenby's Media Interaction Cycle (Leckenby, 2005), web ads can also be described as media hybrids borrowing features from printed media and television among other media. The preference for specific media will probably change over time, and established media will be recombined in new representational forms.

The variable nature of web advertising on the level of graphic appearance, genre characteristics, and media seems to require a broad typology that is not restricted to particular web ad formats such as the typologies usually proposed. Typologies that consist only of a list of ad formats are unable to reveal much about the complexity of the genre- and media-related conditions of web ads. A web ad format such as, for instance, a 'banner,' defined as "rectangular-shaped graphics, usually located at the top or bottom of a web page" (Rodgers & Thorson, 2000), does not supply information about how the ad conveys its message.

Inspired by an empirical study of web advertising in a Danish context, the chapter proposes a typology of web ads that includes the representational strategies of the ad and which is useful in the study of the development of the web ad genre. The first part of the chapter will focus on the methodological implications in the study of genre development and present a method of collecting and documenting unstable content on the web in order to conduct historical text studies. The second part of the chapter describes a typology of the collected material as well as its use in web history.

APPROACHES TO THE STUDY OF WEB ADVERTISING

In web advertising as well as in other text forms on the web, the object of study is not something that is 'just there' to be picked up but is co-constructed during the act of examination. As pointed out by Wakeford (2004), no canonized methodology exists by which to study web phenomena; on the contrary, it seems necessary to select and combine methods from related studies according to the specific purpose. Moreover, the definition of the field of study is decisive for the methodological choices that have to be made (Wakeford, 2004, p. 47). A clear delimitation of the phenomenon of interest is crucial in a field such as the present case of web advertising, which can be studied from a variety of perspectives—as the work of practitioners or designers (e.g., Fourquet-Courbet, Courbet & Vanhuele, 2007), as a text form in itself subject to communicational analysis (e.g., Janoschka, 2004), and, related to studies of media use, as the users' perceptions of web advertising (e.g., McMillan & Hwang, 2005). The different approaches all make up different conceptions of what web advertising is and how it should be studied. From a genre perspective, the choice of approach will inevitably reveal a particular picture of genre development, simply due to the fact that genre cannot be regarded as a 'result' of either the conditions of production, the textual characteristics, or the user oriented aspects of web advertising but must be regarded as an outcome of a complex dynamic between these. In this context, I will, however, focus on the web ad as a text form in itself which is intended for market communication. Given the empirical nature of the study, the chapter presents (inevitably) a somehow limited perspective on web advertising as a genre and will thus not postulate any general assertions on the matter. The presented findings must therefore be regarded in close relation to the particular definition of the phenomenon as well as to the methodological choices that are made.[2]

In order to conduct an empirical study of web advertising, one must formulate an operational definition of the object of interest. As I have just emphasized, the operational definition is not identical to the phenomenon of web advertising as such but only reflects a way of studying it. The operational definition applied here is in line with Janoschka's communicative description of web advertising as a hyperlinked structure of successive advertising messages (Janoschka, 2004, p. 49ff.). For the empirical study, I define web advertising in relation to (at least) two communicational layers. The starting point is the embedded web ad that appears on (or in connection with) a host website as, for instance, a banner on an online newspaper. The embedded web ad includes formats as banners, billboards, skyscrapers, etc. Structurally connected to the embedded web ad is the linked target that appears as another communicative layer in the ad. The target ad typically appears as part of a corporate website (e.g., a product page) or as a provisional microsite hosting a specific campaign. Even if this structural outline of web advertising seems hypothetical because the actual user does not necessarily follow the defined sequence but may enter the structure from alternative entrances (and then perhaps will not regard the target webpage as advertising), I find it useful in a systematic collection of material because it defines the web ad with a 'starting point.'

Having defined the object of study, the next step is to decide from where to collect the material. Table 1 presents an overview of different sources from which to access web advertising. The table illustrates that different sources provide different levels and perspectives of studying the phenomenon as well as different kinds of material.

	Advertising websites	**Ad servers**	**Agencies**	**Private collections**	**Web archives**
Level	publication	distribution	production	exhibition	archives
Perspective	webpage	server	campaign	campaign	web
Material available	observed ads on a specific website	all ads on the server (local or remote)	own productions	ads chosen at random or by value judgment	all ads that are possible to harvest on all harvested websites
Time	synchronous	synchronous	asynchronous	asynchronous	asynchronous
Context	yes	no	no	no	yes
Follow links	yes	yes	yes/no	yes/no	yes/no
Quality	intact	intact	intact	good	variable

Table 1: Sources of web advertising

As a consequence of the operational definition based on ads located on websites, the choice of source for the present study is the 'observational' study of ads on advertising websites. The intention is to record ads on a sample of websites, i.e. to capture ads from diverse ad servers. Principally, a web archive could also be a useful source for that purpose, especially in order to collect material with the aim of studying genre development. However, for the time being, web archiving seems to suffer from technical problems in harvesting, e.g., dynamic content as java scripts.[3] As regards agencies or private collections on the web (e.g., blogs on interactive advertising), they can be useful sources when looking for specific cases or brands, but do not seem to be appropriate in systematic investigations of the genre. In the following, the methodological implications concerning the use of websites as a source of 'observational' study of ads will be discussed in more detail.

THE TRANSIENT OBJECT

In relation to both content and structure (e.g., link structure), it is the task of the web researcher to take into account the dynamic and transient character of the web medium. As regards web advertising, the challenge of continuous updating seems most critical in the case of the embedded web ad, which most often is replaced by another ad from the ad server every time the user enters—or simply updates—the webpage. In other words, it is rarely possible to return to a specific web ad for further examination and analysis even within a very short period of time. The web ad is thus a highly transient object. The preparation of an empirical study of advertising on the web must therefore include methodological strategies to handle the particular media-related conditions in which web ads exist.

Two central methodological challenges concern the transformational character and the infrastructure of the web. These aspects relate to the conditions of selecting and collecting material as well as to the characteristics of the selected material. The transformational character of the web is mentioned by Mitra and Cohen (1999) as an important issue in the analysis of web texts, for instance, in relation to how to manage the material to be analysed. As Mitra and Cohen emphasize, the impermanence of the web text makes it necessary for the researcher to record and save copies of the texts to be examined. Additionally, in relation to the infrastructure of the web, two essential questions need to be considered as regards methodology and the selection of material, namely: Where to begin? And, how much to include? (cf. Mitra & Cohen, 1999, p. 192ff.).

The question 'where to begin' is a consequence of the decentralised infrastructure of the web consisting of an enormous network of interlinked webpages to which the user has 'random access.' On a structural level, there is no 'natural' or evident starting point for the selection of webpages that must be included in the analysis. In their investigation of genres on the web, Crowston and Williams (2000) take the consequences of this fact and use a random sample of webpages selected by a search engine and thereby neglect the placement of the webpage in the structural hierarchy of a website. Even though Crowston and Williams acknowledge the decentralised infrastructure of the web, the methodological procedure does not seem appropriate. As Crowston & Williams self-critically remark, the size of the web makes it inappropriate to base the collection of material on random sampling (Crowston & Williams, 2000, p. 211).

On the contrary, if we follow, e.g., Mitra and Cohen (1999), the researcher must select and argue for the choice of web texts that need to be included in the analysis. It is, thus, important to determine from what to select and what to achieve by the specific study in case. Rather than random sampling, it is therefore relevant, as described by Herring (2004) in relation to studies of computer-mediated communication, to conduct a purposive sampling and from this delimit the study, e.g., "sampling by thread..., by a slice of time..., by periodic intervals..., by event..." (Herring, 2004, p. 9).

In relation to the choices that need to be made, at least two overall practical-methodological questions must be addressed: (1) How can a transient object such as the web ad be collected and documented? (2) From which advertising websites should the material be collected? In the following pages, I will elaborate on these questions in referring to an empirical study of web ads and show how the methodological approach is crucial to historical studies of web texts.

STABILIZATION OF THE TRANSIENT OBJECT

To be able to study the development of web advertising as a genre and not least to document the analytical points, the analysed object must be accessible for examination. Brügger (2005) considers strategies to 'micro archiving'[4] of web texts which are relevant also in relation to the collection and archiving of web ads. Brügger distinguishes between three fundamental ways of stabilising and preserving dynamic objects—document, monument, and imprint (cf. Brügger, 2005, p. 15ff.). The method of archiving depends, however, on the type of object: The document is a transformation of dynamic objects as, for instance, nature or human behaviour into another symbolic form. The object will be reproduced in another medium by means of different techniques: "...by observing, monitoring, posing questions, keeping records and taking notes, interviewing, photographing and filming the concrete object of analysis" (Brügger, 2005, p. 16). In these ways, the researcher creates documents about the object, but the object itself is not preserved. The monument is created when material objects as, for example, tools or media are taken out of their usual contexts and preserved in a museum. And finally, the imprint is a re-creation of immaterial objects such as television and radio by means of transference from one medium to another (a DVD recording of a television programme) (cf. Brügger, 2005, p. 17ff.).

According to Brügger, the web shares characteristics with the dynamics that exist in nature or in society. The web must therefore be stabilised and preserved as a document, partly because of the way it is possible to represent the web, and partly because of the role of the researcher in the process of stabilisation and collection of material:

> Archiving the Internet resembles the document: firstly, because it involves an element of *representation* of the object, insofar as the storage involves some form of "transformation," and secondly, because the actual process of stabilisation involves a relatively high degree of *subjective involvement.* (Brügger, 2005, p. 30)

At the same time, the web differs from traditional documentation, since the web, as an object, is a medium in itself and therefore does not necessarily need mediation. As a consequence, Brügger distinguishes between two fundamental ways of archiving the web: (1) documents *about* the web, and (2) documents *of* the web. Where documents *about* the web are based on observations from quantitative or qualitative methods and thereby create documents which are different from the web, the documentation *of* the web consists of a more 'direct' form of archiving.

In relation to the present study of web advertising, the collected material consists of documents *of* web ads or recordings of the appearance and location of the ads. For the documentation of the web ads, screen-capture and screen-recording software is used. The aim of using this type of software is not to catch the dynamics implied in the navigation between different webpages as such, but to be able to document the movements inside the specific ads, as, for instance, the possibilities of interaction, animation, or moving pictures.

Web ads are dynamic in two ways: (1) they are dynamic as regards the frequency of replacement and updating in relation to their appearance and accessibility to the researcher, and (2) the ads can be dynamic themselves and contain film sequences, animation, navigation possibilities, or other kinds of interactive components. Concerning the interactive ads, in which the user's activity is a prerequisite for the actualization of the ad, it is necessary to record the ad 'in use.' In the present study, the 'use' of web ads consists in screen recordings of my interaction with such elements as games, microsites, or banner ads with a mouse-over effect. In other words, the mouse is directed by me (as a researcher), and thereby I leave my mark on the documents in the collection of material. Thus, the recordings represent a particular reading, perspective or interpretation of the studied object.

As Wakeford points out, in studies of the web the researcher generally takes part in the construction of the examined object (cf. Wakeford, 2004, p. 47). As regards the recordings of the web ads 'in use,' the 'constructedness' is striking. The strong subjective element in the documentation of web ads is inevitable in a medium such as the web as also emphasized by Brügger (op. cit, 2005). The collected material must therefore always be considered as a result of a documentation made by an individual in a given time period, on a specific computer connected to the internet, and with a particular kind of software. In this manner, the collected web ads will only serve as exemplifications, and the analysis of the material will only tell us about a certain appearance of the phenomenon, namely, the one resulting from a given process of documentation.

To sum up, I find it relevant in the process of documentation of web ads to distinguish between photographic and cinematographic modes of documentation (see table 2).

	Technology	**Areas of documentation**
Photographic	copying (if possible) screen-capture software	static web ads context the linked target
Cinematographic	screen-recording software (video recording)	animated web ads interactive web ads (are recorded 'in use') the linked target (eventually 'in use')

Table 2: Photographic and cinematographic modes of documentation

The photographic mode of documentation results in a static document of the (static or dynamic) object by means of screen-capture software (or copied image files). The cinematographic mode of documentation results in a dynamic document of the dynamic object by means of screen-recording software. The use of screen-recording implies that the researcher, as described very precisely by Brügger, becomes a director in a cinematic process of documentation: "...in screen recording, we are nearing a kinematic approach to archiving, where the person doing the archiving becomes his or her own director by treading their own paths in the landscape of the screen" (Brügger, 2005, p. 62).

In addition to the importance of the researcher in the documentation of web advertising, cookies also play a role. Because of their 'invisibility,' it is difficult to calculate with cookies in the process of documentation. As a researcher, you can either choose to accept them as a condition of the web medium or accept them but frequently delete them. It is, however, important to be aware of the existence of cookies as a factor whose significance in the process of documentation cannot be described precisely. The same is true for the significance of the technical system behind the ads, which decides, among other things, how often the ads are replaced, in what geographical areas, or at what time of the day they are displayed.

SAMPLING AND DEFINITION OF NAVIGATION PATHS

A purposive sampling implies that the researcher has to decide the criteria for the selection of the sample: In what media context is web advertising going to be examined? On which websites are the observation and collection of ads going to take place? The present study is based on a genre- and content-related sampling that involves a media context somewhat similar to studies of printed and audiovisual advertising in traditional mass communication media in which ads are located among editorial content in newspapers and magazines or in (between) television programmes. Equivalents to these media exist on the web as online newspapers, online magazines, and websites of television channels, but also web portals, which have a similar mass communicative aim, can be considered as belonging to this media context.

As is customary in web studies within the Content Analysis tradition (cf. McMillan, 2000), the sampling in the present study is based on a website index from which 15 advertising websites were selected within the overlapping genres of portals, information resources, and news media. The following criteria provide the basis for the selection of the websites in the sample: (1) appeal to a broad audience, (2) many services, (3) many visitors, and (4) a broad spectrum of topics. Every website in the sample does not necessarily meet all the criteria. The sample also includes more specialized information resources and news media such as a health website, a computer news website, and a cultural/tourist website. However, taken together, the sample reflects the criteria of broadness and variation in topics.

For each website in the sample, a navigation (or 'reading') path is determined. The navigation paths (cf. the structural example in table 3) are followed in the process of documentation.

front page (update x times)
main article
menu: topic 1
select article
menu: topic 2
select article
sub-topic 1
sub-topic 2
menu: topic 3
select article
menu: topic 4
select article
front page

Table 3: Structure of a navigation path on a news website

Together, the paths in the websites of the sample also cover a wide range of topics—domestic and foreign news, health, sport, culture, business and economy, computer, lifestyle, and entertainment, and services such as search functionalities (search for jobs, cars, and traffic routes) and mail-login are included. The advantage of determining specific navigation paths through each website is that it is possible to repeat the path at every recording (contrary to, say, recording randomly within a certain time period). At the same time, variation in topic is ensured. A disadvantage is when the structure of a website is rearranged so that the determined navigation path cannot be followed. However, in practice, the path can often be reconstructed relatively easily by finding the same topics under new menus.

The purpose of the chosen procedure is to collect all the displayed ads during the researcher's navigation through the defined paths and to record both the ad itself (the embedded web ad) and its target (e.g., a microsite). But to be able to perform this procedure, it is necessary to consider how much of the context should be included in the documentation of the single ad. As to the embedded web ad, it is quite simple to document the context by means of screen capture or screen recording of the particular host website, so that it is possible later in the process of analysis to decide whether the context seems important for the interpretation of the ad. As regards the target ad and its context, the case is more complicated: What are the limits of the target webpage—and are those limits identical to the limits of the ad? Herring reflects upon this methodological problem in relation to content analysis of web texts in a discussion of how many 'layers' of the website should be included (how many links should be followed) and whether it is reasonable to stick to the homepage (front page) as the substantial unit of analysis (cf. Herring, 2004, p. 6).

In the present study of web ads, it does not seem possible to clearly decide where the limits of the target ad are and to formulate precise guidelines in advance. Moreover, there are considerable differences between targets as front pages (the main entrance to the website), as product pages (located decentrally), and as campaign websites (microsites). When collecting and recording web ads, one must acknowledge the subjective element in the process of documentation and in relation to the size of the units in the sample. In other words, it is up to the 'director' who interacts with and records the ads intuitively to decide how much is relevant to archive. This intuition is a matter of genre recognition or the ability to decode which elements of the website that 'belong to' the context of the ad. As mentioned, the recognition of the genre can be rather difficult, because there are no clear boundaries of the target ad. For instance, a webpage will (probably) be conceived differently depending on whether it is entered from a banner ad (where the user is 'tuned in' to be confronted with an advertising message) or from a search engine. During the recordings, it is thus necessary to consider the limits of the ad in relation to what connects the compositional elements: Is there a thematic connection between the elements, or are they connected 'by convention' as standard elements in the corporate website (factual information about the company, contact information, etc.)? Furthermore, the theme of the target ad can be compared to the theme put out in the embedded web ad. However, the task is, in every case and in relation to every type of ad, to consider which linked themes it seems reasonable to follow.

Compared to Herring and the Content Analytical approach, I will not be able to operate with any predefined limits of the target ads during the documentation. Herring raises interesting questions concerning the preparation of the empirical study and proposes a systematisation that can be useful in the process of documentation. However, as regards the delimitation of the ad (what and how much to be included in the archived document), I must be guided by the content and consequently leave the methodological cogency characteristic of the field of Content Analysis. Instead, I will stick to the textual approach illustrated in the metaphor of the director (cf. Brügger, 2005) who by using the navigation paths as a guiding script interpretatively follows the displayed content.

THE HISTORY OF THE COLLECTED MATERIAL

This study of web advertising consists of two collections of material archived from the same sample of websites. The first is a result of a large-scale collection carried out in 2004–2005, and the second is a smaller collection carried out in 2008–2009 in order to be able to test the typology as a means of studying genre development.[5] The systematic recordings of the displayed ads in the navigation paths have resulted in an enormous amount of material—recordings of the embedded ad itself, screen-captures of the context, recordings or screen-captures of the linked target, etc. In the following pages, I will report only on the findings related to the embedded web ads.[6]

To collect material twice within the same sample and to follow the determined navigation paths in preparation for historical analysis obviously raises methodological problems. Due to the dynamic character of the medium, it is not possible to carry out the exact same procedure of collection in a time span of four years. For the second collection of material, I was, however, able to use the same sample; only one website appeared not to carry ads anymore.[7] Yet, I had to reconstruct the navigation paths according to the updated thematic structure of the websites. Actually, this appeared not to be that difficult because I was able to find the same (or similar) topics in the renewed website structures. From a methodological perspective, my estimate is, then, that it is possible to repeat the study of web ads over time within the described framework and that the procedure provides a good basis for comparative studies of genre development.

A TYPOLOGY OF EMBEDDED WEB ADS

The archived material in the collection from 2004–2005 consists of 1,025 documents of embedded web ads, and the collection from 2008–2009 consists of 221 documents of embedded web ads. I will propose the typology as a way of sorting out, mapping, and comparing the collected material. The intention is that the typology should reflect the media as well as the genre-related characteristics of the ads. In many respects, this task is extremely difficult because web ads seem to borrow from a variety of other genres and media. Often, it is also difficult to distinguish genre from media, especially in the digital medium in which older media, when integrated into the digital medium, become genres within the new medium (Finnemann, 2001). Nevertheless, the typology must be clear and consistent as to the distinction between the different types of ads if it is to be useful.

Genre is a way of categorizing texts and is widely used in literature, rhetoric, film theory, and linguistics. In the classic literary tradition, genre is distinguished on the level of enunciation.[8] The level of enunciation is also applied in typologies of advertising: Stigel (2001), for example, differentiates formats of television commercials according to the ways in which the user is addressed.[9] Similarly, in the present typology of embedded web ads, I include the addressing of the user as well as the possible role of the user in the actualisation of the ad. However, the addressing of the user is not a clear-cut dimension in a typology of web advertising because the digital medium is capable of representing modes of address in several media (e.g., print, audio-visual). Moreover, in non-interactive ads, the user is addressed by verbal (spoken or written) utterances accompanied by images, whereas in the interactive ads, the user is also addressed according to what he can do, that is, according to the role that is intended for him and which is manifested in the interactive components of the ads. Principally, there will be a difference between the mode of address in non-interactive fictional forms, in which the user is a spectator to a scene or a story, and the mode of address in interactive fictional forms, in which the user may principally be invited to take part in the 'universe' of the ad. However, an overall distinction between factual and fictionally oriented web ads seems useful.

According to Stigel, the television commercial reflects the formats and modes of expression of the television medium in general. The same situation exists in web advertising in relation to the web medium; however, as it is an inherent characteristic of the digital medium, we find multiple references to or remediation of other media (cf. Bolter & Grusin, 1999). In particular, the web ad seems to remediate advertising formats in other media, for instance, static web ads can resemble the printed poster or the outdoor billboard (cf. Leckenby, 2005), and web ads can use the formats of television commercials. According to Finnemann, these are now genres or sub-genres within the digital medium. The question is now how to handle this media complexity of the web ad in a consistent way as dimensions in the typology. Logically, one solution would be to operate with typologies within the typology.

As mentioned previously, the structure of the web ad is characterised by a hyperlink that connects and leads to the successive messages in the ad. The embedded web ad can thus be described as a graphic link using different compositional strategies (verbal, visual, and audio-visual elements) to attract the attention of the user. The hyperlinked feature of the embedded web ad is the obvious reason why many ads seem to focus (at least at the end, after a sequence of texts and images) on the user's possibility to navigate (using imperatives as 'click here,' 'read more,' etc.). However, in the collected material there are also examples of ads that do not draw much attention to the navigational aspect but appear as a communicational form in themselves, either as something the user should see or read (creating brand awareness) or as something in which the user should actively engage (e.g., a small game). In such cases, the navigational aspect will often be expressed in a more discrete or indirect manner. In fact, even in the cases that accentuate the navigational aspect, there is obviously always something that motivates the possibility of linking. Again, this motivation can be created by means of something the user has to see and read (or, more rarely, listen to) or by means of some kind of activity.

Accordingly, I would like to point to the embedded web ad as a graphic link that reflects the hypertextual features of its surrounding context. More precisely, I consider the embedded web ad as reflecting (though on a small and limited scale) a hypertextual logic implying modal shifts between a reading mode, a link mode, and an editing mode (Finnemann, 2001).[10] In the typology, I will focus on the motivational settings in the ad that motivate the linking as either oriented towards a reading mode or as oriented towards an interaction (editing) mode.[11] Thus, I distinguish between two modes of orientation in the embedded web ads:

1. Web ads that are oriented towards the usual ways of receiving ads, as something you see, read, and/or listen to, and that directly or indirectly refer to the link possibility

2. Web ads that are oriented towards the possibility of interaction and that directly or indirectly refer to the link possibility

With reference to the collected material, the considerations outlined above result in the typology presented in table 5. To sum up, the web ad is investigated in relation to two dimensions of orientation—an overall distinction between factual and fictional forms (and a hybrid form in between) and a distinction between ads oriented towards reading (and then eventually navigation) or interaction (and then eventually navigation). The form of the link (direct or indirect) is not clearly connected to specific dimensions in the typology (e.g., fictionally or factually oriented ads), rather it seems that particular types of links can enter into combinations in all the dimensions, e.g., a fictionally oriented audio-visual ad followed by an explicit link.[12]

As regards the reading-oriented ads, I find it necessary to operate with typologies within the typology and make distinctions between non-sequential verbal/visual forms, sequential verbal/visual forms, and audio-visual forms (film/video).[13] In relation to the interaction-oriented ads, a similar distinction is not (yet) necessary, nor does it (yet) seem evident.

An overview of the types (sub-genres) of web ads in the typology is presented in table 4:

Factual forms (didactic, informational)	
The signature (n-s)	Name or logo of the sender or brand, eventually with a slogan.
The simple composition (n-s)	Text dominated and often with direct address to the user, typically by imperatives. The product is often seen isolated from a context. Images serve as illustrations of the text.
The advertorial (n-s)	Imitates the surrounding context, appears as editorial content.
The simple animation (s)	Text dominated and often with direct address to the user, typically by questions. The product is often seen isolated from a context. Images serve as illustrations of the text.
Presenter (a-v)	Direct address by a speaking presenter, illusion of 'eye contact.'
Testimonial (a-v)	The testimony of a reliable person, often without 'eye contact.'
Voice-over (a-v)	An off-screen voice connects the events displayed on the screen. Direct address.
Contact and order forms	Online forms to initiate contact with the sender.
Tools and menus	E.g., search facilities, menus with predefined options.

Continued on following page

Factual/fictional forms (hybrids)	
The compound composition (n-s)	The image is dominating and is often associative. The product is shown in a certain context, and there is a transference of qualities between context and product. Hard facts are presented in the text.
The compound animation (s)	Interplay between text and images. The text is mainly factual but may also consist of figurative language. The product is not seen as isolated. Images are primarily associative.
Voice-over+ (a-v)	Combination of direct and indirect address. Interplay between the visual and the spoken.
Try-outs	Experimental forms, e.g., 'create your own postcard'
Gimmicks	Hidden point to be revealed by the user or simply 'something the user can do.' Often playful, explorative, or thematised interaction.
Games	Small games, rule and/or turn based.
Fictional forms (associative, emotional)	
The complex composition (n-s)	The image is dominating and highly associative and emotional. It is not possible to distinguish product from context. Lifestyle oriented and thematised. Text can be included, e.g., the brand.
The complex animation (s)	Dominated by highly associative or emotional images. Product and context are merged. Lyrical compositions or montage by means of juxtapositions of image and/or text fragments.
The dramatised animation (s)	Displays a short story within a unified whole, without sound.
Drama (a-v)	A story with a plot where the characters are speaking for themselves. The user is not addressed directly (until the pay-off), but is an implied spectator.
Montage (a-v)	Lyrical composition with a looser connection of the elements. Associative.
* (n-s): non-sequential verbal/visual form, (s): sequential verbal/visual form, (a-v): audio-visual form	

Table 4: Types of embedded web ads in the typology

The types of web ads in the typology reflect the collection of material and cannot be regarded as an exhaustive picture of how web ads are made or can be made. Other types of ads can be included in the overall frame of the typology.

Examining the typology, it is obvious that the embedded web ad is strongly oriented towards the factual and didactic modes of expression and implies a direct appeal to the user. The embedded web ad is most often conceptualised as something to read or watch and eventually click on. From the typology, it is also evident that the embedded web ad is quite 'simple' as regards the compositional structure. Only a minority of the embedded web ads appear innovative (but not necessarily more appealing).

THE TYPOLOGY AND THE STUDY OF GENRE DEVELOPMENT

The typology is a way of mapping large collections of embedded web ads. Table 5 displays the different ad types in the two collections. As regards the development of the genre, the second collection of material has a relatively low degree of reference and will only serve as a test of the typology. Also in relation to time, the span is relatively short. However, the typology can be a useful tool in the study of the genre development as it includes the representational strategies of the web ads. The typology and the proposed sub-genres seem to work, not least because it proved rather easy to map the new collection of material from 2008/2009 within the overall framework of the typology. The use of similar methodological approaches makes it possible to compare the two collections and to a lesser extent the number of ads in the material.

ORIENTATION		Factual	Factual/Fictional	Fictional	→ Direct or indirect appeal →	Link
Reading (94.7%) (84.6%)	Non-sequential verbal/visual forms (33.6%) (23.5%)	Signature (8.6%) (0.5%) Simple composition (21.1%) (21.7%) Advertorial (1.4%) (-)	Compound composition (2.3%) (1.4%)	Complex composition (0.2%) (-)		
	Sequential verbal/visual forms (59.5%) (58.9%)	Simple animation (55.7%) (56.6%)	Compound animation (3.4%) (1.8%)	Complex animation (0.1%) (-) Dramatised animation (0.3%) (0.5%)		
	Audio-visual forms (1.7%) (2.3%)	Presenter (0.5) (0.5%) Testimonial (-) (0.9%) Voice-over (0.7%) (-)	Voice-over + (0.1%) (-)	Drama (0.1%) (-) Montage (0.3%) (0.9%)		
Interaction (5.3%) (15.4%)		Contact/order forms (1.7%) (1.8%) Tools and menus (2.5%) (8.6%)	Gimmicks (0.9%) (2.3%) Try-outs (0.1%) (-) Games (0.1%) (2.7%)			

Table 5: A typology of embedded web ads. The left brackets represent the share of the ad type in the material from 2004/2005, the right brackets represent the material from 2008/2009, cf. (2004/2005) (2008/2009).

In the material from 2004/2005, the embedded web ad appears as a predominant factual genre, and most of the ads are not oriented towards interaction. This is also the dominant picture in the material from 2008/2009. However, the share of the interaction-oriented ads has increased significantly (from 5.3% in 2004/2005 to 15.4% in 2008/2009) at the expense of the signature and the advertorial. Tools and menus constitute the major part of the interaction-oriented ads, but contact and order forms as well as games and gimmicks are dominant within this category. This means that the user is increasingly addressed according to something he or she can do yet still in a highly factual setting. In relation to the typology, this change is not only a question of 'new' or 'innovative' use of technology; it should be regarded as a change in representational strategy and in what constitutes the motivation to link. Due to the increased share of interaction-oriented ads, future research may need to sharpen the distinctions of the ad types in this category, not only on the basis of empirical material (as it is the case in the present typology) but also theoretically.

Compared to typologies such as lists of web ad formats mentioned in the introduction, one of the strengths of the proposed typology is that new sub-genres can be added within its overall framework. So, when the typology is used historically in order to study genre development, the overall framework of the representational strategies will remain intact, and new types of ads can be added according to the tendencies in the empirical material.[14] Accordingly, the typology can include possible innovative sub-genres, not with reference to the graphic format but with reference to how the user is addressed or is supposed to act, and with focus on the organization of verbal, visual, and auditory modes of expression. This means that the typology can also take into account that genre development may occur within particular web ad formats.

CONCLUSION

How can the typology contribute to web history? First and foremost, the typology is a way of handling large collections of documents in relation to what can be identified as the media aesthetic characteristics of the embedded web ad. But I also consider the typology as a useful tool to compare collections of web ads over time and to study the development of the genre. As a framework to study genre development, the typology will present general mappings of the representational strategies of embedded web ads on the basis of the chosen methodological approach (the definition of the object, the sampling, the documentation process, etc.), and consequently it does not rest on a criterion of 'innovative' or 'interesting' cases. The typology reveals a rather grounded perspective on the history of web advertising, which is naturally limited by its empirical foundation but avoids tendentious stressing the importance of new ad formats that might turn out to be appealing and effective but only tell us part of the story of how web ads actually look. It might be objected that the typology only reveals a partial story of web advertising. The point is, however, that it does so on the level of procedure. Accordingly, what the typology illustrates about web advertising and genre development must be viewed in connection to the implemented methodological approach. In addition, the overall framework of the typology can be applied to other kinds of samplings and other ways of documenting and collecting embedded web ads.

NOTES

1. An exception is McMillan's typology of internet advertising (based on literature reviews and interviews with practitioners) in which identified formats are mapped according to four communicative purposes (McMillan, 2007, p. 20).
2. A focus on the textual characteristic of web advertising might (from a general view) exclude the potential importance of the distributive and social aspects of the genre, cf. the sharing of commercial videos on social network sites.
3. In a Danish context, the netarchive.dk is responsible for a systematic 'harvesting' of Danish websites (cf. http://www.netarchive.dk/). Besides the technical problems with the dynamic content, following links in the archived material will sometimes result in a jump in time, since the webcrawler cannot always capture all link destinations within the same period of time.
4. 'Micro archiving' is carried out by individuals in relation to a limited area of study (Brügger, 2005, p. 10).
5. The ads in the first collection of material are recorded in October 2004, December 2004, February 2005, and in April 2005. The ads in the second collection of material are recorded in December 2008 and in January 2009.
6. The function of the embedded web ad is not restricted merely to create click-throughs; it has a communicative effect on its own and can therefore rightly be regarded as an advertising form in itself.
7. Based on the findings from the 2004–2005 collection, which show that this particular website only resulted in a very low number of ads, the website was just left out. Alternatively, it could have been replaced with a similar website.
8. Cf. the first-person utterance in lyric, the switching between first-person narration and utterances from other characters in epic, and the dialogue of the characters in drama.

9. Stigel proposes the following television commercial formats ranging over a spectrum from factual (direct address) to fictional formats (indirect address): presenter, testimonial, voice-over, voice-over+, drama, and montage (Stigel, 2001, p. 332ff.).
10. According to Finnemann, the reading mode is reading "as usual," the link mode includes navigating and browsing, and the editing mode is "interactive behaviour changing the future behaviour/content of the system" (Finnemann, 2001, p. 43).
11. It is, of course, possible to find examples of ads that are oriented towards both modes, but, in my view, one of them will often appear more remarkable than the other.
12. This combination illustrates the undeniable factual reference of advertising, cf. "advertising as a basically persuasive, informative and referential act" (Stigel, 2001, p. 338). Even in the fictional forms a reference to the product, brand, or sender is necessary (at least at the end).
13. The sub-typology of the non-sequential as well as the sequential verbal/visual ads is partly inspired by the categorization of Dyer (1982, p. 88ff.), who outlines a distinction between simple, compound, complex (and sophisticated) ads according to their ways of referring to and presenting the product by means of text and images (on a scale from the factual and denotative reference to the fictional and connotative/associative reference). The sub-typology of the audio-visual forms is the categorization proposed by Stigel (2001).
14. In the present context, I do not consider how many ads it takes to constitute a sub-genre (cf. Foot & Schneider, 2006, p. 177). The strategy is to map all the documents in the material.

REFERENCES

Bolter, J. D. & Grusin, R. (1999). *Remediation. Understanding new media*. Cambridge, MA & London: The MIT Press.

Brügger, N. (2005). *Archiving websites. General considerations and strategies*. Aarhus: The Centre for Internet Research. Retrieved January 30, 2009, from http://www.cfi.au.dk/publikationer/archiving/archiving

Cook, G. (2001). *The discourse of advertising* (2nd ed.). London & New York: Routledge.

Crowston, K. & Williams, M. (2000). Reproduced and emergent genres of communication on the World Wide Web. *The Information Society, 16*(3), 201–215.

Dyer, G. (1982). *Advertising as communication*. London & New York: Routledge.

Faber, R. J. & Lee, M. & Nan, X. (2004). Advertising and the consumer information environment online. *American Behaviorial Scientist, 48*(4), 447–463.

Finnemann, N. O. (2001). The internet—A new communicational infrastructure. *Papers from The Centre for Internet Research*, no. 02. Retrieved January 30, 2009 from http://www.cfi.au.dk /publikationer/cfi/002_finnemann

Foot, K. A. & Schneider, S. M. (2006): *Web campaigning*. Cambridge, MA: The MIT Press.

Fourquet-Courbet, M.-P. & Courbet, D. & Vanhuele, M. (2007). How web banner designers work: The role of internal dialogues, self-evaluations, and implicit communication theories. *Journal of Advertising Research, 47*(2), 183–192.

Herring, S. C. (2004). Content analysis for new media: Rethinking the paradigm. *New research for new media: Innovative research methodologies symposium, working papers and readings* (47–66). Retrieved January 30, 2009, from http://ella.slis.indiana.edu/~herring/newmedia.pdf

Janoschka, A. (2004). *Web advertising: New forms of communication on the internet.* Amsterdam/Philadelphia: John Benjamins.

Leckenby, J. D. (2005). The interaction of traditional and new media. In M. Stafford & R. Faber (eds.), *Advertising, promotion, and new media* (pp. 3–29). Armonk, NY: M.E. Sharpe.

McMillan, S. J. (2000). The microscope and the moving target: The challenge of applying content analysis to the World Wide Web. *Journalism and Mass Communication Quarterly, 77*(1), 80–99.

McMillan, S. J. (2007). Internet advertising: One face or many? In D. W. Schumann & E. Thorson (eds.), *Internet advertising: Theory and research* (pp. 15–35), Mahwah, NJ & London: Lawrence Erlbaum.

McMillan, S. J. & Hwang, J.-S. (2005). Measures of perceived interactivity: An exploration of the role of direction of communication, user control, and time on shaping perceptions of interactivity. In M. Stafford & R. Faber (eds.), *Advertising, promotion, and new media* (pp. 125–147). Armonk, NY: M.E. Sharpe.

Mitra, A. & Cohen, E. (1999). Analyzing the web: Directions and challenges. In S. Jones (ed.), *Doing internet research: Critical issues and methods for examining the Net* (pp. 179–202). Thousand Oaks, CA & London & New Delhi: Sage.

Rodgers, S. & Thorson, E. (2000). The interactive advertising model: How users perceive and process online ads. *Journal of Interactive Advertising,* 1(1). Retrieved January 30, 2009, from http://www.jiad.org/article5

Stigel, J. (2001). The aesthetics of Danish TV-spot-commercials: A study of Danish TV-commercials in the 1990s. In F. Hansen & L.Y. Hansen (eds.), *Advertising research in the Nordic countries* (pp. 327–350), Frederiksberg: Samfundslitteratur.

Wakeford, N. (2004). Developing methodological frameworks for studying the World Wide Web. In D. Gauntlett & R. Horsley (eds.), *Web.Studies* (2nd ed) (pp. 24–48), London: Arnold.

CHAPTER 12

Web Archiving in Research and Historical Global Collaboratories

Charles van den Heuvel

> Thus the moving image of the world would be established—its memory, its true duplicate. (Paul Otlet, 1935, p. 391)

WEB ARCHIVING IN THE HUMANITIES AND SOCIAL SCIENCES

Given the amount of information on the World Wide Web that people all over the globe produce on a daily basis, it is rather surprising to see, how seldom web archives are used as a historical source to study recent developments in our society. Berners-Lee describes, in *Weaving the Web* (1999), his future vision of the World Wide Web as a dream in two parts. In the first part, the web becomes a more powerful means for collaboration among people. Berners-Lee imagines information space as something to which everyone has immediate access, not just to browse in, but also to create in. In the second part of the dream, collaborations extend to computers. Machines become capable of analyzing all the data on the web—the content, links and transactions between people and computers—resulting in a "Semantic Web" (Berners-Lee, 1999). The fulfillment of the latter part of Berners-Lee's dream still seems far away. This knowledge construction is time-consuming and requires technological expertise. However, the first part of the dream is becoming reality. Given the success of Wikipedia, YouTube, Flickr, Facebook and other Web 2.0 applications, it is fair to say that people are augmenting the World Wide Web with data and information and that they use it more and more as an instrument to collaborate. As such it is an increasing source of what people do with information and how they create an Information Society. Brewster Kahle created the Internet Archive in 1996 to prevent the web's content from disappearing into the past. Three years later he decided to include books, television programs, movies, and music in this large digital repository. There is no doubt that the Internet Archive's collection and the various web archiving initiatives of national libraries and archives are already of value for individual and collaborating researchers in the humanities and social sciences. However their potential is not fully exploited yet for various reasons. Legal and policy issues are often used as arguments. In some countries web archives can be stored but for copyright reasons cannot be published. Other libraries, especially national libraries, only archive websites with domain names of their own country, either by law or out of tradition. We leave these legal and policy issues further aside but want to focus here on a different possible explanation of why web archives are not fully exploited by researchers, in particular in the humanities and social sciences.

In the report of the American Council of Learned Societies' Commission on Cyber Infrastructure for Humanities and Social Sciences: *Our Cultural Commonwealth,* one argument, amongst technological, economic, legal, and institutional ones that would explain the limited use of such a digital infrastructure in these groups of disciplines, stands out in particular and addresses "the conservative culture of scholarship" (ACLS, 2006, p. 21). However, the report does not explain, or only implicitly explains, where this "conservatism" is coming from (Wouters, 2007). It might be explained as the result of a mismatch between the scholar in the humanities and social sciences and the provided tools and services. Tools and services will only be used if they serve researchers in their daily work. The problem requires a better understanding of the scholarly practices of researchers. The first experiences with ethnographic research of scholars who use digital services and tools (DEFF, 2006) and the analysis of historical collaboratory projects (VKS-IISH) are promising but still scarce, especially within the humanities. Therefore, we have to limit ourselves for the moment with a general exploration of what is required in order to study web histories for research. In some studies small-scale solutions are explored to make web archiving suitable for research (Brügger, 2005; Schneider, Foot & Wouters, 2009). Other test out the robustness of large infrastructures by adjusting the research questions accordingly.[1] Another possible middle ground to explore is the set of requirements needed for a large infrastructure that allows web archiving for research in a flexible way by acknowledging the various individual practices of scholars as much as possible.

THE IN-BETWEEN MACHINE: REQUIREMENTS FOR WEB ARCHIVING IN RESEARCH

In 2008, a collaboration between the Virtual Knowledge Studio in Amsterdam and Maastricht University (The Netherlands) started a project to develop an infrastructure for web archiving in research in the humanities and social sciences called The In-Between Machine. The In-Between Machine initiative builds on the experiences of various projects and tries to find new ways for preserving web archives together with their contexts of creation and use. Cataloging activities of curators and web-based annotation activities by researchers will be kept together as part of the aim to create a collective memory. The project is called The In-Between Machine for various reasons. The infrastructure aims at mediating between durable and ephemeral data, between databases and web spheres, between data and annotations, between various disciplines (in particular, the humanities and social sciences) and their methods, between curators and researchers, between experts and lay experts, between individual and collective memory. To make web archives more suitable for research The In-Between Machine needs to meet following requirements: a better search interface to web archives, facilities to annotate them, individually and collectively, and finally an infrastructure to preserve these annotations together with the web archives.

Searching: Beyond 9/11 and Political Campaigns

The way we use web archives for research is limited by the design of the interface of the Wayback Machine. One types in a URL and clicks on a date to see a previous state of a website. Clicking on a date can be very useful for some forms of research in the humanities and social sciences. Researchers that would like to study the impact of September 11, 2001, for instance, could easily work from there. The interface of the Wayback Machine in its present form is also suitable for the study of historical events (if the data are present, which is certainly not always the case) between specific begin and end dates. This might explain the success of the use of web archives to analyze political campaigns. Political parties, candidates, and the public are producing a great amount of information on websites in a fairly brief period, which can be compared in retrospect with other sources of information. We have very good examples (Foot & Schneider, 2006; Kluver et al., 2007), but it would be a pity not to go beyond this use of websites as historical sources of research. We might wish to analyze, for instance, how websites portrayed the change of public opinion regarding the wars in Iraq or Afghanistan over a longer period. Such questions do not begin or end with a fixed date; URLs referring to dates are in such cases completely arbitrary. Moreover, it is not all about dates. Websites can be seen as dynamic cultural creations. Art or media historians might want to study whether some trends can be recognized in the way artists use websites for their work. Websites can be seen as large data sets for linguists who use the web as a corpus to study changes in language or as focal nodes of social networks studied by sociologists, ethnographers, psychologists, etc. In short, we need an interface that can serve an extremely diverse group of researchers in the humanities and social sciences who sometimes follow a path of links for in-depth studies using databases and on other occasions hover above web spheres to get a glimpse of groups around certain nodes of shared interests.

Structuring, Annotation and Contextualization for Individual Research and in Collaboratories

Website curators and users, individually and collectively, need to be able to contextualize web archives for research in the form of labeling or marking texts and other media (sound, images, films, etc.). This way expertly classified and annotated collections of archived Web objects can be created and opened up to non-experts structuring their knowledge in folksonomies and personal tags (Dougherty, 2007).

Preservation of Web Archives with Annotations for Collective Memory

Various projects are aimed at overcoming the problems related to the quantity and ephemeral, dynamic nature of digital data and improving the standards of web archiving. The In-Between Machine project focuses primarily on archiving for and annotating by researchers. The In-Between Machine not only allows researchers to collaborate and to annotate collectively but can also be seen as a selection tool to establish which archives to preserve for our collective memory. The web archiving policies of most national cultural heritage institutions to select the domains of their own country are almost never in line with the needs of researchers. Perhaps the distinction could be made on the criterion of data enrichment. For example, if one had to choose between a copy of Newton's *Principia* with and without Einstein's notes, both heritage institutions and researchers would probably keep the former for posterity. The assumption is that for annotated web archives, this choice would be similar.

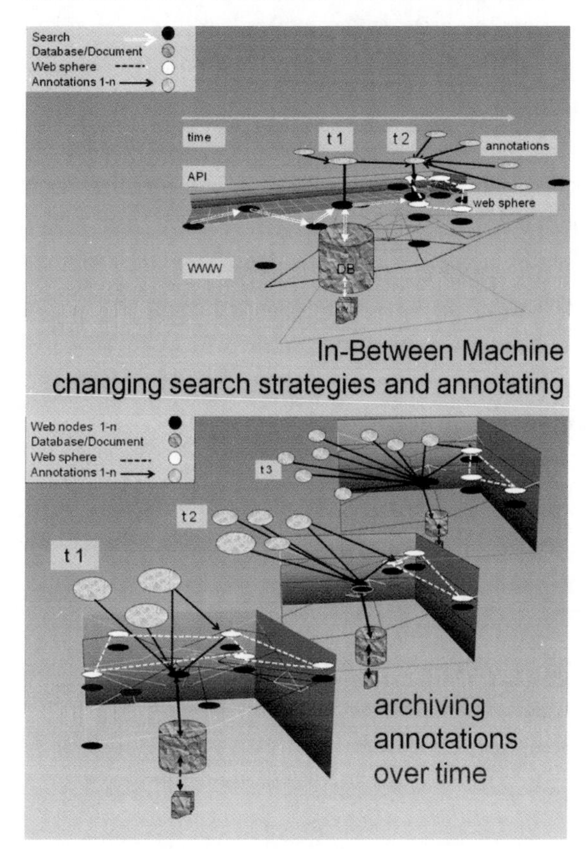

Figure 1: Concept of the In-Between Machine: archiving websites with annotations for collective memory

The proposed In-Between Machine is a complex application on top of the web and web archives. However, it is not the first time that an infrastructure that could handle static and dynamic, durable and ephemeral material was considered. Moreover, its main use was in collaborative research.

A HISTORICAL EXPLORATION OF ARCHIVING DURABLE AND EPHEMERAL MULTIMEDIA ON A GLOBAL LEVEL: KAHLE-OTLET

The mission of the founder of Internet Archive (1996), Brewster Kahle—to give the whole planet access to human knowledge—stands in a long tradition. When

Kahle decided not to limit himself to saving websites but to include books, movies, and music—in short to archive everything—he called his digital repository "Library of Alexandia, v. 2." So why start this historical expedition of web histories with Paul Otlet? Why not follow the historical examples Kahle himself mentions as his sources of inspiration, such as Benjamin Franklin, Thomas Jefferson, or Andrew Carnegie? (Feldman, 2004; Mills, 2006).

Paul Otlet (1868–1944), who was raised in the bourgeois milieu of Brussels and studied law, had a different background than the internet entrepreneur with a degree in computer science and engineering. Nevertheless, we claim that Otlet's experiments to implement a global knowledge system expressed in his *Traité de Documentation* of 1934 and in other publications and visualizations might still be relevant for a further development of the digital infrastructure of Kahle's Internet Archive and its interface, the Wayback Machine, to make a better fit between the needs of scholars and this important source of collective memory.

Otlet was on a continuous mission similar to Kahle's plan to develop the Internet Archive into a global repository. There is a similarity in their dreams to make all knowledge production in all formats accessible to everyone, all over the world. Moreover, they use similar strategies—financial, organizational, technical, epistemic and the creation of political will—in fulfilling their dreams. By the end of the nineteenth century, Otlet had taken various actions to embody the knowledge of the world. First he created a bibliographical database, the Universal Bibliographical Repertory, that soon was augmented by a Universal Iconographic Repertory and an Encyclopedic Repertory of Dossiers (Rayward, 1975). To order these databases, Otlet developed the Universal Decimal Classification (UDC) on the basis of Melvil Dewey's (1851–1931) Decimal Classification System. It was typical for Otlet to organize and materialize such activities both in institutions and buildings. In 1895, the year of the start of the bibliographical repertory, Otlet founded the International Institute of Bibliography (IBB). Fifteen years later, Otlet, together with Henri La Fontaine (1854–1943), conceived the project of the "Palais Mondial" that would bring all their initiatives for knowledge organization on a global level together. Later Otlet dubbed the Palais Mondial "the Mundaneum," which had a turbulent history of closures and re-openings until it more or less died with the outbreak of World War II and Otlet's death in 1944 (Rayward, 1975). It is important to note for our historical exploration how Otlet's plans of organizing global knowledge might be relevant for preserving and annotating website archives for collective memory in the future. He conceived the Mundaneum not just as a conglomeration of buildings but primarily as an infrastructure for a networked knowledge-based global society, consisting of material and virtual components. The Mundaneum was conceived as a hierarchical network of buildings around a global center, and the word was used as an architectural metaphor for knowledge organization and dissemination on a global level (Heuvel, 2008). In *Monde: Essai d'universalisme* (1935), Otlet wrote: "The Mundaneum is an idea, an institution, a method, a material body of work, a building and a network" (Otlet, 1935, p. 448).

This material and virtual Mundaneum was the driving force for reorganizing the documentation of all knowledge in a planned manner; integrating technical and epistemic strategies. How fundamental this re-organization would be becomes clear from Otlet's description of the ultimate problem of documentation:

> Man would no longer need documentation if he were to become an omniscient being like God himself. A less ultimate degree would create instrumentation acting across distance which would combine at the same time radio, x-rays, cinema and microscopic photography. All the things of the universe and all those of man would be registered from afar as they were produced. Thus the moving image of the world would be established—its memory, its true duplicate. From afar anyone would be able to read the passage, expanded or limited to the desired subject that could be projected on his individual screen. Thus, in his armchair, anyone would be able to contemplate the whole of creation or particular parts of it. (Otlet, 1935, pp. 390–91, transl. Rayward, 1990, p. 1)

This quote is interesting for our quest of web archiving for collective memory in relation to other resources, because it integrates the documentation of all knowledge of the world, including the preservation of ephemeral knowledge recorded in stable and dynamic media, to enable humanity to use and produce information in whatever format. This integration was visualized in its ultimate form in one of the final sketches that Otlet made of the Mundaneum in the year 1943 (see figure 2). The Mundaneum is depicted as a transmitter of knowledge by sound (radio-telephone) and by image (radio-television)—Otlet uses the term Thinking Machine—that allows people all over the world to participate actively.

However, consideration of what the inclusion of all these various media implied on a technical and epistemic level goes much further back. As early as the end of the nineteenth century, Otlet started to think about a fundamental change of bibliography by involving researchers all over the world to describe and purify information. Otlet envisioned different technical means and media for transferring knowledge leading to "substitutes for the book" and new forms of documentation.

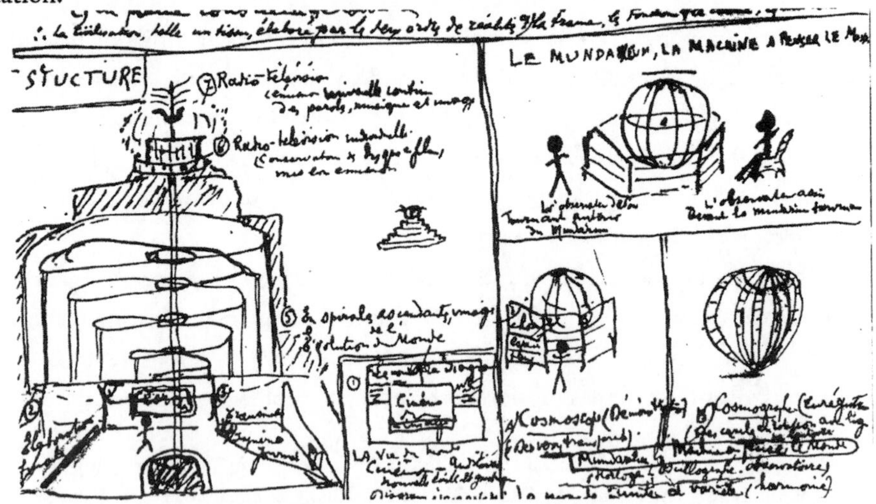

Figure 2: Mundaneum transmitter of knowledge (1943) (detail). EUM 14-120, nr 112. Mons, Mundaneum ©

SUBSTITUTES FOR THE BOOK AND NEW FORMS OF DOCUMENTATION

Fundamental for an understanding of Otlet's knowledge architecture is the notion that the book is nothing more than a container of ideas that might be conveyed in a more efficient way.

Otlet's search for different ways of transferring knowledge was directed on the one hand at developing alternative (but related) book forms such as visual encyclopedias or atlases, and on the other hand at exploring technical possibilities for retrieving, organizing, and disseminating information by means of other media such as film (especially microfilm), radio, and television. As early as 1906 Otlet expressed his views together with Robert Goldschmidt in the publications *Sur une forme nouvelle du livre: le livre microphotograpique* to be followed by their publication: *La conservation et la diffusion international de la pensée: le livre microphotique* (Goldschmidt & Otlet, 1925; Rayward, 1990). The first idea behind it was to reduce the enormous quantity of books. However, Otlet was convinced that the miniaturization and standardization in microfilm and other media would also allow an efficient interchange of alternative publication formats such as educational material, atlases, museum displays, and ephemeral material. This efficiency was based on what Otlet called the Monographic Principle. It means that texts (but also other forms of information such as formulas, charts, images, schemes, etc.) should be dissected into their basic elements and recorded on standardized cards or sheets of paper. These chunks of information could then be reassembled over and over again in new combinations of publication formats. Otlet was thinking of mechanical ways to dissect and reassemble information dynamically in standardized formats, comparable to the way the browser builds up a web page after a search. Although Otlet could never have foreseen the World Wide Web, he was convinced that the latest media of his time would transform radically, probably even replace, the book in its role of disseminating knowledge. He especially believed in the potential of wireless (radio) to become "a universal network that would permit the dissemination of knowledge without limitation" (Otlet, 1909, p. 29; Heuvel, 2008, p. 140).

Otlet realized that these alternative book forms and substitutes for the book required for new and more efficient ways of organizing knowledge. This implied the integration of the latest technology for processing and publishing information with new concepts of documentation. Before addressing the question of the relevance of Otlet's concepts of documentation for web archiving, we will briefly explore what is involved in this particular form of archiving. The World Wide Web is an ephemeral and dynamic form of publication. "Web documents," if that term can be used at all, are (1) published and available (mostly without charge) from any place connected to the Internet, (2) structured in a non-linear way as hypermedia using direct and actionable links between content pieces (3) contain not only text but combinations of images, sounds, and textual content as well, and (4) can be the result of distributed (sometimes open) authorship (compare Masanès, 2006, pp. 16–17). The characterization of the web as distributed hypermedia permanently authored at a global scale, entails, as Masanès points out correctly: "that Web archiving can only achieve preservation of limited aspects of a larger and living cultural artifact" (Masanès, 2006, p. 19).

It is interesting to read this notion of web archives as "cultural artifacts," in relation to a recurrent theme in the history of information sciences, that of "information as thing" (Buckland, 1991a, 1991b; Hjørland, 1997; Frohmann, 2009). Despite the fact that Buckland describes this information as "physical objects such as data and documents" (Buckland 1991b, p. 43), he proposes to discuss the "digital document" from a historical perspective as a special case of an answer to the question "What is a 'document'?" Although this question can be answered in many ways and Frohmann (2009) in his recent study "Revisiting 'What Is a "document"?'" challenged its validity, we will follow Buckland's reasoning as to how far the definition of "document" could be extended to "digital documents" in that it has explicit references to Paul Otlet. Buckland observes that for Otlet documents could be three-dimensional, including museum objects, for instance, and more importantly that objects themselves (not only representations hereof) can be regarded as documents (Buckland, 1998, p. 216). We shall return to the three-dimensionality of documents when discussing Otlet's interfaces. Here it suffices to say that Otlet's notion of multi-dimensional, non-linear documentation of multimedia is important for our comparison with web archiving. However, Buckland recognizes another 'evolving' notion in the work of Otlet that is important for understanding the nature of digital documents: Whatever functions as a document is a document, even if it is not in the traditional form of a document. Apart from recognizing the relevance of describing documents in terms of function rather than physical format, it is important to note that Otlet's main concern is not the object but the process of documentation. With this we do not just imply the process of documenting knowledge communicated as facts, subjects, or events (compare Buckland, 1991a, p. 351) but also future transformations of documentation of dynamic and ephemeral information.

This is crucial for web archiving. Several authors describe web archives as things: "artifacts" (Masanès, 2006), stabilized dynamic objects "documents, monuments and imprints" (Brügger, 2005) or propose an object-oriented approach (Foot & Schneider, this volume).[2] However, it is important to realize that a web archive just freezes an instant and provides a web memory that is part of the World Wide Web. In order to access the information of the web archive, it has to be reconstituted in the infrastructure of the Internet and the application of the World Wide Web on top of it, which changed since the moment of capture. This means for our historical exploration of web archiving that not only infrastructures should be considered that focus on the preservation of documents that have dynamic, ephemeral, non-linear, multimedia, and distributed characteristics but also on forms of documentation in which information is heavily dependent on, is actually inseparable from, the technologies that produce and preserve it.

In order to study transformations in web archives over time in relation to these highly dynamic documents, it will be useful in the future to include research on event-ontologies (Shaw & Larson, 2008) or instantiation (Smiraglia, 2008) while for changes in content the web philology approach of Brügger (2008) is very promising. However, for forms of documents that are inseparable from the instruments that reproduce them, once again the historical views of Otlet about the future of documentation will be of interest.

Not all the phases that Otlet describes in this process towards "hyper-documentation" can be discussed here. The sixth stage of this process, indicated by Otlet as a "sens-perception document," in which texts, images, and sound documents are combined with other tactile (smelling and tasting) ones, would certainly be of interest within the context of current explorations on synesthetics and of recent challenges to handle the fuzzy documents of embodied/tangible computing. However, the previous phase in this process of hyper-documentation, the one of the "document-instrument," might be more relevant for web archiving: "In the fifth stage, the document intervenes again to register directly the perception created by instruments. Documents and instruments at this point are linked in such way that they are not two distinct things, but one single" (Otlet, 1934, p. 429). This stage, in which documents are no longer separable from the instrumentation that produces them, in which the Monographic Principle is linked with the process of hyper-documentation, seems to come close to the way in which web pages are built up on the web over and over again and also to the way in which web archives have to be re-created over and over again. So far, Otlet's views on alternatives and substitutes for the book and hyper-documentation were discussed to explore possible similarities with the ephemeral, dynamic, and non-linear characteristics of web documents. However, Otlet is also interesting for his anticipation of another of the above-mentioned characteristics of Web publications, distributed authorship. Otlet's views on the collective use of documents via telecommunication will now be explored.

Multimedia substitutes for the book play a crucial role in Otlet's later images of knowledge dissemination for collective use in networks. Two related drawings, both with the title Documentation and Telecommunication (circa 1937), intended for the unpublished *Encyclopedia Universalis Mundaneum* show an early variant of multimedia in which telephone, radio, gramophone, film, and television are combined and transmitted as part of courses and teleconferencing. While the first drawing focuses on the telecommunication of knowledge in ephemeral non-book formats; the second one visualizes the linkage of users to various repositories of books, films, disks in a "network of universal documentation" (see figure 3).

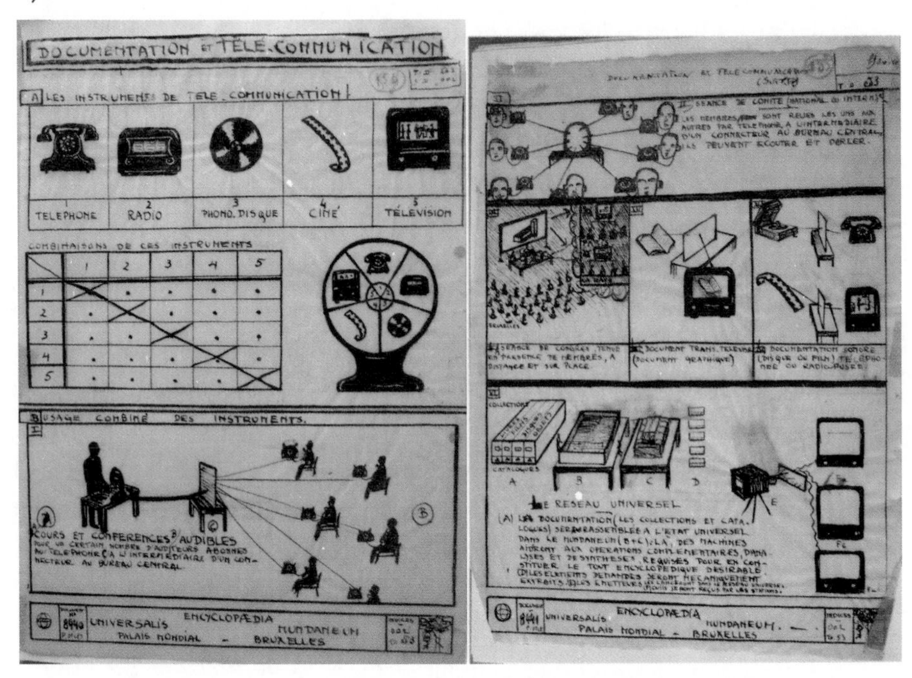

Figure 3: Documentation and Telecommunication 1 and 2
EUM 3-14-132 and 133 Mons, Mundaneum ©

In this network, documentation (collections and catalogues) is composed in a universal format in the Mundaneum and connected with other bibliographic repertories. Then machines assist in complementary operations of analysis and synthesis of the encyclopedic whole and extract desired elements mechanically. Finally transmitters send the desired knowledge elements through the universal network, where they are received by stations. In order to streamline the process and to avoid high transfer costs, Otlet designed a protocol with telegraph codes that stood for the operations that followed such requests for information. There were letter codes for the methods of transport (phone, plane, or mail), for the format of copy (typed, photographed, microfilm), for the kind and amount of information (title, full text, abstract and contextual information) and for various divisions in countries, periods, and languages.[3] However, Otlet envisioned this network of universal documentation as enabling to use but also to navigate and contribute collectively valuable information to enhance the knowledge of the world.

THE WAYBACK MACHINE AND HISTORICAL MULTIDIMENSIONAL DYNAMIC INTERFACES

Although the potential value of the Internet Archive for research can hardly be overestimated, we have argued that the original interface to get access to its web archives, the Wayback Machine, does not facilitate the search and annotation of relevant information by researchers in the humanities and social sciences. Moreover, we claimed that an exploration of historical notions of new forms of documentation and of historical global collaborations might support a better fit between user requirements and this important resource of collective memory. Here, we argue that Paul Otlet's visions of future possibilities might be useful once again to create a better interface than those available to us now.The interface of the Wayback Machine—that combines a URL with a date to one instant of one website—is one-dimensional and static. However, Otlet had already explored ways to provide access to a synthesis of the entire corpus of the ever-growing knowledge of the world in a multi-dimensional and dynamic form. Although Otlet was hindered by the flat, static character of paper, he experimented with a dynamic application of a multi-dimensional interface in at least one of his visualizations of the *Plan Mondial* (Heuvel, 2008) (see figure 4).

Otlet's axes are not gliding scales but in reality subdivided into structured administrative levels (from local to global), into the six continents, and into distinct historical periods. Nevertheless, his mobile representation seems to come close to modern computer simulations and would provide more semantic operability than the interface to web archives by the Wayback Machine. Moreover, the ability of this dynamic visualization to let the user move along an axis that expresses degree of reach (degré d'étendue), albeit here in a rigid, structured way, meets in principle the requirement for the proposed interface that researchers can alternate between general and in-depth searches. In this visualization Otlet's focus was purely on navigation; however, in other studies he explored ways in which researchers can contribute to the enrichment of global knowledge.

144

LE PLAN MONDIAL

I- Toute formation sociale est une synthèse (combinaison) des divers éléments suivants dégagés par l'analyse sociologique (distinction).

① Les domaines	② Les secteurs	③ L'instrumentation.
11) Le physique 12) L'économique 13) Le social 14) Le politique 15) L'intellectuel 16) Le religieux	21) Public 22) Associations 23) Personnes.	31) Le but. (C.U) 32) Science mondiale. 33) Principes mondiaux. 34) Plan mondial. 35) Fédération mondiale 36) Constitution mondiale 37) Survey mondial. 38) Cité Mondiale.
④ Degré d'étendue.	⑤ Espace:	⑥ Temps.
41) Local. 42) Régional. 43) National. 44) Continental 45) Mondial.	51) Europe. 52) Asie. 53) Afrique. 54) Amérique. 55) Océanie 56) Polanie.	61) Préhistoire. 62) Antiquité. 63) Moyen-Age. 64) Renaissance. 65) Temps contemporains

On peut figurer ces six ordres de données par un cube aux trois dimensions (1,2,3) et mobile suivant trois grands axes (4,5,6).

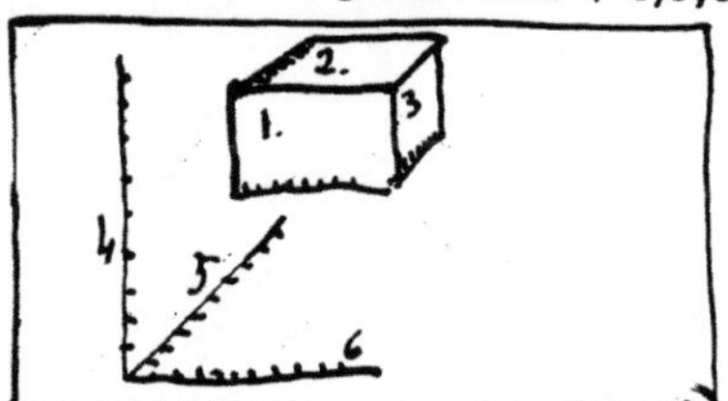

I- Quel que soit l'aménagement qui interviendra après la guerre, la Cité Mondiale aura une raison d'être dans cet aménagement et il sera inévitable quel qu'il soit d'instaurer un plan mondial. Ainsi le voudra cet ébranlement du monde, les exigences nouvelles, les nouveaux objectifs dès qu'on voudra les fonder sur des bases doctrinales.

II- Ce plan mondial intéresse le plan même de la Cité (plan idéologique et urbanistique) et il est désirable qu'on le trouve inscrit dans le dit plan de la Cité.

Figure 4: Dynamic visualization of the Plan Mondial

The three visible sides of a cube represent (1) the domains of knowledge, (2) the organizational sectors (public, private, associations), and (3) the instruments (world knowledge, world plan, world federation, world constitution, etc.) that had to be embraced within the *Plan Mondial*. This cube is shown as moving along three axes: 4 (degree of reach), 5 (space) and 6 (time), which would continuously change the relationship among the data. EUM 3-14-119 (old number 144) Mons Mundaneum ©

HISTORICAL EXAMPLES OF USING DYNAMIC AND EPHEMERAL INFORMATION FOR RESEARCH

For Otlet technological and epistemic strategies to process information, such as the Monographic Principle, were not just aimed at mastering glut, but also at a purification of knowledge by means of abstraction to enhance the quality of knowledge with as an ultimate goal the creation of a new kind of civilized universal society. As early as the end of the nineteenth century, Otlet pleaded for a systematic recording of facts, statistical data, and interpretations in his essay: *Un peu de bibliographie* (Otlet, 1891–1892). Frohmann (2008) recently explored the meaning of Otlet's systematic quest for "writing the facts," for modern concepts of knowledge organization in what he calls "the free play of electronic signifiers." We will try to bring this discussion further by linking historical views of pre-web initiatives with designs for interfaces and protocols of collective data enrichment with current debates around web-based infrastructures for collective annotating by scholars.

In 1912 Otlet and the Scottish sociologist and town planner Patrick Geddes proposed the preparation of an *Encyclopedia Synthetica Schematica* by an international group of collaborating scientists presenting scientific results in the form of charts and diagrams (Heuvel, 2008). These visualizations would themselves become objects of study that focused on the properties necessary for them "to represent schematically different notions and points of view of different scientific questions."[4] This encyclopedia went no further at the time than its printed title page and a short introductory note. However Otlet continued to pursue the idea of scientists collaborating together in a search for "objective truth."

In the view of Otlet scientific truth is impersonal, controllable, and verifiable for all. This requires an organization of the research results that is built on the unification and coordination of methods in combination with standardized and calibrated instruments. The book plays a role in the codification of knowledge, but Otlet uses it in a broader sense as "all sorts of registration of thought," including personal notes for research. He suggests replacing the term "book" with "document intellectuel" (Otlet, 1913, p. 383-384). This intellectual document is not limited to text but also includes registrations of sound and images on gramophone disks, photographs, and film for scientific communication. To bring all these views and formats of intellectual documents together, everyone needs to work together on the creation of *Le Livre universal de la Science* (The Universal Book of Science), that Otlet describes as "an unlimited work, always up-to-date, constantly growing, concentrating, absorbing, synthesizing, systematizing every intellectual product from the moment it is born" (Otlet, 1913, p. 385). At first sight, it seems an early form of Wikipedia, but Otlet believes too much in objective science to leave knowledge production to the wisdom of the crowd without control. In meetings leading towards the World Documentation Conference in 1937 (Rayward, 1983), a Universal Network of Documentation was set up, in which public and private documentation centers and intellectuals would work together. Its basic principles: "a—cooperation and exchange, b—centralization and decentralization, c—private and official activities, d—freedom and discipline to meet practical actions and e—open access and paid usage" (Otlet, 1937, p. 14) suggest at first sight a certain degree of openness in the distributed network, similar to the web. However, Otlet had already discussed streamlining the implementation of the network in much detail (Otlet, 1937). He prescribed standardized sizes for cards and publications and designed a complex protocol for the development of the UDC under the control of an international commission. For our comparison with web archiving and collaboratories in research, it is important to note that the control system of the UDC had features in common with hypertext and shared databases.

Boyd Rayward compared the UDC to hypertext in the following way:

> The monographic principle applied to standardized cards and sheets represented one of the major components of modern hypertext systems—nodes. The other, links and navigational systems, is reflected in the transformation by Otlet and his colleagues of Melvil Dewey's Decimal Classification system into the Universal Classification System (Rayward, 1998, p. 71).

Different from purely topical classification schemes, the UDC did not just order subjects or topic in classes by numeric codes but also allowed for linking to additional facets such as place, language, physical characteristics via its auxiliary tables of connector signs, a system of related parts that by numeric codes and connectors, such as "+, / and :" provided "the links, the genealogy even, of ideas and objects, their relationships of dependence and subordination, of similarity and difference" (Rayward, 1998, p. 71). On a technical level the classification system made it possible to link annotations to specific documents, or parts of (interrelated) documents that, following the Monographic Principle, were (re)composed around a classification number. The linkage characteristics of the UDC would not only allow connecting various classification systems (see figure 5) but also creating a space of contributors around documents. The latter could revise documents in the form of annotations, ranging from additions to various points of view (see figure 7).

These links were made manually, but Otlet also studied mechanical ways to create a system of ideas, a mechanical brain. This mechanical brain should not just serve the intellectual work of collaborating scholars but, in principle, it should serve everyone: "Like the technical machine allows not qualified workers to make perfect products, the intellectual machine does not require a specific education of the one that uses it" (Otlet, 1935, p. 238). Alex Wright states that Otlet's vision allows marrying top-down classification systems such as the UDC with socially constructed information spaces, such as MySpace.com, Flickr.com and del.icio.us with their own folksonomies and tags (Wright, 2007, p. 192). This statement is interesting for our exploration of the preservation of website archives with annotations for collective memory and suggests the need for a closer look at the infrastructure that Otlet proposes for personal and collective data enrichment. Moreover, it brings us to the ongoing debate on the authority of experts and lay experts in Web 2.0.

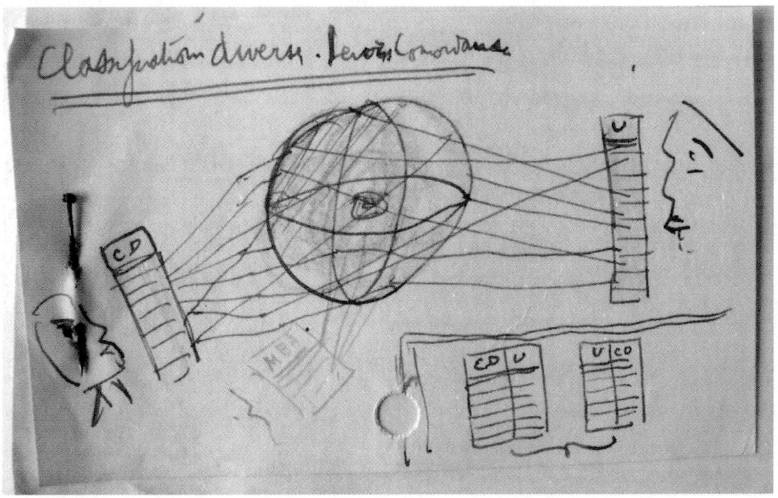

Figure 5: Otlet linking various classification systems
EUM 1-4-164 Mons, Mundaneum ©

PERSONAL ANNOTATIONS AND COLLECTIVE DATA ENRICHMENT

For Otlet the process of documentation not only involved the creation of a knowledge system but a social system aimed at creating a better society. "One can imagine a social state that makes progress in its whole by an instrumentation based on very high levels of abstraction that would be made available to everyone" (Otlet, 1935, pp. 238–239). But where is the individual in all this?

Otlet's sketches make clear that he was struggling to position the individual in his knowledge systems. Sometimes the observer stands in the center (see figure 6), but in most cases he watches from outside towards global knowledge representations. This tension between the personal and the universal does not only have implications for the way users are supposed to navigate but also for the manner in which they are supposed to contribute information to his knowledge constructions. When Otlet visualizes a person in the center, he is foremost an end-user of documents that are already organized and filtered by the UDC in the documentation apparatus around him. Otlet's concept of personal knowledge organization is strongly related to, is actually a microcosm of, his universal classification system.

Otlet recognizes the value of extracting personal notes from documents:

> Preserved, classified, revised, continuously enriched with other notes derived from other sources, they could become a real book: a particular book for each person of which one could say: "My Book," "My Encyclopedia"...an artificial memory of everything one desires to recall. (Otlet, 1934, p. 319).

He also believes in the value of preserving these annotations for collective memory. By classifying and storing the notes together with bibliographical descriptions of documents, "One could avoid new transcriptions often subject to errors, keep up with facts and ideas annotated at various moments. The confirmations by others that may also express different aspects of the same thing" (Otlet, 1934, p. 319). The process seems at first sight similar to Wikipedia in which the involvement of more people adding and editing certain lemmas leads to the improvement of those lemmas in particular and the digital encyclopedia in general. However, for annotating documents Otlet has eminent scholars in mind, who, regulated by protocols for intellectual work, would further develop the Universal Network of Documentation and the UDC to reach ever-higher levels of scientific objectivity.

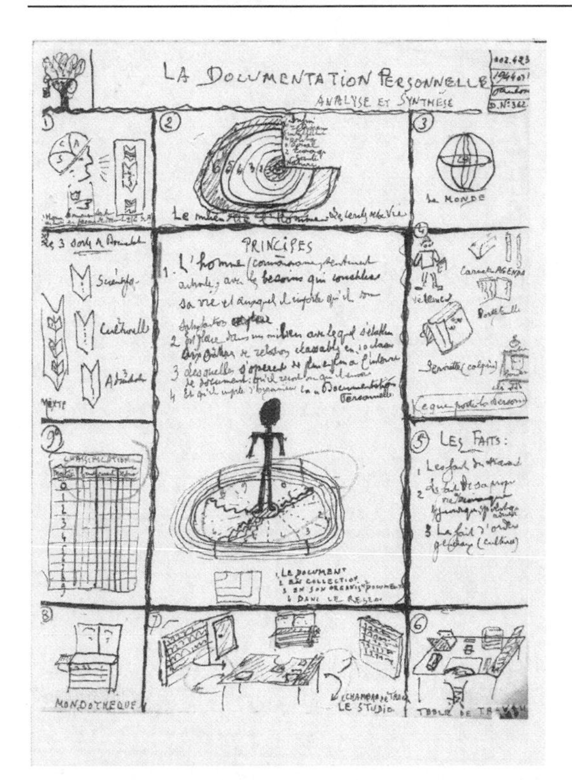

Figure 6: Otlet—Personal documentation EUM 15-228 Mons, Mundaneum ©

Yes, when Alex Wright states that Otlet's vision allows for linking classification systems such as the UDC with socially constructed information spaces, he is right in a technical sense. However, it would be wrong to read Otlet's Universal Network of Documentation simply as a wiki or his "personal classifications" as folksonomies. For Otlet the producer of the knowledge is foremost an outsider of the system whose contribution would only be recognized after a long process of editing by what we would call nowadays "domain experts." Compared to Wikipedia, Otlet's knowledge system and collaboratory is more top down but at the same time also more transparent. Edits and annotations do not merge directly with the information but remain visible in an ordered way, describing the provenance and intention of the proposed data enrichment of the Universal Documentation Network. Therefore Otlet's views on data enrichment might be taken into consideration as part of the proposed infrastructure to preserve website archives and annotations as a collective memory for research.

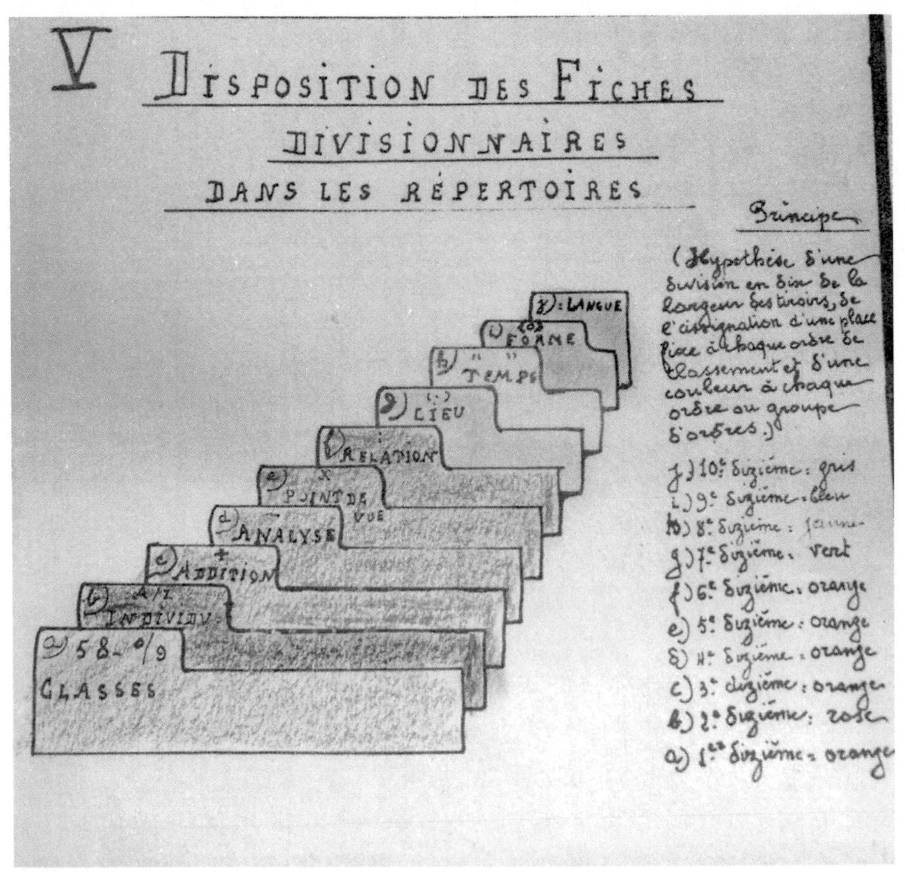

Figure 7: Provenance and intention of annotations in relation to knowledge class in repertory

Card box with color code for tabs. Knowledge class in relation to name of the annotator whose input is classified as addition, analysis, point of view or relation to other subjects, to space, to time, to form, i.e., medium, and to language. EUM Affiches Table de classification 58 Botanique (detail) Mons Mundaneum ©

BRINGING IT TOGETHER: OTLET AND THE IN-BETWEEN MACHINE

We have argued that large infrastructures that preserve web archives together with other digital repositories, such as the Internet Archive, do not have the necessary interfaces and annotation systems to exploit them fully for research. Existing systems such Hanzo Archives or I BreadCrumbs allow researchers to annotate website archives, but such annotations are not kept together with these archives to create a collective memory. Meghan Dougherty has made clear in her dissertation that tags by users are not just means to enrich data but also can become instrumental in navigating and giving sense to shared paths in web archives (Dougherty, 2007). Building on her ideas for an interface tool, Wayfinder, we intend developing an infrastructure, The In-Between Machine, that covers the middle ground between collective memory and annotations for research. Furthermore, we have argued that apart from future ethnographic studies of the scholarly use of web archives, it might be useful to explore historical infrastructures of preserving and annotating dynamic, ephemeral material for research. To this end Otlet's epistemic and technical strategies were analyzed in relation to web archiving. We concluded that his Monographic Principle, in which information chunks expressed in various media (text, image, static and dynamic, durable and ephemeral) corresponded very well with the multi-dimensional, non-linear character of web archives. Moreover, we claimed that Otlet's views on the future of documentation, in which documents cannot be separated from the instrumentation that produces them, are crucial for our understanding of web archives as instant memory parts of an ever-changing World Wide Web. In that sense Otlet's statements of 1934 come closer to the nature of web archiving than recent interpretations of specialists that approach web archives as objects similar to books and other forms of documentation. Finally we tried to demonstrate that Otlet is interesting for web archiving because of his futuristic views on forms of documentation that require reconsideration of knowledge organization and production and also from a technical point of view. Otlet's multidimensional interface that we have analyzed in some detail allowed access to objects on various levels. As such it is coming closer to the requirement that researchers must be able to change strategies in navigation rather than employ the rather flat system of URLs and dates of the Wayback Machine. The same applies to issues of annotation. Following the work of Wright, we claimed that Otlet's system of auxiliary tables allows keeping together dynamic annotations with changing content. However, we have challenged Wright's idea of reading Otlet's Universal Documentation Network as a social space such as Wikipedia. The nature of Otlet's knowledge system and the collaboratory he sees developing around it are too hierarchical for that. However, the knowledge system's transparency might be usefully investigated in discussions of the role of authority in distributed authorship.

The attempt by Otlet to uphold scientific authority by designing protocols for scholarly collaboration and developing classifications or typologies of annotations comes close to recent attempts to differentiate in forms of expertise in Web 2.0. Larry Sanger, co-founder of Wikipedia, built a new user-generated encyclopedia, Citizendium, reviewed by domain experts. Software developers try integrating top-down classification with bottom-up tagging (Facetag) and try to give identity to links by visualizing their provenance (HarvANA) or by assessing their value (GenTech Datamodel). The question of whether the future lies in structuring information by experts, in the collective wisdom of the crowd, or in an "in between solution" is ideological rather than technical. Infrastructures envisioned to preserve, navigate, and enrich web archives as documents for research will shape that future. However, not just technical solutions to preserve, disseminate, and enrich information on a global level but also the way knowledge is organized and used by experts and non-experts all over the world creates the cultural history of mankind. It requires a historical approach of knowledge organization in relation to practices of use.

ACKNOWLEDGMENTS

I wish to thank Bernd Frohmann for sharing his text "Revisiting 'What Is a "document"'?," unpublished at the time that I wrote this chapter and the Mundaneum in Mons for the use of images that illustrate this publication. I am indebted to Meghan Dougherty, Richard Smiraglia and Sally Wyatt, who read earlier versions of this chapter. I am particularly grateful for the extensive comments made by Paul Wouters and Boyd Rayward on the final version.

NOTES

1. Meyer, Madsen and Schroeder, "The World Wide Web of Humanities: Archives for researching the Web", paper for Web_Site Histories Conference, Aarhus University. Centre for Internet Research, October 7, 2008. See, further event Humanities on the Web. Is it working? Oxford Internet Institute, March 19, 2009: http://www.slideshare.net/etmeyer/wwwoh. Retrieved June 5, 2009.
2. See, the contribution of Foot and Schneider, chapter 2.
3. Mons, Mundaneum, MB 225.141 IIB (2004–5)—Nota 915, 1926. 01. 30.
4. Mons, Mundaneum, EUM, I 4 Généralités, Orbis Encyclopedia Synthetica, Printed title page *Encyclopedia Synthetica Schematica* 'Matériaux pour l'élaboration d'une méthode et d'une synthèse à l'aide de schemas et de diagrammes, publiés par une collaboration interscientifique et

international sous la direction de Patrick Geddes et Paul Otlet', with an introductory note in typescript of September 17, 1912, p. 1.

REFERENCES

ACLS (2006). *Our cultural commonwealth: The report of the American Council of Learned Societies' Commission on cyberinfrastructure for humanities and social sciences.* New York: American Council of Learned Societies. Retrieved August 30, 2009, from http://www.acls.org/cyberinfrastructure/OurCulturalCommonwealth.pdf

Berners-Lee, T. (1999). *Weaving the web. The past, present and future of the world wide web by its inventor Tim Berners-Lee with Mark Fischetti.* London: Orion Business.

Brügger, N. (2005). *Archiving websites. General considerations and strategies.* Aarhus: The Center for Internet Research.

Brügger, N. (2008). The archived website and website philology: A new type of historical document? *Nordicom Review,* 29(2), 155–175.

Buckland, M. (1991a). Information as thing. *Journal of the American Society for Information Science,* 42(5), 351–360.

Buckland, M. (1991b). *Information and information systems,* Westport, CT: Praeger.

Buckland, M. (1998). What is a 'document'?, In T.B. Hahn & M. Buckland (Eds.), *Historical studies in information science* (pp. 215–220). Medford, NJ: Information Today Inc. Reprint *Journal of the American Society for Information Science,* 48(9), 804–809.

DEFF (2006). *The Hybrid library from the users' perspective.*

Dougherty, M. (2007). *Archiving the web: Collection, documentation, display and shifting knowledge production paradigms.* Seattle: University of Washington Ph.D. dissertation.

Feldman, S. (2004, June). A conversation with Brewster Kahle. *Security,* 2(4). Retrieved July 6, 2008 from http://www.acmqueue.com/modules.php

Foot, K.A. & Schneider, S.M. (2006). *Web campaigning.* Cambridge, MA, London: MIT Press.

Frohmann, B. (2008). The role of facts in Paul Otlet's modernist project of documentation. In W.B. Rayward (Ed.), *European modernism and the information society* (pp. 75–88). London: Ashgate.

Frohmann, B. (2009). Revisiting 'What is a "document"?' *Journal of Documentation,* 65(2), 291–303.

Goldschmidt, R. & Otlet, P. (1925). *La conservation et la diffusion internationale de la pensée: Le livre microphotique.* Publication No. 144. Bruxelles: IIB. [*The preservation and international diffusion of thought: The microphotic book*] (Rayward, 1990, paper 15).

Heuvel, C. van den (2008). Building society, constructing knowledge, weaving the web: Otlet's visualizations of a global information society and his concept of a universal civilization. In W.B. Rayward (Ed.), *European modernism and the information society* (pp. 127–153). London: Ashgate.

Hjørland, B. (1997). *Information seeking and subject representation: An activity-theoretical approach to information science.* Westport, CT, London: Greenwood Press.

Kluver, R., Jankowski, N., Foot, K. & Schneider, S. (Eds.) (2007). *The internet and national elections: A comparative study of web campaigning*. New York: Routledge.

Masanès, J. (Ed.) (2006). *Web archiving*. Berlin, Heidelberg: Springer Verlag.

Mills, E. (2006). *Brewster Kahle's modest mission: Archiving everything*. Post on ZDNet News: June 23, 2006 11:00:00 AM. Retrieved July 7, 2008 from http://news.zdnet.com/2100-9588_22 6087167.html

Otlet, P. (1891–92). Un peu de bibliografie, Palais, Organe des Conférences du Jeune Barreau de Belgique, pp. 254–271. In Rayward, W.B. (1990) 11–24.

Otlet, P. (1909). *La function et les transformations du livre: Résumé de la conférence faite à la Maison du Livre 14 November 1908*. Bruxelles: Musée du Livre 11.

Otlet, P. (1911). *L'Avenir du livre et de la bibliographie IIB Bulletin* (16), 275–296.

Otlet, P. (1913). *Le livre dans les sciences, Conférence faite à la maison du Livre par M. Paul Otlet.* Bruxelles, Musée du Livre 25–26, 379-389.

Otlet, P. (1934). *Traité de documentation: Le livre sur le livre: Théorie et pratique*. Bruxelles: Editiones Mundaneum, Palais Mondial.

Otlet, P. (1935). *Monde, essai d'universalisme: Connaissance du monde, sentiment du monde, action organisée et plan du monde*. Bruxelles: Editiones Mundaneum/D.van Keerberghen & Fils.

Otlet, P. (1937). Réseau universel de documentation: Documentatio universalis (rete), *IID Communicationes*, (4)1, 13–16.

Otlet, P., & Goldschmidt, R. (1906). *Sur une forme nouvelle du livre: Le livre microphotograpique* [On a new form of the book: The microphotographic book]. In Rayward, W.B. (1990), 204–210.

Rayward, W.B. (1975). *The universe of information: The work of Paul Otlet for documentation and international organization*. Moscow: FID.

Rayward, W.B. (1983). The International Exposition and The World Documentation Congress, Paris, 1937. *Library Quarterly*, 53 (July): 254–268.

Rayward, W.B. (1990). *International organisation and dissemination of knowledge: Selected essays of Paul Otlet, translated and edited with an introduction by W. Boyd Rayward*, FID 684. Amsterdam, New York, Oxford, Tokyo: Elsevier.

Rayward, W.B. (1998). Visions of Xanadu: Paul Otlet (1868–1944) and hypertext. In T.B. Hahn & M. Buckland (Eds.), *Historical Studies in Information Science* (pp. 65–80). Medford NJ: Information Today Inc. Reprint *Journal of the American Society for Information Science*. 45 (1994): 235–250.

Rayward, W.B. (Ed.) (2008). *European modernism and the information society*. London: Ashgate.

Schneider, S.M, Foot, K.A., & Wouters, P. (2009). Taking web archiving seriously. In Jankowski, N.W. (Ed.) *E-research: Transformations in scholarly practice*. New York: Routledge: 205–221.

Shaw, R. & Larson, R.R. (2008). Event representation in temporal and geographic context. In *Research and advanced technology for digital libraries*. Berlin, Heidelberg: Springer: 415–418.

Smiraglia, R. P. (2008). A meta-analysis of instantiation as a phenomenon of information objects. In *Culture del testo e del documento IX, 25, gennaio-aprile*, 5–25.

VKS-IISH – http://www.virtualknowledgestudio.nl/projects/globalhubs.php

Wouters, P. (2007). Forging new academic identities: Key dilemmas and challenges in digital scholarship. Simpson Center, University of Washington, Seattle, May 1.

Wright, A. (2007). *Glut: Mastering information through the ages*. Washington, DC: Joseph Henry Press.

CHAPTER 13

Collecting and Preserving Memories from the Virginia Tech Tragedy: Realizing a Web Archive

Brent K. Jesiek and Jeremy Hunsinger[*]

INTRODUCTION

For countless thousands worldwide, the plurality of perspectives surrounding the events of April 16, 2007, at Virginia Tech ruptured their everyday lives. The shock of that day's rupture demanded new understandings of the world, the renegotiation of the commonplace, and the realization of new and old mnemotechnical practices. These emergent places and practices provide foundations for innovations in web history.

The first steps toward these foundations were taken two days after the tragedy. Tending to still-fresh emotional wounds and trying to nurture a sense of community, a discussion occurred that would lead to the digitization and preservation of memories from around the world. Enabling the inscription of shared experience and shared grief opened new pathways for documenting and re-imagining past events. Conversation over beers in a pub in downtown Blacksburg, Virginia, gradually drifted to the subject of preserving and sharing memories. In the collegial atmosphere of the pub, a need was recognized and a response began to crystallize. We sought out a set of tactics and tools that could capture memories, allowing anyone to commit memories from April 16th to a system, in the process constructing a new commonplace. It was agreed that a new web site and archive should be launched by the Center for Digital Discourse and Culture (CDDC) to collect the many digital artifacts qua memories created in the wake of the tragedy.

We collected our first items from the general public on April 24, and over the next few days more items trickled in. Confident with the stability and functionality of the system, on Monday April 30 both the CDDC and Virginia Tech's College of Liberal Arts and Human Sciences did formal press releases announcing the launch of the April 16 Archive (Elliot, 2007). Within less than two weeks of the tragedy, the archive was being realized.

Web archives with a memorial mission, such as the April 16 Archive, are pragmatic and outwardly social archives that do not overwrite the memories of visitors and contributors. Instead, these memory repositories allow for open and social re/constructions of both commonplaces and related mnemotechnical practices surrounding those memories. The memory that such events demand is more than formalized archival memory; these memories must have elements that serve the memorial missions surrounding the event, and they must be inclusive of the plurality of interests of those affected. Such an archive must strive to enable the re/negotiation of both individual and social memories.

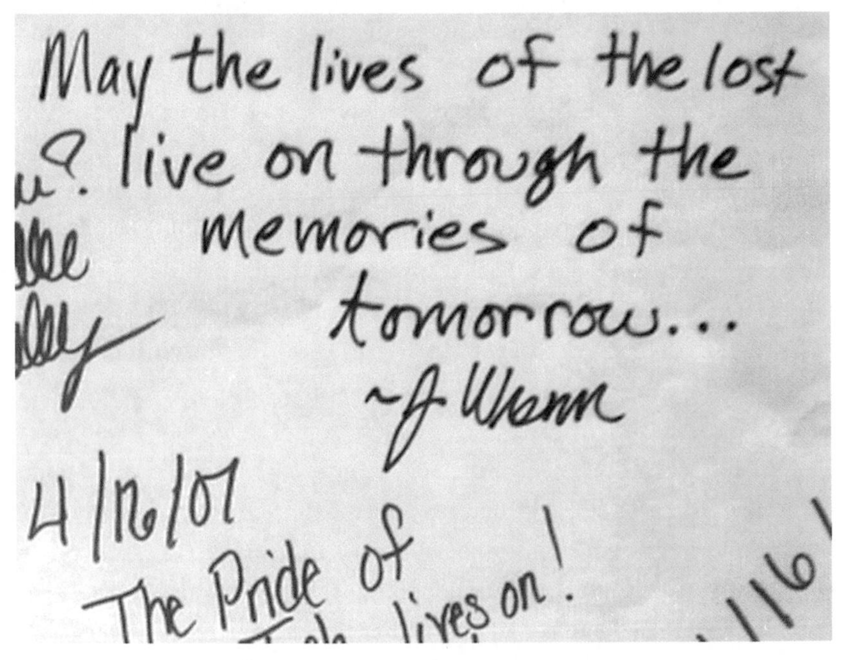

Detail of memorial board at Virginia Tech's Squires Student Center.
Photo taken April 19, 2007, by Jinfeng (Jenny) Jiao.
(http://www.april16archive.org/object/928)

MNEMOTECHNICS: FROM PHYSICAL ARCHIVES TO DIGITAL MEMORY BANKS

The memory bank takes its place among a plurality of archival technologies that map onto our mnemotechnical practices. Virginia Tech's archival practitioners, for example, are building more traditional physical collections from the countless memorial items and gifts that arrived on campus after the events of April 16 (Vargas, 2007). But as these and other physical archives are finalized and constructed, they become less an active remembrance of events and more a concretization of positionalities and primary narratives, which over time develop toward the histories that the majority of people will share. And as these positions in turn become less inscriptions of memory—especially as they are recognized as official markers of history and begin to provide (re)mediated (re)interpretations of events—they become abstracted from the memories that were their original foundations, the social meanings in which they are embedded, and the memorial actions for which they were intended. It is in this abstraction from the social that the personal and minoritarian meanings of the archive slough off over time, and the canonical, majoritarian histories arise and overwrite memories.

Like all movements toward the co-creation of majoritarian perspectives, professional archivists are keenly aware of the need to capture primary sources, which then inform the development of fundamental narratives and their respective majoritarian positions. The understanding of this abstractive co-creation is inscribed in and through their mnemotechnical practices and archival technologies. Capturing and writing these positions are determined by the intrinsic limits of memorials and archives, limits that are often built into the physical containers of meaning and the physical representations of events. Given their limitations, such physical archives cannot reflect a wide plurality of positions. Nor are they very good at capturing or valuing the ephemeral or personal, because their principles of accession often focus on gathering what archivists recognize and value most, namely, the documents behind the primary narratives.[1] At the same time, source materials representing minoritarian positionalities and memories remain largely invisible and marginalized.

Bound by their physical and economic constraints, physical archives are also unable to archive everything contributed. By mid-August of 2007, for example, officials at Virginia Tech had documented more than 60,000 items related to the events of April 16 but admitted they were not yet done counting (Vargas, 2007). One library administrator added that she had "probably just seen 1 percent of what there is." By August, the school had provided a scant 800 cubic feet for storing archived goods. Such limitations suggested that archiving even 5 percent of all physical objects collected and received in the wake of the tragedy would be difficult to achieve, despite the fact that this target was recommended to Virginia Tech by consultants from the Library of Congress. It is often the most ephemeral and personal items that are first to be cast off when archivists face such brutal constraints. And as such items are disposed, so too are their histories.

"Memorial at the top of the drill field"

Photo taken April 21, 2007, by Ross Catrow. Licensed under Creative Commons Attribution-Share Alike 2.0 Generic (http://www.april16archive.org/object/1566)

In the past, debates about inclusion centered on provenance, space, and economics. These constraints led archivists to strategically and selectively pursue the restriction, maintenance, and/or expansion of collections. Yet the debates surrounding what is to be included in a given archive are transformed by the politics surrounding the technologies and practices of memory, even as these debates are simultaneously (re)constructed by and through these same technologies and practices. And without spatial limitations, the politics of the archive become partially unbound. This unboundedness reflects back on the mnemotechnical systems that represent the next generations of digital collections and digital histories.

With a digital collection, the archivist's traditional physical frame of reference starts to be replaced by the digital spaces of bytes and bits. Admittedly, web-based archiving systems are still constrained by the physical limits of storage inherent in digital storage devices such as hard drives, flash memory, or related technologies. But even with current technologies, our rapidly expanding ability to collect and preserve ever-more expansive arrays of digital information breaks the spatial and economic metaphors of the archive and transcends the structures that previously informed accession policies.[2]

By pursuing a project based on the expression of contributed memories and taking a highly social and pragmatic approach to questions of provenance, we sought to limit our subjection to some of the difficulties that have long vexed archivists. Opening our principle of accession to everything related to the April 16 event allowed us to better enable our users to focus on the immediacy of their memories and their expression of personal knowledges and experiences related to the event. This type of approach makes it far more likely that web history projects will help uncover other narratives—including more personal, even minoritarian narratives—that can provide counter-points that may support or possibly undermine the primary narratives that become central to the shared majoritorian history.

Having an archival system that encompasses every related expression of memory that anyone wishes to commit—and that curators and collectors elect to gather—provides a very different constitution of the possible histories of an event, especially as compared to what is usually provided by archival missions that are constructed and constrained in a top-down manner. This conception of memory is further transformed by systems that allow diverse collections of people from all over the world to openly access and contribute to an archive. The technology that we adopted for our project matched digital ethos to digital archive, thereby motivating and informing how we approached the building and management of our system.

The April 16 Archive embodies one conception of communal memory—the commonplace or shared reference point—called the "digital memory bank." This type of web application is being used to preserve the richness of the present as it transitions to the past, thereby ensuring that digital records related to a given historical subject or period are readily accessible and carefully preserved for future access. George Mason University's Center for History and New Media (CHNM) is a leader in this area, and kindly provided us with an early version of their Omeka software platform to launch our archive.[3]

Extensively informed by the work of archivists, technical experts, historians, and other scholars, this application is the product of numerous years of experience and many iterations of development (Cohen and Rosenzweig, 2005). In fact, the software has been refined through a series of ambitious projects, including the September 11 Digital Archive, which in 2003 became the first digital acquisition of the Library of Congress (Library of Congress, 2003), and the Hurricane Digital Memory Bank, which collects stories and digital objects related to hurricanes Rita and Katrina. Given such a successful and esteemed track record, we were pleased to accept CHNM's gracious offer of software as we realized another member in this family of well-known memory banks.[4]

MEMORIES, LIVES, SOCIAL WORLDS, AND HISTORIES

Funes the Memorious exemplifies the problem of perfect living memory (Borges, 1999). He could not act, because whenever he would seek to remember how to do something, every detail would come to his mind, occupying it to the exclusion of future action. Digital memories from a perfect recorder would act much the same way, absorbing the whole of the time of experience and reordering our personal narratives in relation to it. In absorbing and reordering time, perfect memory can overwhelm one's life, to the point where the story takes longer to tell than the lives and events actually lived. Funes' perfect memory shows that history demands a limit to inscription. The completeness of his memory cannot be stated or even known, because it overwrites the living of life and the shared world the living creates, leaving only the private world of his memories.

This imagination of perfect memory is illustrated differently in Calvino's essay for the 50th anniversary of IBM (Calvino, 1995). In this story, Calvino's digital archivist is an adulterer and eventually a murderer. He is capable of the most heinous of acts, yet because of his complete power of overwriting the past, he can erase his actions including the existence of his victims. Calvino illustrates the terror of both the archives specifically and history in general: namely, that history can be 'cleansed' or erased. The erasure of shared social memory by archivists, historians, and memorialists plays a part in the knowledge-politics of the archive (Brown and Davis-Brown, 1998).

Part of the knowledge-politics of the archive is the control of memory via processes of archiving, historicizing, inscribing, or memorializing. Archivists and historians can remove memory from the social sphere and institutionalize the memories in archives, books, and memorials. This is what Calvino was illustrating about the control of the memory, or the capacity to bring about the virtual murder of thousands of people by the systematic exclusion of their memories and thus the exclusion of their lived lives and social worlds. Although it is often necessary and important to exclude some facts in the construction of histories and anthropologies, opening up a wide space for the future development of diverse narratives, stories, and interpretations demands that we capture and preserve what we can (Augé, 2004). One possible way of opening up the spaces of interpretation is exemplified by Fabian's claims about the migration toward commentary on archives, instead of the continual reproduction of archival contents (Fabian, 2008).

Capturing memory by digital contribution avoids the control of accession and de-accession and thus the problem of exclusion of memories. Public contribution of memories allows for a plurality of social worlds to manifest themselves in the memory bank. In that our project was centered on memories—and not conducting research or constructing narratives from the materials in the archive—our archive is no longer centered on the archive, archivist, or writers of primary narratives. Our power over the physical hardware and hence content of the archives is minimal, given that copies of the archive live on elsewhere in mirrors and search engines.

Instead, the archive is partially decentered across real and virtual territories, and then reconstructed through the memories of a plurality. Further, the gross weight of the individual differences inherent in a multiplicity of contributions problematizes the experiences of time and space through which we tell narratives. In reading the contributed memories, one gets a sense of time and space as plural, yet shared, in relation to the materials contributed. This reading gives people a new way to make sense of how they relate to their memories, their spaces, their senses of time, and their possible histories.

The ability of the archive to become a space of shared memories also creates aporetic spaces and times, where people can get lost—hopefully fruitfully and beneficially—in the relations between their own memories and the contents of the memory bank. As a result, one's own personal narrative of events is partially displaced. Relating to the contributed memories of others, our memories are reconstituted with a new sense of being, with a different history, and with an expanded sense of both the significance of an event and our own relation to it.

The question of personal significance should not be overlooked, as it is central to the idea of a memory bank. Contributions to memory banks help both break down the subjective experience of personal memories and project the memories into a social landscape. The sharing of memories allows people to perform their significance in these social spaces, thus breaking the confines of their local community. The value of this social re/construction of memory and its commonplaces is in its capacity to overcome pre-existing social and physical barriers, such as the tired dichotomy of direct participant versus outside observer, curator versus contributor, or designer versus user. By promoting contribution based on a model of decentered sociality, people can see how archival contributions from others are also about themselves, in both the singular and plural.

Memory banks often tend strongly toward an inclusive, heterogeneous, and non-hierarchical character. This improves the ability of people to create multiple and differential relationships to a given archive or subset thereof. These relationships are the foundations of multiple and overlapping narratives that ground social relations. As the memory bank becomes a repository for personally significant memories and digital objects, the relations between the memory bank and the contributors do not simply involve a separation of memories from their point of origination but rather become part of a shared social world.

Memory bank architectures thus replicate less the repository values of the traditional archive and instead embody the relationships of participatory cultures and social memories that are exemplified by next-generation web technologies such as blogs, wikis, and social network sites such as Facebook and MySpace. On these sites, memories are frequently shared with other participants, and histories are re/constructed on the fly. And while the items in a digital memory bank are not mutable like the text in a wiki, they do allow recontextualization through tagging items with keywords, and this functionality enables the construction of new relationships of knowledge and memory.[5] The types of participatory contribution and reconstruction of memories enabled by memory banks can help people make sense of the importance of their own narratives, and those of others.

The problem of Funes' inaction is not a problem for the creators or participants in these new types of archives, as the memories that are shared leave adequate spaces for continued action and participation. Rather than paralyzing us with either overwhelming knowledges or a fear of erasure, social memories provide commonplaces for discussion, experience, and action. The social memories constituted by an emergent community of memory bank contributors and users are thus not a cause of inaction but rather an inspiration for action, for creating new identities and histories. The memory bank provides for living our lives in the context of memories that transcend but do not overwrite individual experiences.

As we have discussed, a perfect memory is not the goal of a memory bank.

Virginia Tech memorial on Info Island in Second Life®.

Image captured April 17, 2007, by Wagner James Au. Reproduced courtesy of nwn.blogs.com (http://www.april16archive.org/object/124)

THE APRIL 16 ARCHIVE: REALIZING A WEB ARCHIVE AS COMMONPLACE MEMORY BANK

Memory bank applications like Omeka represent a democratization of web archives and web history, lowering the barriers that previously limited the ability of committed individuals or small groups to establish their own digital memory banks. With an early version of the software and a little help from the CHMN, we used our pre-existing systems and technical expertise at the CDDC to respond rapidly in the face of tragedy.

By mid-August of 2007 the archive consisted of more than 1,000 discrete digital objects, by early December of 2007 that number had reached 1,400, and by December of 2008 the archive featured well over 2,000 items. Our staff, including six part-time workers who assisted with the project, collected a large share of these items, but many others were contributed directly to the site. The archive includes numerous stories, images, poems, and tributes available nowhere else on the web. Our accession policy for submissions has been as inclusive and agnostic as possible. By mid-December of 2008, we recorded more than 30,000 unique visits to the site and over 200,000 page views. The archive has also been mentioned or profiled by more than thirty major online and traditional media outlets (Grant, 2007; Goff, 2008; Guess, 2008; Haegele, 2007; Hutkin, 2007; Vargas, 2007).

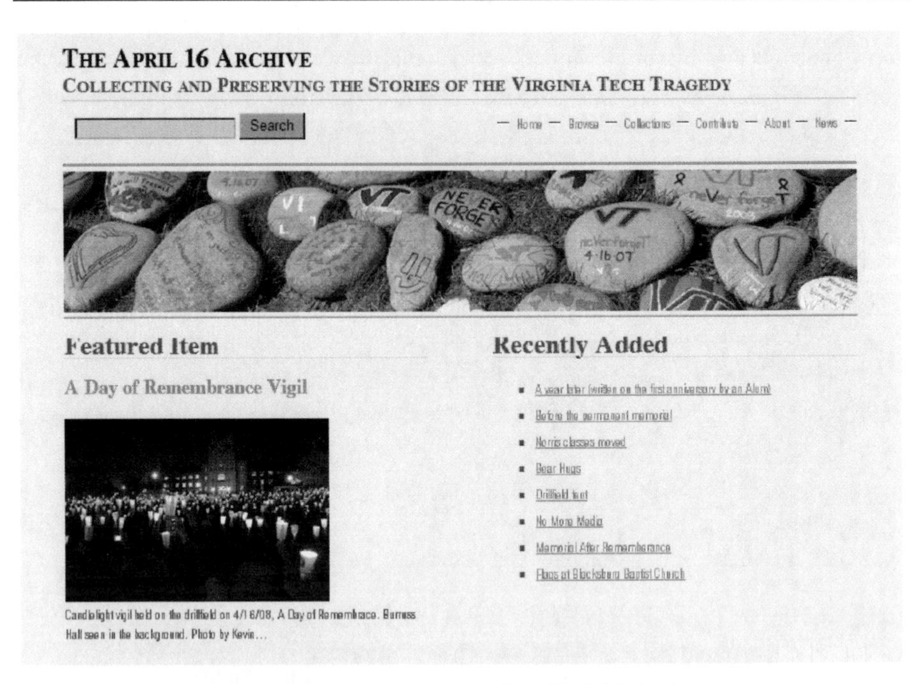

Main entry/index page for the April 16 Archive.

Captured by Brent K. Jesiek on December 12, 2008.
(http://www.april16archive.org/)

We have mostly focused our own collecting efforts on materials that: (1) are most likely to get lost with the passage of time, (2) provide unique, local, and/or original viewpoints rather than repeating widely known facts or knowledge, and (3) are not well indexed or preserved elsewhere. To provide further perspective on this wide variety of content, the archive now features:

- A collection of official university e-mails sent on April 16 (http://www.april16archive.org/object/62).
- Hundreds of photos and stories in special collections dedicated to memorial and other related events at Virginia Tech and beyond, including collections for the April 16th Memorial Dedication, A Concert for Virginia Tech, and the Anniversary Commemoration (http://april16archive.org/collections).
- Relevant articles and editorials from more than sixty college and university newspapers and media outlets, mainly in North America.

- Commentaries and reflections on topics as diverse as gun control, violence, Asian- and Korean-American identity, mainstream media coverage of the event, new media and citizen journalism, privacy laws, and mental health.
- Reactions to the events of April 16 from more than a dozen countries all over the globe and in languages as diverse as Romanian, Spanish, Korean, Chinese, and French.
- Documentation of memorial structures and events in the Second Life® virtual world.
- Numerous audio files, including talk show excerpts, musical tributes, and podcasts.
- Official follow-up reports produced by the university and state and federal governments.
- Many pieces of digital artwork and poetry.

Some of our work has also focused on collecting Creative Commons (CC) licensed content, including text, photos, and audio. Given the nature of the project, we can archive this material without having to obtain author/creator permission, so long as we follow and reproduce the associated license provisions. To date, more than 20% of the items in the archive are archived under CC licenses.

MEMORY BANK LIMITS AND CHALLENGES

Despite the successes outlined above, we also recognize the limits of our efforts. For instance, we were disappointed by an almost complete lack of unpaid volunteers willing to help curate or collect materials, despite many early expressions of interest. This problem was likely compounded by the typical ebb and flow of the academic calendar, continued emotional fall-out from the tragedy, and a desire among faculty and graduate students to get back to neglected work. Similar factors likely played a role in the rapid decline of other promising initiatives such as *Hokies 4/16: A Memorial Project* (http://hokies416.wordpress.com/), a blog-based archive started by student staff of *The Collegiate Times*, Virginia Tech's student-run newspaper. Independent memorial and tribute sites launched by the school's alumni were also quite impressive but similarly lost momentum over time.[6]

Perhaps for similar reasons, most visitors to our archive browse without contributing reflections or artifacts. And those who do contribute materials are faced with significant demands on their immaterial labor. At present, the system requires that users submit items individually, provide ample associated information (description, keyword tags, identity of creator and contributor, etc.), and agree to certain consent provisions. In addition to the labor required to submit each item, contributors may feel pressured to edit or enhance the objects they submit, especially in anticipation of public display.

In light of these challenges, having paid full-time staff (in the form of co-author Brent Jesiek) and part-time workers proved extremely valuable for curating and growing the archive. But involving these other individuals in the project brings into further relief the emotional impact of this kind of archival work. For some, helping on the archive serves as a coping mechanism, but for others it is simply traumatic. This diversity of emotional responses helps us realize the purpose of the archive in respect to sense making and creating a common understanding. The process of forgetting is an important part of overcoming trauma. And forgetting, like remembering, takes work (Olick and Robbins, 1998). Both forms of memorial work are important for recovery from traumatic events. Inasmuch as our labors were both constituting new sets of memories and causing the reprisal of forgotten ones, one can clearly see that working on a memory bank of this type is emotionally charged labor.

Another related difficulty centers on maintaining a balance between our memorial and archival missions. While first and foremost framed and positioned as an archive, such a site is also necessarily a memorial. In fact, we explicitly noted this aspect of the project in an initial press release: "Through this archive, we aim to leave a positive legacy for the larger community and contribute to a collective process of healing, especially as those affected by this tragedy tell their stories in their own words" (Elliot, 2007).

To date, we have received two complaints about materials in the archive. One of the items in question—a graphic yet realistic comic-style drawing of the perpetrator opening fire in a classroom—was temporarily removed from public view because we decided it was simply too raw, visceral, and immediate. But we decided that a second troubling item—an anonymous "confession," taken from a blog site, where a self-described "loner" expressed sympathies with Seung-Hui Cho's actions—should remain online. Our decision was strongly informed by our commitment to make visible minoritarian views that are frequently marginalized, made invisible, and disappeared by mainstream media, traditional archives, dominant public discourses, and majoritarian histories.[7]

But as we troubled over the inclusion and visibility of individual items in our collection, we also quickly realized that our archive was only designed to hold certain kinds of content. For example, our platform was not designed to capture or display entire web sites—although few systems actually do, and none do it perfectly (Brügger, 2005). We faced related problems trying to capture large collections of commercial/copyrighted content, including content from big media outlets. We responded by partnering with the Internet Archive (www.archive.org) to "scrape" relevant web content using their Archive-It service (www.archive-it.org) for about six weeks. This collection eventually grew to include partial mirrors of more than forty memorial and tribute sites, commercial and non-commercial media coverage, and other relevant content (http://www.archive-it.org/collections/694).[8] The historical dimension of this archive was expanded around the one-year anniversary of the tragedy, when many of these same sites were scrapped again. We also led the development of a similar Archive-It collection dedicated to the Northern Illinois University Shooting (http://www.archive-it.org/collections/970).

As our ambitions and archives expanded, we found that we were challenging pre-existing assumptions about who owns and controls institutional memories. In a university, library staff, university relations workers, individual department and program representatives, and trained historians may play various official and unofficial roles in ongoing efforts to collect, preserve, or even purge certain memories. When there are deviations from these scripts, tensions can emerge. As word about our work spread on campus, specific concerns were expressed about our lack of engagement with relevant faculty in other departments, the possible encroachment of our project onto the turf of other university units and research groups, and the extent to which we were exploiting the tragedy for financial gain. In summary, our impulse to archive—which we ultimately felt was altruistic—revealed some potent politics of memory, likely exacerbated by a post-traumatic institutional climate.

An appropriate division of memorial labor eventually emerged as we worked out these challenges, and today a small number of online archives hold thousands of relevant items. The formation of the Prevail Archive, for example, was driven by a mix of legal requirements and enthusiastic student volunteers (http://www.prevailarchive.org/). It now features some 6,000 pages of official university documents and communications related to the events of April 16. Virginia Tech's Digital Library and Archives, on the other hand, has posted scans and photos of more than 2,000 memorial items, including banners, cards, memory books, and posters (http://scholar.lib.vt.edu/416_archive/). And finally, Virginia Tech's Digital Library Research Laboratory maintains an index of many hundreds of relevant photos and videos, many culled from sites like Flickr and YouTube (http://www.dl-vt-416.org/resources.html). We are also working on a spin-off project that will feature thousands of newspaper front pages with coverage of the April 16 tragedy, from both the United States and around the world (http://april16archive.org/frontpages/).

The issues and challenges described here are mainly practical, political, and economic. Ultimately, our desire to preserve a social memory outside of normal archival institutions and practices helped move the archive from start-up phase to more routine types of maintenance, collection, curation, and promotion. But the sociality and materiality of digital memories—including the people, the servers, and disk drives—are not permanent. The foresight of pursuing a project as quickly as we did should not be overlooked, because establishing early access to memories helped constitute the actuality and thus the value of the contributions to the memory bank. It is now clear that the value of the archive is self-legitimizing as reflected in the wide range of memories, knowledges, and sentiments it now contains.

FROM CONTRIBUTING MEMORIES TO WEB HISTORY

Memory banks provide an alternative contribution to web history. Contrary to web archiving, which seeks to archive the pages and other artifacts found on computers connected by the hypertext transfer protocol (http), memory banks supplement the collection of those artifacts with another set of artifacts, including items that diverse individuals find, perhaps digitize, and contribute. The items in memory banks should not be viewed as the products of either institutionalized production processes or web-based regimes of production. Instead, these artifacts represent memories and narratives, and as such the histories they tell need not be universally considered necessarily true; they are reflections of individual memories.

The difference is not only in the mode of production of memorial digital artifacts as compared to the production of other artifacts and pages on the web. The April 16 Archive includes screen captures, poems, digital photos, newspaper articles, academic papers, e-mails, and wide variety of other material that a diverse assortment of individuals and groups have produced. Many of these artifacts have a non-Internet origin and their provenance is easily questioned. Many are also derivative of materials with clear provenance, but their derivation may have aesthetic or other memorial value to the contributors. This does not remove their value to web history, but we do have to recognize that they will not necessarily provide the legitimation or justification toward all versions of the history. These digital artifacts qua memories ultimately provide a plurality of possible narratives for historians of the web to follow and further research.

Yet these artifacts are now necessarily a part of the web mode of production because they have been contributed to, and integrated with, the database that generates the archive's web pages. And copies of these web pages now live elsewhere, including on search engines like Google and on Archive-It site. And their value as a reflection of memory, experience, and perspective can provide insights into web history, giving us deeper insights into the arena of digital products and production that surrounds the web.

The story of the development of the April 16 Archive is itself a small chapter in the larger web history of the tragedy. The items within the archive, on the other hand, provide a rich yet necessarily incomplete representation of the complex and troublesome reality of April 16. In an important sense, many of these same artifacts provide further evidence for some of the other—and as of yet largely unwritten—chapters of the web history of April 16.

We already know from experience that the archive works well for finding specific pieces of information and evidence, including from text and photos. In fact, researchers have already incorporated information and photos from the archive into journal articles and book chapters. We have also found that searching for specific keywords and tags can reveal interesting patterns and relations in the data set. And we feel strongly that many of the blog posts, photographs, and pieces of digital art in the archive reveal some of the most personal and least mediated responses to the tragedy, especially as compared to more commercial and official sources. Along similar lines, our collection of college media coverage on

April 16 is extensive and unique, especially in revealing how university administrators, faculty, and students responded to the tragedy, including on the web.

But in using such materials, researchers face a host of challenges. First there is the question of coverage, since it is often difficult to tell how comprehensive our holdings are, or to know what decisions were made about including or excluding certain items. In some cases, for example, our staff archived the comments associated with blog posts and news articles, but often they did not. Even more generally, those of us who made the effort to contribute our memories and other content often selected what we thought were especially unique or representative items, but these decisions were scarcely documented. Verifying sources can also prove thorny because some items lack attribution; others are attributed to pseudonyms, and still others list names of contributors but no contact details.

Items in our archive frequently point to other relevant sites, sources, and themes, and as researchers go elsewhere they face similar challenges. Links to sources may go dead, making it difficult or even impossible to view and verify original content. Or, one might wonder about the methodology used to select the 6,000 documents in the Prevail Archive, or the process used to pick "representative" items for inclusion in the library's archive of physical memorial items. Those probing the web mirrors hosted at Archive-It, on the other hand, may find that some sites have not been completely scraped, or that mirrors of sites are not available for specific desired dates or date ranges. Those following links from the 4/16 Digital Library may find that photos and videos posted on sites like Flickr and YouTube are no longer available for viewing, whether due to disabled user accounts or conscious decisions by individual users or site administrators. Given such difficulties, research strategies and methods are being rewritten for digital archives and sources. In fact, one might conclude that doing web history demands that we first and foremost grapple with the ephemeral nature of the web.

CONCLUSION

The plurality of the experience of memory grounds the motivations for building and maintaining the April 16 Archive. We cannot escape the history of the event, its social implications, nor the memories that were created as a result of it. We may forget some things, but we should strive to preserve those memories that are worthy of collecting and preserving, and the process of valuation and provenance should be opened up to include users and contributors as this type of archiving is a profoundly social process. Achieving the goal of an open, social, and pragmatic memory bank has demanded that we grapple with a range of difficult issues, including as related to accession and de-accession, balancing archival and memorial missions and organizational politics.

Work on the archive continues, albeit at a slower pace given our limited resources and other active projects. We are pleased, however, to report that the archive is now running on a newer version of the Omeka platform, which promises additional features. A new plug-in structure, for example, will likely make it easier to add new features and enhancements, including those developed by an emergent community of individuals and groups who are adopting Omeka as part of their own mnemotechnical systems.

And then there are the countless items that remain unarchived, some scattered about the web, others residing on phonecams and hard drives, memory sticks, and CDs. Surely, items will continue to trickle in over coming months and years. But very early on, we came to terms with an important reality of our undertaking: namely, that we will only ever capture a small fraction of the digital ephemera related to the events of April 16. Even if we aspired to a perfect machinic memory of April 16, this goal is a practical impossibility. Yet if the items we have archived provide some measure of healing and recovery for those touched by the violence of that fateful day, if they stand as a memorial to the victims, if they provide insights for researchers, then this project will be a success. By these measures, perhaps we have already succeeded.

When Brent first chronicled the events of April 16 on his web log, he closed with the following:

> The remainder of the semester will not be easy here. I know not what lies ahead, but hope that this community can come together and find some way forward, out of this mess. For those outside of the community, please be thinking of us. We need all of the positive energy you can spare. (Jesiek, 2007)

The archive represents one way forward—among many—out of the mess of April 16, 2007. But even more generally, we intend that memory bank projects like our own can help reveal the increasing importance of the social re/construction of archival memory, including as it relates to the re/writing of web history.

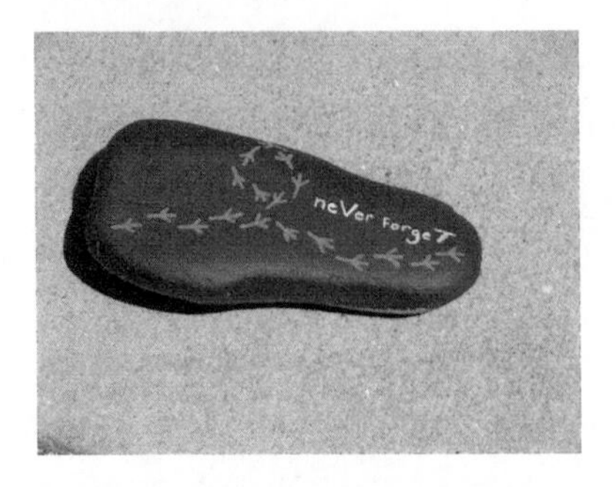

Memorial stone painted on the one-year anniversary of the tragedy, as part of a "Remembering Through Art Creation" event on the Virginia Tech campus.

Photo by Kacey Beddoes. (http://april16archive.org/items/show/2336)

ACKNOWLEDGMENTS

This chapter is dedicated to all who have been touched by the April 16 tragedy at Virginia Tech. We also extended our appreciation to the organizers of the 2007 48th Annual RBMS (Rare Books and Manuscripts Section of the American Library Association) Preconference, who invited us to speak about the April 16 Archive in a seminar session. Thanks also to Ben Agger and Timothy W. Luke, who invited us to submit a chapter about the archive to their edited volume titled *There Is a Gunman on Campus: Tragedy and Terror at Virginia Tech* (Jesiek and Hunsinger, 2008). The conference presentation and chapter provide some of the basis for this chapter. And finally, we thank all who so graciously contributed to and/or helped out with the April 16 Archive. Without their support and encouragement, we would never have been able to collect and preserve so many important memories.

NOTES

* Brent K. Jesiek and Jeremy Hunsinger are equal co-authors on this chapter.

1. Commentators such as Young (2003) and Koelsch (2007) have challenged this tradition by promoting the archival value of ephemera and realia. Of course, defining "value" is a dicey proposition, and Young also suggests skepticism about the prospect of collecting "even more extremely transient types of information" from the web (p. 25).
2. It is telling, for example, that the current size of the April 16 Archive—which consists of more than 2,000 digital objects—represents just a few gigabytes (GB) of data. By contrast, the web server on which the archive is hosted has a total storage capacity that is many hundreds of times larger (about 1.4 terabytes).
3. Other projects attempt to realize this type of mission through the use of other software platforms. Memory Archive (memoryarchive.org), for example, uses the MediaWiki application to create a wiki-styled interface that allows for the submission, editing, and preservation of user-generated memoirs on virtually any topic.

4. Other innovative sites based on this platform include *Mozilla Digital Memory Bank* (http://mozillamemory.org/), which represents the history of the Mozilla open source software community, and *Gulag: Many Days, Many Lives* (http://gulaghistory.org/), which features vignettes and biographies of prisoners who spent time in the Soviet Gulag system. This software was also used to create *Thanks, Roy* (http://www.thanksroy.org/), a web site dedicated to collecting memories of Roy Rosenzweig, the recently deceased founding director of GMU's CHMN. For still more examples, see the Omeka showcase page (http://omeka.org/showcase/).
5. Memory bank developers are also exploring other possibilities for recontextualization and social interaction, such as rating and commenting features.
6. Sites by VT alums include "We Are Virginia Tech" (http://frankandsharon.net/tribute/), "The Virginia Tech Online Memorial" (http://vt-memorial.org/about_website.html), and "Virginia Tech Tribute" (http://virginiatechtribute.blogspot.com/).
7. As Lilly Koltun has noted, one often finds a startling level of homogeneity in the digital realm, despite possibilities otherwise. But she goes on to argue that recognizing the endless repetition of messages and content in the digital landscape "obliges us to look actively to find the occluded, but nevertheless present, divergent marginal voices" (Koltun, 1999, p. 120).
8. However, this approach did not solve the problem of trying to capture materials related to April 16 from social networking sites like Facebook, which block access to web-crawling and severely restrict users from capturing and reproducing scraped content. This is unfortunate, as Facebook was an important site for online activities related to April 16. Given such policies, who will tell the web histories of Facebook, and how?

REFERENCES

Augé, M. (2004). *Oblivion*. Minneapolis, MN: University of Minnesota Press.

Borges, J. (1999). *Borges: Collected fictions* (translated by A. Hurley). New York: Penguin.

Brown, R. and B. Davis-Brown. (1998). The making of memory: The politics of archives, libraries and museums in the construction of national consciousness. *History of the Human Sciences,* 11(4), 17–32.

Brügger, Niels (2005, January). *Archiving websites: General considerations and strategies*. Århus, Denmark: The Centre for Internet Research. Retrieved 11/27/07 from http://www.cfi.au.dk/publikationer/archiving/guide.pdf

Calvino, I. (1995). World memory. In *Numbers in the Dark and Other Stories* (translated by Tim Parks, preface by Esther Calvino). New York: Pantheon Books.

Cohen, D. J., and R. Rosenzweig. (2005). *A guide to gathering, preserving, and presenting the past on the web*. Philadelphia, PA: University of Pennsylvania Press. Also available at http://chnm.gmu.edu/digitalhistory/

Elliot, J. (2007, May 1). Center for Digital Discourse and Culture launches April 16 Archive. *Virginia Tech News*. Retrieved 11/29/07 from http://www.vtnews.vt.edu/story.php?relyear=2007&itemno=261

Fabian, J. (2008). *Ethnography as commentary: Writing from the virtual archive*. Durham, NC: Duke University Press.

Goff, K. (2008, January 10). Stories in the electronic age. *The Washington Times*. Retrieved 11/14/08 from http://www.washingtontimes.com/news/2008/jan/10/stories-in-the-electronic-age/

Grant, E. (2007, May 2). Archiving tragedy, promoting healing. *American Historical Association blog*. Retrieved 11/29/07 from http://blog.historians.org/resources/208/archiving-tragedy-promoting-healing

Guess, A. (2008, February 20). New tool for online collections. *Inside Higher Ed*. Retrieved 11/14/08 from http://www.insidehighered.com/news/2008/02/20/omeka

Haegele, K. (2007, May 6). The April 16 Archive tells a new kind of Virginia Tech story. *The Philadelphia Inquirer*. Retrieved 11/29/07 from http://www.thelalatheory.com/dig7.html

Hutkin, E. (2007, June 24). Tech to archive shooting mementos. *The Roanoke Times*. Retrieved 11/29/07 from http://www.roanoke.com/vtmemorials/wb/121923

Jesiek, B. (2007, April 16). VT Tragedy—My Account. Blacksburg, VA: The April 16 Archive. Retrieved December 10, 2007 from http://www.april16archive.org/object/11

Jesiek, B. & Hunsinger. J. (2008, forthcoming). The April 16 Archive: Collecting and preserving memories of the Virginia Tech Tragedy. In B. Agger & T.W. Luke (Eds.), *There is a gunman on campus: Tragedy and terror at Virginia Tech*. New York: Rowman & Littlefield.

Koelsch, B. (2007, April 9). *Research and instructional uses of ephemera and realia in academic library archival collections*. Master's thesis, University of North Carolina at Chapel Hill School of Information and Library Science. Retrieved December 11, 2007 from http://etd.ils.unc.edu:8080/dspace/handle/1901/380

Koltun L. (1999). The promise and threat of digital options in an archival age. *Archivaria*, 47(spring), 114–135.

Library of Congress. (2003, August 15). Library accepts September 11 digital archive, holds symposium (News Release). Retrieved 11/21/07 from http://www.loc.gov/today/pr/2003/03-142.html

Olick, J., & Robbins, J. (1998). Social memory studies: From 'collective memory' to the historical sociology of mnemonic practices. *Annual Review of Sociology*, 24, 105–141.

Vargas, T. (2007, August 19). Preserving the outpouring of grief: Va. Tech Archives 60,000 Condolences. *Washington Post*, Sunday, August 19, 2007, A01.

Young, T. (2003) Evidence: Toward a library definition of ephemera. *RBM: A Journal of Rare Books, Manuscripts, and Cultural Heritage*, 4(1), 11–26.

CHAPTER 14

Research-based Online Presentation of Web Design History: The Case of webmuseum.dk

Ida Engholm

BACKGROUND

With the advance of the WWW, the internet has become not only a mainstay of society's communication infrastructure but also perhaps the most important cultural medium of our time. For libraries and archives, the advance of the internet has led to a change in conservation obligations with associated strategic and technological challenges, as it is now necessary to establish acquisition strategies for dynamic materials as well as technologies for conserving and presenting them. Some countries maintain national web archives, and international private initiatives, such as webarchive.org collect, store and present websites on the internet. However, there are as yet only sporadic attempts by museums to conserve and document web history and culture.[1] This may seem surprising, given the importance of the internet as a culture-producing medium. But how can web museums be established, what are the inherent challenges, and how do web museums differ from traditional museums?

The core responsibility of any museum is the research-based acquisition, registration, storage, and presentation of material and immaterial cultural legacy.[2] These tasks are essential for any museum, and in establishing a museum for the web, they form the natural point of departure for a definition of purpose. However, these tasks must be rethought to some extent in relation to the digital medium and the web technology context that the web museum is embedded in. With the core museum tasks as its point of departure, the following discussion highlights some of the challenges that must be addressed in establishing the institutional setting for a web museum. These challenges will be exemplified through a case story, which forms the basis for a discussion of the ways in which the web museum may facilitate future studies of web history.

ACQUISITION, REGISTRATION, AND CONSERVATION

The primary and most important challenge that must be considered in establishing a web museum may be what the collection should include and how the objects should be acquired and conserved.[3] The mainstay of any museum is its collection, which is based on a selection process in relation to the museum's subject field and communication efforts.

Unlike archives and libraries, museums are not required to maintain a broad and representative archive. Archiving decisions are defined by the museum's topical, temporal and geographical area of responsibility. Accordingly, the acquisition process must rely on a more selective practice than that applied by libraries and archives.

The museum's selection procedures must ensure that relevant sites within the subject area are archived and that relevant knowledge about the selected sites is documented and made available. This requires doing research into and establishing knowledge about the museum's subject area to ensure that archiving and presentation are based on a high degree of knowledge about the acquired objects.

The acquisition process involves challenges conditioned by the museum's digital and dynamic subject field. Unlike traditional museums, the collection in a web museum is not based on objects of a concrete or finite nature. The museum's objects have no physical form that one can walk around or pick up but are determined by 'invisible' technological components and codes, distributed on one or multiple servers. Furthermore, they are conditioned by the computer through which they are represented, including transmission speed, screen resolution and browser version.

A museum that has the internet as its object of study and acquisition must be able to 'stabilise' the object, to some extent, in order to study and archive it. In that sense, the challenges surrounding web archiving are situated as an intersection between research and archiving in the broadest sense (Brügger, 2005, p. 9).

These challenges include questions concerning what constitutes the museum object. What is its extent? What does it consist of? Which versions should be included? This delimitation involves considerations of a content-related, academic, and technical nature, which in the following pages will focus on two issues: (1) a definition of websites as objects of analysis and archiving as well as considerations concerning the object/work concept applied by museums, and (2) technical issues concerning archiving strategy for the acquisition of websites.

THE OBJECTS IN THE COLLECTION AND THE 'WORK CONCEPT'

Until a few years ago, internet research focused mainly on analyses of the internet and the web as a media form and on socio-cultural analyses of online activities (e.g., Benoit & Benoit, 2000; Bolter & Grusin, 1999; Hine, 2000; Howard, 2002). Relatively few attempted studies of websites as objects of analysis and archiving in their own right (e.g., Schneider & Foot, 2004; Brügger, 2007). One possible point of departure for a definition of websites with a view to museum acquisition is Niels Ole Finnemann's definition, presented with a view to specifying websites as objects of analysis and archiving. Finnemann proposes the following definition:

> A website can be defined as a site, 1) i.e., a delimited set of addresses on the Internet, which is delimited insofar as 2) they are subject to an overall editorial control of their content, which is 3) freely accessible to the general public either through payment or free of charge and with or without user indication and a password. (Finnemann, 2005, quoted in Brügger, 2005, p. 11).

Thus, a site is defined by its URL and by virtue of having coherence (i.e., representing the end-result of a series of editorial and design decisions).

This definition makes sense in relation to the study and acquisition of specific websites; however, it only goes part of the way toward addressing a museum's need to analyse and archive broader trends and semantically related events across sites. Steven Schneider and Kirsten A. Foot (2004) propose a strategy for analysing (and acquiring) websites, called 'web sphere analysis' (p. 188ff). Here, the object of analysis is defined as a broader 'semantic object,' which is thematically oriented and involves multiple sites and their interconnectedness. This object definition addresses the need for documentation beyond a specific website with a view to including related sites and more of the WWW network; as such, it is useful for museums that focus on socio-cultural aspects of internet usage and interaction.

The definitions shed light on issues that web museums must address in their acquisition processes and the establishment of a collection; these issues go beyond what constitutes a website and on to discussing the distinction between work and relevant context as well as the line between relevant contexts and the rest of the web.

These issues are linked to a more fundamental discussion about the work/object concept that a museum uses as the basis for its collection. The work/object concept is determined by the principles of research and historical study that govern the museum's purpose definition and activities. What is selected, how much, and how frequently it should be documented all depend on this purpose definition. For example, an art-oriented museum may aim to document expressive stages in the development of the internet by archiving individual exemplary sites, while a media museum wishing to document news dissemination would find it essential to archive news stories from the same news site over several days.

The collection's work concept is the key to academic delimitation, and for a web museum it is the key to the acquisition of digital web material. Like traditional museums, a web museum has to locally decide what acquisitions are relevant. The work concept also affects the technical archiving strategy that the museum bases its collection on.

ESTABLISHING AND CONSERVING THE COLLECTION—TECHNICAL CHALLENGES

The most common strategies for internet archiving are *macro* and *micro archiving*.[4]

Macro archiving is archiving on a large scale (of a large number of websites and, in principle, infinitely). This form of archiving is typically applied by government and private institutions, both nationally and internationally, that have the necessary computer power, storage capacity, and technical expertise (Brügger, 2005, p. 11). The acquisition process is automated through the use of search robots, perhaps supplemented with event harvesting and selective automated acquisitions of specific domain or site types, and it aims to secure the national and international cultural legacy.

By contrast, *micro archiving* is carried out on a small scale, both with regard to space (a limited number of websites) and time (a limited, isolated period). It is typically carried out by individuals who have considerable computer power and storage capacity and whose technical knowledge of archiving or of the subsequent treatment is either lacking or on an amateur level. Archiving is often done on the basis of an immediate, here-and-now need to preserve an object of study (Brügger, 2005, p. 10).

Because the purpose of web museums is not to provide exhaustive documentation and because museums rarely have the computer and storage capacity or technical expertise that, for example, national or international libraries and archives have access to, it would seem most appropriate to base the acquisition process on micro archiving. The following section outlines some possible strategies for micro archiving that web museums may use in establishing a collection.

Archiving in Cooperation with National Libraries or Archives

One possible archiving strategy for the museum is to cooperate with a national library or archive that is already engaged in digital archiving and the ongoing conservation of harvested sites. Thus, to the extent that the institutional collaboration and legal restrictions allow, the museum can select and link to sites in the archive.[5] Thus, the museum becomes an interface that structures and presents the selected sites. The collection is defined by the sites that the museum has selected from the archive, which are 'physically' situated in and conserved by the archive but pulled into the museum's interface through links.

The benefit of this strategy is that the acquisition process takes place within the framework of cross-institutional cooperation, which, in principle, gives the museum access to a larger number of sites, and the technical handling and resource requirements do not have to be shouldered by the museum alone. The drawback is that the museum does not have its own collection but relies on cooperation with other institutions in the acquisition process.

Micro Archiving

Another possible strategy is for the web museum to have its own collection and do its own micro archiving of the sites it wants to acquire. Various types of software enable the download of dynamic sites and the selection of sections that are deemed conservation worthy. It takes an ongoing delimitation process based on research and archiving principles to determine what should be archived, how 'deep,' and how often.[6]

To achieve optimal archiving and subsequent playback, however, micro archiving often requires a certain degree of processing, where image files, links, animations, etc., are 'attached' to enable the archived site to be played back in relevant browsers and ensure that users do not encounter inactive links and incomplete layouts. The latter, for example, is a problem in webarchive.org, which uses automated harvesting and takes no subsequent steps to address the fact that links and websites contain files, animations, or other 'loosely associated' elements that eventually 'die,' rendering the sites incomplete.

Another form of micro archiving that the museum may use as a supplement is the 'manual recording' of 'screen films' through the computer's built-in recording facilities. This makes it possible to capture and document interactive actions and is particularly suitable, for example, for games, quizzes, and animations that are only activated through user interaction.

A simpler form of micro archiving requires fewer resources and involves screen shots of websites, including the download of static images of websites. This form of archiving documents the layout and design of websites but fails to capture any interactive, auditive, or animation elements.

The benefit for museums in managing their own acquisition is that it gives them full control over what is acquired and how. The drawback is the resources it requires and the fact that most museums, unlike libraries, do not have the resources and expertise necessary for the ongoing conservation and presentation of the archived material on new platforms and browsers.

Acquisition Based on Hardware as Well as Software

Some museums may choose to supplement the digital archive with the acquisition of contemporary hardware and the ongoing download of browser versions, software, and support programs. This form of archiving contributes to the historically correct recreation of web sites and their content, appearance, interaction and usage situations at given times in web history. In terms of presentation features, this strategy lets the museum's visitors access sites through period-specific computer platforms and browsers.

However, it is a demanding approach in terms of the acquisition and maintenance of both hardware and software (not just relevant browsers but also the plug-ins and additional programs needed to play back the sites in the form in which they were originally developed).

The strategy also has limitations in relation to presentation, as the web museum or parts of it can only be accessed from the physical museum where the hardware and software are available and not, for example, online. In the following section, the focus will be exclusively on the challenges implied in establishing online web museums, not on hardware-related issues.

Metadata Treatment

Like traditional museums, web museums must establish practices that provide formalised descriptions of the acquired objects with data about their content, form, and nature. These metadata are essential, as they not only contribute to the documentation of the objects but also to their ongoing assessment in the collection as well as in the communication efforts.

In most cases, metadata can be pulled directly from the documents, but in some cases it is a complex and demanding task to achieve adequate resource descriptions. This is certainly true of older web materials, which have not been archived in a museum or a professional archive but have to be pulled in from private archives, for example. Because systematic archiving strategies for the internet did not really get off the ground until the late 1990s and the early 2000s, much of the early history of the web has already been lost.[7] Thus, documenting the first years of the medium's development often requires that museum professionals do 'archaeological' work in private archives and develop alternative methods of acquisition, such as pulling images of websites from printed user manuals, how-to-books or inspiration books.

Due to the rapid development of the internet, it is often a challenge to get the desired information. Websites are constantly changing or disappearing, and subsequently it may be difficult to determine the senders or factual information about the release date, year, or—if only screen dumps are available—information about digital material type: software, plug-ins, etc. This is not least a challenge in relation to copyright issues. As part of their registration procedures, like traditional museums and regardless of acquisition method, web museums must comply with regulations for copyright clearance for the website that is being archived. If the copyright holders cannot be determined, it may prove difficult to achieve the clearance necessary to exhibit a site.

As outlined above, acquisition, registration and conservation constitute a complex task. In some ways, a web museum's tasks resemble those faced by traditional museums, but in important ways, a web museum faces new technological and media-specific conditions. This is also the case for the major museum task of presentation and communication, which must also address the web-specific context of a web museum.

PRESENTATION AND COMMUNICATION

In our globalised and postmodern world, the role of museums has increasingly changed to being a part of the experience economy. The transformation of the economy from industry, revolving around production, to an orientation toward symbols and consumption has led to changes in the activities of cultural institutions (Lash & Urry, 1994). Previously, a museum's primary role was to conserve and present the cultural legacy through informational and educational activities, while today's museums are also required to highlight and strengthen experiential values and the focus on the user. Museums navigate between the goals of providing education and experiences, and the entertaining and audience-engaging elements in museums' communication efforts are taking on growing importance. In physical museums, the communication takes place within the concrete architectural setting that is available to the museum. A museum that has the web as its subject field presents its medium through the medium itself, which places certain specific demands on presentation and communication.

Academically and in terms of content, the presentation approach must be anchored, first of all, in the research that takes place in relation to the museum's collection and which is explicitly reflected in the themes, structure and presentation of exhibits and their context in the communication efforts. The communication requirement frames and facilitates research and the ongoing knowledge production, and contributes to information and experiences aimed at the audience through the utilisation of the attraction potentials of the museum and the medium.

One unique communication challenge faced by an online web museum is that access to the museum is computer mediated and web based. The museum's usage context is the user's home, educational institution or workplace, from whence the user may visit the museum alone or together with others. Thus, the first communication concern that a web museum will have to consider is what target groups to address, including which types of user equipment and user competencies the museum is going to accommodate. If the museum chooses to aim for the lowest common denominator in terms of experience and equipment, that clearly places certain restrictions on the communication approach. On the other hand, the museum will exclude potential target groups if it sets high standards for user competencies and equipment.

As a web-based exhibition form, web museums have rich opportunities for creating informative and experience-oriented presentations, because these presentations, unlike those in traditional museums, unfold exclusively in the digital medium. Art curator Steven Dietz (1998) calls this form of museum site an *online exhibition*. According to Dietz, the online exhibition is designed for the medium itself and thus characterised by a freer form of communication than, for example, traditional museum sites, which are bound to 're-present' existing exhibitions in physical museums. The communication form, design and interaction of the online exhibition are solely required to create exhibitions that work within the setting provided by the medium.

This allows online museums to rethink presentation and communication forms in relation to the medium itself and the museum context they enter into. To a large extent, the communication form can be characterised by the approach that Jay David Bolter and David Grusin (1999) call *hypermediacy*. In their book, Bolter and Grusin argue, with the concept of *remediation* as their point of departure, that old media are not replaced by but rather remediate existing media, and they point out various forms of remediation. *Hypermediacy* is characterised by media visibility and a reinvention or transformation of existing media forms, unlike *immediacy*, which is characterised by media transparency, and which reflects the urge to copy the functions and modes of expression of existing media, thus, in principle, hiding the medium's own characteristics.

The online communication form invites less recycling of existing media forms and more innovation in relation to content as well as interface.[8] On a more specific level, in developing a presentation and communication strategy, it is essential to address certain aspects, specifically the content-related, technical, user-related and aesthetic dimensions of the website.[9]

The content considerations have to do with the museum's purpose and involve considerations in relation to communication form and target group (including decisions concerning language level, text structure, the balance of text versus visual, dynamic, and auditive content, etc). The technical issues involve decisions concerning what technology and software the site should be based on (accessibility in relation to computer platforms and browsers and whether users will have to download support programs and plug-ins to access the site). User considerations relate to the use of the site and involve issues concerning information architecture, navigation and interaction, which—in combination with the technical elements—enable users to interact with the site. The aesthetic dimension concerns the visual and any auditive appearance aspects, including graphics, colours, ambience, etc.

Naturally, these concerns are inherently contextual, as they are only expressed through the specific use by museum visitors realising the content and 'acts' that the museum enables based on their varying equipment, interests and purposes.

A web museum faces the challenge of managing the core areas of acquisition, registration, conservation, research, and communication while also making the museum an attraction with a high degree of visibility and impact. Like physical museums, web museums must mediate between the two poles and face the challenge of optimising communication in the spirit of the experience economy without selling out with regard to the museum's archiving and research principles. These requirements to a new communication practice are not just a general demand in society, they are inherent in the developments within cultural and postmodern reception forms that involve the user as a participant in collaborative sense-making. The digital technologies promote a user-engaging and experience-based communication form, with inherent demands for user-friendliness and recognisability. Thus, in developing communication strategies for web museums, it is important to utilise the potential of the medium and contribute to experiential goals while ensuring a certain degree of 'convention preservation,' as the site must appear comprehensible and useable and make cultural sense to the target group.

As evident from this discussion, the issue of establishing a web museum involves certain considerations concerning the materials and work concept the collection should be based on. How are the objects and their context delimitated? In addition, this issue involves certain technical decisions about the archiving form (types of micro archiving and extent) and about the communication approach, including decisions about target groups, user equipment, communication form, and the balance between usability and attraction.

In the following section, these issues will be addressed through a specific case, webmuseum.dk, a Danish online museum for the design history of websites, which has been developed in a collaborative effort by the Danish Centre for Design Research and the Danish Museum of Art & Design. The presentation will focus a discussion of the ways in which museums may serve as the basis for future web history studies.

WEBMUSEUM.DK

Webmuseum.dk is an online museum for the design history of websites. The museum will open in the spring of 2010 and will then be available via the internet and through a physical access point at the Danish Museum of Art & Design.

As far as I know, this is the first museum for website history that documents the development of the web from the point that the internet went graphic in the early 1990s to future web developments. The academic focus is on the *design* history development of the web, and through its association with the Danish Museum of Art & Design, the museum embeds itself in an arts & crafts context. Thus, the museum's purpose is to collect, conserve, and present web material to document web design development.

So far, the documentation of web design development has been sporadic. In the literature, publications about web design have been characterised mainly by normative and pragmatic how-to introductions on website design that offer guidelines for the development of websites. A few publications outline a historical perspective on website design, where the main focus, however, is less on documentation and more on establishing a basis for a presentation of new guidelines and offering inspiration for the production of websites (e.g., Siegel (1997); Fleming (1998); Cloninger (2002); Rosenfeld & Morville (2006); Niederst (2007)). Within design history, Philip Meggs (1998) and Ida Engholm (2003, 2005) have described aspects of web history with a focus on the graphic development of the medium. At this time, however, there are virtually no attempts at assembling a comprehensive historic documentation like the ones we see in art, design, or media history.

In a web context, the only known attempt to document web design development is the site DigitalCraft.org. It was established in 2003 by the Museum für angewandte Kunst in Frankfurt as a collection and online exhibition venue for website design, computer games, and social communities. As an exhibition venue, it differs from the archives because it has a curated and structured presentation practice. The site focuses mainly on contemporary website design, however—"fast-moving trends in everyday digital culture" as the introduction states (Nori, 2005a). There has been no preservation of older materials although that is one of the future goals of the collection (Nori, 2005b).

The sporadic efforts at assembling a historical documentation of web design development in the literature and online as well as the lack of access to useful website design material in the archives formed the background for the decision to establish Webmuseum.dk.

Acquisition and Conservation

The initial considerations addressed the basic establishment of the collection. Should the museum have its own collection or establish a partnership with an existing archive? Initially, the possibility of culling websites from the national archive, *netarkivet.dk*, was explored. This Net Archive harvests Danish web material through an automated procedure based on broad, inclusive criteria. Here, the web museum would find examples of historical and future web material for its presentation and communication activities. Unfortunately, existing Danish legislation precluded this approach due to legal issues in relation to national legislation involving the Legal Deposit Act, the Copyright Act and the Data Protection Act. This legislation complex restricts access to the Net Archive, in practice rendering the bulk of the material accessible only for research purposes and subject to application. For the museum, this meant that material from the Net Archive could not be presented or republished.

Therefore, it was decided to base the museum on in-house micro archiving. The collection is managed by a team of editors who assess the sites critically and contribute to the museum's knowledge production and presentation activities by recording the selected items and their context and writing about them. As most of the museum's material is no longer available online, it is stored in the form in which it was available. For early stages of web development, the exhibits are mostly in the form of screen dumps and scanned images from printed user manuals and how-to books. In addition, micro archiving is carried out with the software program WebDevil, which makes it possible to archive entire sites and to specify how many pages should be included. In most cases, only representative components have been selected, and few sites are stored in full.

In addition to micro archiving, the museum also archives short screen films of selected sections of sites. As part of the research prior to the creation of the museum, the editorial team interviewed researchers, programmers, web designers, and concept developers; some of these interviews provided background material, while others are included in the museum as video interviews that serve to frame specific websites or to introduce a theme in the museum.

The Material Culture Concept of the Collection

The museum's association with a design and arts & crafts-oriented context means that its theoretical foundation rests on that section of design history whose academic matrix is art history. Accordingly, the design concept is qualitatively oriented toward certain object types, which may be viewed or canonised as exemplary on the basis of normative perceptions. Thus, in a museum context, acquisition and presentation require that the objects are at an exemplary level in terms of technical properties, materials, and aesthetic qualities. Of the aspects that constitute form and meaning, the main descriptive focus is on the design process/design moment and on the artist/designer personality behind the 'work' as they are manifested in the given form.

In the context of recent design history, however, with inspiration from anthropologically oriented and material-cultural design studies, the descriptive focus in recent years has been increasingly on the context surrounding the works and on the social, cultural, and user/consumer-related structure of the design process and object. In design research, this has led to new approaches to design history, in which the work-oriented angles have been replaced by studies of specific cultural spheres concerning specific products (e.g., du Gay et al., 1997; Clarke, 1999) or more context-based design history documentations with an emphasis on material culture and everyday objects and the way people relate to them rather than an identification and analysis of exemplary works (e.g., Miller, 1987; Attfield, 2000).

In developing webmuseum.dk, the goal has been to combine a prescriptive work-oriented approach with an approach anchored in material culture studies. From a historical perspective, it is seen as essential to select websites that can be considered exemplary from a normative point of view in terms of technical features, user aspects, and/or graphic-aesthetic properties. As a design history museum, webmuseum.dk has an obligation to document prominent website design from various periods and to present key examples of artistically experimental and original website design to a contemporary audience.

The rationale for an approach based on material culture studies springs from the unique character of the web and on the desire for a broad and contextualised presentation of design history. Unlike other media, the web is an open publishing medium, where both professional and private contributors provide content and design websites. A web museum must reflect this diversity of senders and appearances. Thus, the web museum not only includes websites that are professionally exemplary from an aesthetic 'work-oriented' normative point of view; it also seeks to capture the conditions governing web development and distribution as well as web production and consumption in popular culture. The material culture perspective is reflected in the fact that the web museum not only presents websites as isolated objects but also highlights the cultural forms that sites assume from the time of their conception until the point of distribution and use. Hence, the museum's collection not only contains websites but other objects as well in the form of video, image, and audio files that help contextualise and frame the story. Thus, in qualitative normative terms, the work concept is centred partly on exemplary sites and partly on a historical category of objects that manifest aspects concerning technology, function, industry, economics, and culture at a given time in web history.

Figure 1. Layout of the frontpage of webmuseum.dk with an entrance to the collection, "Samling," and the user driven section, "Forum." The website will later be available in English.

The Interface Level of Presentation and Communication

In establishing webmuseum.dk, there was an emphasis on creating an interface that was both user friendly and experience oriented. The communication strategy is based on the fundamental clash between 'web' and 'museum.' The web museum has both a curated museum mode and a free, user-oriented—and user-produced—mode.

The curated museum mode offers the visitor a mediated insight into the development of the internet through structured thematic exhibitions. Here, as in other sections of the web museum, the user is offered a thorough presentation of a subject area in the form of text material and visual content (and, in some cases, audio files) of relevant examples of web material. The content is presented in its most ideal form as thumbnails or screen shots. This presentation form is closely associated with traditional museum communication.

The free, user-oriented mode lets the user move flexibly between presentation levels and shift independently between the web museum's many presentations. Users can zoom in and out on the museum material and shift their view sideways as well as up and down. This provides access to different levels of communication: From any thematic presentation, the user can zoom in on a particular exhibit, move directly from there to another exhibit from a different exhibition context, and then zoom out to the thematic presentation level of this new context. The presentation that the individual user experiences is determined by his or her individual interests or desires. The user may choose to stick to a coherent narrative thread or move freely through the exhibition. This flexibility is also possible in physical museums today, but in a physical context the user does not have access to all the exhibits simultaneously. The user's behaviour affects the appearance of the web museum, and the user's choices affect the location and context of materials in relation to each other. With this presentation form, the web museum seeks to take full advantage of the hyper-flexibility of the internet.

Exhibition Activities

As in any museum, the core is the collection, the content and extent of which is based on a selection process that is related to the research and communication activities. Thus, visitors to the web museum will not find an exhaustive representation of material from the first days of the internet until the present day. Instead, the collection contains the exhibitions that the editors have chosen to display based on a requirement of historical representation. As more and more communication activities are initiated, the collection will grow.

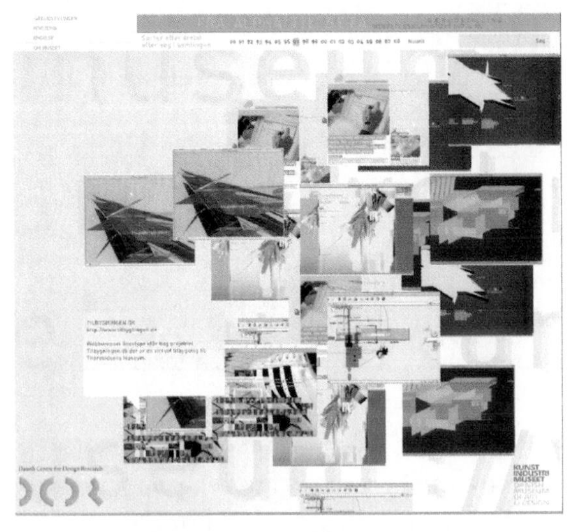

Figure 2. The user can move flexibly between different presentation levels. Here a search result from the archive of the webmuseum.

In the first version of the museum, users can explore the collection in three different modes:

- In a permanent exhibition where the collection is curated and placed into thematic contexts.
- Through free or thematic search functions within the collection.
- In a user-driven section where social technologies enable users to create their own exhibitions and share files and opinions with other users.

The permanent time line exhibition. This exhibition serves as a 'time line corridor,' which is structured around particular years with entrances to various main and subordinate narratives that follow the corridor chronologically from the museum's first year, 1990, until today. Structurally, the exhibits are placed in 'side galleries' organised as independent main narratives. The presentation uses text, images, and video clips that support the works and their context. The exhibitions are intended as sources of learning and knowledge and tell stories that offer insights into web history. The narratives focus on technological, social, economic, and aesthetic factors that affect web design development and on the particular factors that affected the development of the individual site.

There are seven main narratives, which visitors can follow in a linear fashion or jump into and out of, as they wish.

Side Gallery 1 addresses the time when *"The internet goes graphic" (1990–93)*. This exhibition describes some of the factors that made the web a mass medium. The essential claim is that the expansion of the web is mainly due to the introduction of a graphic user interface. With the launch of WWW and the first browser, Mosaic, in the early 1990s, the internet went from being a text- and code-based instrument to a platform that features images, sound, and graphics that enables ordinary people to use the medium. The exhibition presents the first graphic websites and describes the technical background for their development and use. Due to limitations in the html-code and the browsers, websites were very similar in their appearance and interaction—text on a white or grey background and purple links.

Side Gallery 2 focuses on the two-year period of 1994–95; this narrative describes the economic and business aspects that gave the web its commercial power and the impact of these factors on the design idiom. Entitled *"The web as a commercial medium"*(1995-97), this exhibition describes the earliest examples of e-commerce and new forms of interaction between companies and users as well as among users. Technically, the possibilities were expanding in terms of layout, navigation, and graphics. The first fashion trends arose: Black became the preferred background colour on the web, and the blue links encountered competition from designed 'buttons' that made for a more varied design expression.

Side Gallery 3, entitled *"A new profession: web designer" (1995–97)*, describes how the new medium creates the new profession and explores the web designers' influence on design developments. Graphic designers and interface designers enter the stage; web design becomes a professional discipline; art and design schools begin to offer web design as a subject, and the first web design award is handed out. In companies and organisations, the IT department is no longer in charge of developing websites, and an entire new industry of web designers and web agencies sees the light of day.

Side Gallery 4 deals with *"The gold rush period—the web designer as star"* (1997–1999); it addresses the economically successful period when established companies and organisations go online, and web agencies become successful as the mediators that secure companies an online presence and thus increase success and earnings. The exhibition illustrates how the increased competition between browsers and a growing number of plug-ins created new possibilities for differentiation through graphic elements and interaction features. This development often led to high levels of creativity, possibly at the cost of user-friendliness.

Side Gallery 5 describes the *"Dot.com crisis"* (2000–2003), when the faith in the new dot-com companies failed, and the stock market plunged. Experimentation and creativity with regard to graphics, animation, and interaction were replaced by a growing emphasis on user-friendliness and viable business concepts. This narrative assumes that the dot-com crisis and the subsequent economic crisis after the attacks on 9/11 caused the slowdown in design creativity as well as the stereotyping of design solutions and navigational conventions that remain evident to this day.

Side Gallery 6 looks at the many new possibilities and interaction forms that arise in website design with the *"Fusion of media platforms"* (2003–), including film streaming on the web, web-based news, and live TV. The main focus is on the technical and media-related factors that affect the design and use of the web, factors that are becoming ever more dynamic and interactive.

Side Gallery 7 looks at recent web trends with an emphasis on so-called *"Web 2.0 technologies and social software"* (2004–) such as MySpace, Flickr, and FaceBook, on which providers do not send information or sell content but instead act as gatekeepers for community formation and user-to-user interaction. In this context, the users are the new designers.

The exhibitions reflect the museum's presentation and communication strategy, which is not focused exclusively on works but rather on describing web history through cross-sectional synchronous perspectives that reflect curated decisions about what is considered historically essential with a view to offering a diachronous presentation of design history.

Free or thematic searches in the collection. In addition to visiting the permanent exhibition, users can also search the collection. The museum offers two search options: a free-text option that offers direct access to individually available, specific data and a timeline that frames particular periods and offers thumbnails representing the individual items in the collection; users may choose a particular site or an object, video, image file, etc., within the given period.

User-created websites and exhibitions. An equally important part of the web museum is the community section, where users can upload websites, links, and files and create their own exhibitions subject to certain predefined conditions.

This section contrasts the conventional museum presentation format in several ways: It is open to anyone, and the exhibition of websites and participation in the debate about web design and web history are not reserved for academically trained connoisseurs. It also acts as a sort of acquisition site for the web museum, as the sites exhibited by users in principle function as submissions for the collection. If they are accepted into the collection, they are registered and provided with metadata, and they may also enter into the museum's own exhibitions. Thus, professional expertise joins forces with the audience in establishing reservoirs for the documentation of web history.

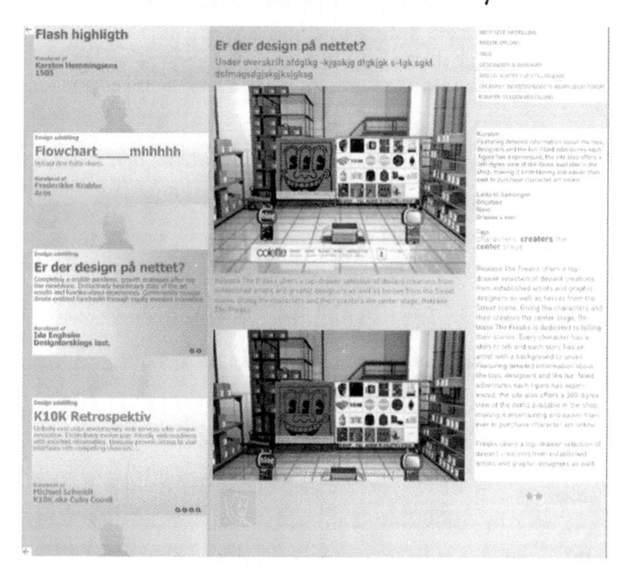

Figure 3. Examples of user created websites and exhibitions in the community section of the webmuseum.

In connection with the museum opening, the Danish Museum of Art & Design set up a room with computers, where the web museum's virtual architecture is projected on the wall to allow more visitors to get an impression of how they may use the museum. This option links the physical and the digital museum in what might be called, in the words of Alisa Barry from the Natural History Museum in London, "a virtuous circle between the virtual and physical space" (Barry, 2006). Webmuseum.dk aims to serve as an independent entity as well as provide a link to the Danish Museum of Art & Design as part of this museum's permanent collection of twentieth and twenty-first-century design history.

THE WEB MUSEUM AS A PLATFORM FOR WEB HISTORY STUDIES

Apart from offering presentation and experiences facilities to the public, web museums should also be venues for historians who study and write about web history. The collection makes material available along with knowledge about the acquired digital objects. Metadata treatment makes it possible to apply search criteria to the material, which lets researchers carry out specific searches in the collection. In addition, the web museum's exhibitions offer thematic documentation that adds to the knowledge about the objects and selected sections of the web and its history. These exhibitions add context and perspective to the material in the collection and supplement the basic sources with interpretative perspectives.

The internet technology also gives web museums new possibilities for creating new types of connections between experts (academics, artists, designers, programmers and others) and the public as well as among museums and educational, and research institutions. This provides additional avenues for knowledge production and for interdisciplinary debates and evaluation. Here, the museum's role is to mediate between various expert discourses and the participating public. This requires ongoing experiments in relation to establishing collections, registration, communication, and research and in relation to the development of formats for documentation, exchange, and cooperation.

At webmuseum.dk, the collection, the exhibitions and the community section are intended as a reservoir for the development of web design history. The museum includes sites from periods that have very limited archival representation in other contexts and which would have been lost had they not been acquired and registered.

Thus, in the web museum, design researchers and web historians will find material documenting the earliest years of the web as well as material reflecting contemporary and future developments. In the collection, researchers will find information about the design, functionality, technical structure. and development of web materials. In addition, the museum's exhibitions contribute to the documentation of the development of the internet as a design-based medium and the technical, social, economic, and cultural conditions that have determined web design production and usage through the brief but turbulent web history of the internet.

CONSERVING OUR CULTURAL LEGACY

The European Community has put the digital conservation of our cultural legacy on the political agenda; at the level of national culture policies, this bolsters the efforts of libraries and museums in pursuing these goals. For years, researchers have been arguing the importance of preserving the fleeting internet culture and establishing contexts for the objects, events, and concepts that this media-expanded form of human communication offers.

The many libraries and archives emerging in current years are testimony to the growing awareness of the need to collect and convey internet history to ensure that this important part of our cultural legacy is preserved and made available to researchers and the public.

It is, however, also important for museums to establish online institutions to document web history and offer professional perspectives as a contribution to the documentation of specific areas of digital history, society and culture. Webmuseum.dk is an early attempt at establishing a museum for web design history. One hopes that other museums with other purposes will follow, contributing to our common cultural memory.

NOTES

1. Apart from the case discussed, webmuseum.dk, as far as is known knowledge the only other web museum currently in existence is digitalcraft.org, cf. below.
2. See, for example, the International Council of Museums (ICOM): http://icom.museum/definition.html. Obviously, the role of the museum is being defined on an ongoing basis to match the general development of society. For example, the museum's tasks are the subject of ongoing debate within the field of museology or museum studies, which is the study of the organisation and management of museums and museum collections, and which also involves theoretical and methodological reflections on the practice and tasks of museums.
3. This chapter uses the term 'digital objects' about websites and web documents; in the following section, the issue of defining and delimiting websites as museum objects will be discussed.
4. In his book *Archiving Websites: General Considerations and Strategies* (2005), Niels Brügger describes the two forms of archiving with an emphasis on micro archiving. The book offers an in-depth discussion of the possibilities and challenges inherent in micro archiving, tests of various types of archiving, software and links to resources on net archiving. The discussion of

micro archiving approaches is based on Brügger as well as experiences with establishing an archiving strategy for webmuseum.dk.

5. In Denmark, legislation—specifically the Legal Deposit Act, the Copyright Act and the Data Protection Act—restricts access to the national web archive, which hampers cooperation with museums and others on the presentation of sites. The archive is closed to the public, and access requires the approval of a formal application.
6. See also Niels Brügger (2005), who, in the book mentioned on Note 4, offers an in-depth description of challenges and possibilities in relation to micro archiving.
7. In Denmark, the first national internet archive was established in 2004. The International Internet Preservation Consortium (IIPC) was established in 2003, and the world's currently largest web archive, archive.org only goes back to 1996.
8. See also Sophie Løssing (2008), whose Ph.D. dissertation addresses various museums' communication online as well as various types of museum sites.
9. These points were suggested by Ida Engholm (2004) as the basis for an analysis of websites; they may also be used as a checklist of aspects that should be addressed in a development context.

REFERENCES

Attfield, J. (2000). Wild things: The material culture of everyday life. Oxford, New York: Berg.

Barry, A. (2006). Creating a virtuous circle between a museum's on-line and physical spaces. In J. Trant & D. Bearman (Eds.), Museums and the web 2006: Proceedings. Toronto: Archives & Museum Informatics. Retrieved May 2, 2009, from http://www.archimuse.com/mw2006/papers/barry/barry.html

Benoit, W.J. & Benoit, P.J. (2000). The virtual campaign: Presidential primary websites in campaign 2000. American Communication Journal 3(3). Retrieved July 20, 2009, from http://www.acjournal.org/holdings/vol3/Iss3/rogue4/benoit.html

Bolter, J.D. & Grusin, R. (1999). Remediation: Understanding new media. Cambridge, MA: MIT Press.

Brügger, N. (2005). Archiving websites: General considerations and strategies. Aarhus: Centre for Internet Research.

Brügger, N. (2007). The website as unit of analysis? Bolter and Manovich revisited. Northern lights: Film and media studies yearbook, 5, 75–88. Bristol: Intellect.

Clarke, A. (1999). Tupperware—The promise of plastic in 1950s America. Washington, DC: Smithsonian Institution Press.

Cloninger, C. (2002). Fresh styles for web designers. Indianapolis, IN: New Riders.

Dietz, S. (1998). Curating (on) the Web. Retrieved August 2009, from http://www.archimuse.com/mw98/papers/dietz/dietz_curatingtheweb.html.

du Gay, P. et al. (1997). Doing cultural studies: The story of the Walkman. London, Thousand Oaks, CA, New Delhi: Sage.

Engholm, I. (2003). Www's designhistorie—Website udviklingen i et genre—og stilteoretisk perspektiv. Ph.D.dissertation. IT University of Copenhagen.

Engholm, I. (2004). Webgenrer og stilarter—om at analysere og kategorisere websites. In I. Engholm & L. Klastrup: Digitale verdener—de nye mediers æstetik og design. Copenhagen: Gyldendal.

Engholm, I. & Salamon, K. L. (2005). Webgenres and -styles as socio-cultural indicators—an experimental, interdisciplinary dialogue. In In the making: Proceedings, Copenhagen, April 29th–May 1st 2005.

Fleming, J. (1998). Web navigation: Designing user experience. Beijing, Cambridge, Cologne: O'Reilly.

Hine, C. (2000). Virtual ethnography. Thousand Oaks, CA: Sage.

Howard, P. (2002). Network ethnography and hypermedia organization: New organizations, new media, new myths. New Media & Society 4(4), 550–74.

Lash, S. & Urry, J. (1994). Economies of signs and space. London: Sage Publications.

Løssing, S. (2008). Internettet som udstillingsramme. Workingpaper, no. 3. Center for Digital Æstetik-forskning, University of Aarhus.

Meggs, P.B. (1998). A history of graphic design. New York, Singapore, Toronto: John Wiley & Sons, Inc.

Miller, D. (1987). Material culture and mass consumption. Oxford, UK: Blackwell.

Niederst, R. J. (2007). Learning web design: A beginner's guide to X(HTML), style sheets, and web graphics. Beijing: O'Reilly.

Nori, F. (2005a). Management of meanings—annotations on the curatorial work realized by digitalcraft.org. Digitalcraft.org. Retrieved May 2, 2009, from http://www.digitalcraft.org/index.php?artikel_id=552

Nori, F. (2005b). A Decade of Webdesign. DigitalCraft.org. Retrieved May 2, 2009, from http://www.digitalcraft.org/index.php?artikel_id=550

Rosenfeld, L. & Morville, P. (2006). Information architecture for the World Wide Web. Cambridge, MA: O'Reilly.

Schneider, S. M. & Foot, K.A. (2004). The web as an object of study. New Media & Society 6(1): 114–122.

Siegel, D. (1997). Creating killer websites—The art of third-generation site design. New York: Hayden Books.

EPILOGUE

The Future of Web History

Niels Brügger

As the present volume has shown, web history can be considered a field of study as well as a discipline in its own right within internet studies. However, there is still work to be done in establishing this emerging field. Web scholars should engage in discussions about the theoretical and methodological foundations of web historiography just as they should be encouraged to launch concrete studies of the history of the web to create a solid basis for our understanding of the web of the past. This future work could follow a variety of paths, only three of which shall be outlined below.

BROADENING THE CONCEPTION OF WEB ARCHIVING

There is no doubt that web archiving is an important issue in the writing of web history—and with good reason. Archived web material is different in many ways from other well-known archived media types; in addition, the very process of preserving this object of study differs significantly from the process of preserving other types of objects. But if we aim to develop the full potential of web historiography, we may have to broaden our conception of web archiving.

I would suggest that web archiving be defined as the deliberate and purposive preservation of web material. This is a broadening of the more limited understanding of web archiving as web harvesting, which basically means that a web crawler downloads files from a web server by following links from a given starting point. Web harvesting is usually conceptualized as web archiving, but if we limit the conception of web archiving to web harvesting, we exclude valuable web material which may be the only existing sources of information about what the web actually looked like in the past. The following examples of what is excluded serve to illustrate this point.

First, identifying web archiving as web harvesting excludes two types of web material: (1) pre-public material such as design outlines, dummies, beta versions, and so on, which may be used to shed new light on what was actually made public on the web; and (2) public material that has been published in other media such as print media (newspapers, magazines, books, journals) or television (commercials, TV spots, films, etc.). Especially with regard to the early period of web history, in many cases the only preserved sources are these indirect, non-digital pieces of evidence. For instance, possibly the oldest existing version of whitehouse.gov can be found in a Danish television newscast from 1994. In comparison, the oldest existing version in the Wayback Machine (see archive.org) is from 1996, and it has no graphics, images, or sound, while the televised version has both images and Bill Clinton's speech.

Second, identifying web archiving as web harvesting excludes two forms of archiving. The transformation of html text into images is one form; one of the simplest ways of preserving web material is to capture what is present on the screen by saving it either as a static image or in the form of moving images by recording a screen movie. Especially screen movies can be important, for instance, in archiving streamed web material, or user experiences on personalized websites, in web-based computer games, and the like. The other form that is excluded is the delivery of web material. In contrast to web harvesting, where the web material is retrieved from the 'outside' by contacting the web server, the material can be delivered from the 'inside'—that is, directly from the producer. There are pros and cons to these other ways of web archiving, but at least they should be considered as in many cases they constitute the only sources for the writing of web history.

In attempting to find and preserve one of the most valuable sources for the writing of web history—the web as it appeared in the past—we should not only look in web archives based on web harvesting; we may have to broaden our search to include other media types, just as we may have to use material archived in other ways than by web harvesting, and we may have to contact the producer of the web material.

WEB HISTORY BEYOND NATIONAL BOUNDARIES

Up until now, the very few studies of the history of the web have almost exclusively been based on a national web sphere. However, it may be relevant to supplement this national approach by a broader understanding of web history in terms of geographical extension. The rationale is twofold. On the one hand, more and more events in our age have to be understood and explained on a global scale. Take, for example, the histories of national financial crises, of a number of local wars of the world, or of great sports events. On the other hand, for some years now the web has played an important part in these global phenomena by virtue of its integral role in the communicative infrastructure of our societies; these events simply cannot be analyzed and explained exhaustively if the web is not part of the analysis.

Therefore, because many events are global in scope and, to a large extent, an in-depth explanation of these involves the web, it may be relevant to extend the scope of web history beyond national boundaries. Web historians who acknowledge the importance of collaborating across national borders will be better able to study these global events.

The major national and transnational web archiving institutions have been collaborating for years. In 2003 they formed the IIPC (The International Internet Preservation Consortium), and other transnational organizations have been formed among web archiving institutions. However, global collaborative historical research projects among web scholars remain the exception. An event that might be a testbed for a transnational historical study of the web is the Olympic Games in London in 2012. Web scholars and web historians could cooperate across national borders as well as with the different archiving institutions. The web historians could provide the research questions and do the historical analyses, and the archiving institutions could initiate a global archiving project to create transnational and coordinated web corpora and set up procedures for integrating already archived web material from previous Olympic Games. For instance, Australia's national web archive, Pandora, holds collections of websites in relation to the Olympic Games in 2000, 2004, and 2008; my personal web archive contains a substantial part of the Danish web activity in relation to the Olympic Games in 2000; the national Danish web archive made an event web harvest of the Olympics in Beijing in 2008, and similar corpora probably exist in other archives.

RESEARCH INFRASTRUCTURE

If we are to transform the emerging field of web history into a research tradition in web history in its own right, the required research infrastructure needs to be defined.

First, research fora which can persist over time need to be created in order to facilitate the exchange of ideas and discussion of results among web historians. These could be conferences, seminars, roundtables, email lists, international peer-reviewed journals, and the like. A small contribution to this part of the infrastructure is the email list webhistory@imv.au.dk (established in November 2008).

Second, it may prove to be vital to the growth of web history as a discipline within internet studies that researchers initiate collaboration with the national and transnational archiving institutions and organizations. This may be a way of making visible the specific needs of the web historian with regard to the quality and the nature of the web material he would like to study (including the constant development of adequate archiving methods). It may also be a way of getting the body of historical web material to grow and thus gradually creating more objects to study.

*

The three above-mentioned points are nothing but fragments of what remains to be addressed in future discussions of web history. It is hoped that the present volume has contributed to an increased interest in web historiography by setting the tone for future studies of the history of the web.

CONTRIBUTORS

Megan Sapnar Ankerson

Megan Sapnar Ankerson is a Ph.D. candidate in Media and Cultural Studies at the University of Wisconsin, Madison. Her dissertation examines the historical dimensions of web style, production practices and new media cultural industries within the social and economic context of the dot-com boom. Her work has been published in *Convergence Media History* (2009), *NMEDIAC: The Journal of New Media and Culture* and *The Electronic Literature Collection*, vol. 1 (2006). She also co-founded and edited the online literary journal *Poems That Go* (2000–2004).

Niels Brügger

Niels Brügger is an Associate Professor at The Centre for Internet Research, Aarhus University, Denmark. His primary research interest is website history, and in 2007 he started writing a history of the Danish Broadcasting Corporation's website entitled "The History of dr.dk, 1996–2006." He has published a number of articles, monographs, and edited books, including *Archiving Websites: General Considerations and Strategies* (The Centre for Internet Research, 2005) and *Media History: Theories, Methods, Analysis* (ed. with S. Kolstrup, Aarhus University Press, 2002).

Ida Engholm

Ida Engholm is an Associate Professor at the Danish Centre for Design Research, Royal Academy of Fine Arts—School of Architecture in Copenhagen. Her main research area is design history with a special focus on web design history. She has published several monographs, edited books, and articles and is founder and co-editor of the journal *Artifact*, published by Routledge, see also: http://www.re-ad.dk/research/engholm_ida(46212).

Charles Ess

Charles Ess is Professor of Philosophy and Religion and Distinguished Professor of Interdisciplinary Studies, Drury University (Springfield, Missouri, USA). Recent publications include *Digital Media Ethics* (Polity Press, 2009), *Information Technology Ethics: Cultural Perspectives* (IGI Global, 2007, co-edited with Soraj Hongladarom), and *The Blackwell Handbook of Internet Studies* (co-edited with Mia Consalvo, 2010). Dr. Ess co-chairs the biannual conference series on "Cultural Attitudes towards Technology and Communication" (CATaC—<www.catacconference.org>). During 2009–2012, Ess will serve as a professor MSO (*med særlige opgaver*) in the Department of Information and Media Studies, Aarhus University, Denmark.

Vidar Falkenberg

Vidar Falkenberg (MSc) is a Ph.D. fellow at the Department of Information and Media Studies at the Aarhus University, Denmark. His dissertation examines the history and development of online newspapers in Denmark in the perspective of media history. He is a member of the Centre for Internet Research and co-organized the conference "Web_Site Histories: Theories, Methods, Analysis" with Niels Brügger in October 2008.

Kirsten Foot

Kirsten Foot is an Associate Professor of Communication and adjunct faculty in the Information School at the University of Washington. She has studied online activism in elections and in the fair trade movements in the U.S. and the U.K. Her current research focuses on anti-human trafficking coalitions. She is coauthor of *Web Campaigning* (MIT Press, 2006) and coeditor of *The Internet and National Elections* (Routledge, 2007). She is an editor of the Acting with Technology series at MIT Press, and as codirector of the WebArchivist.org research group, she develops methods for studying social and political action on the web over time.

Alexander Halavais

Alexander Halavais is an Associate Professor at Quinnipiac University, where he teaches in a graduate program in interactive communication. His research examines the relationship between social media and informal learning and the social outcomes of that relationship. He is the author of *Search Engine Society* (Polity Press, 2009).

Charles van den Heuvel

Charles van den Heuvel studied Art History at Groningen University (Ph.D., 1991). He worked as a senior researcher for several Dutch universities, as a librarian in Florence, and as a map curator of Leiden University Library. At the moment he is senior researcher for the Virtual Knowledge Studio for the Humanities and Social Sciences of the Royal Netherlands Academy of Arts and Sciences (KNAW), where he is involved in various research projects, including web archiving for research. His recent research interest is history of information science. He is initiator of the project: Architecture of Knowledge. European Antecedents of the World Wide Web.

Ken Hillis

Ken Hillis is Professor of Media and Technology Studies in the Department of Communication Studies, The University of North Carolina at Chapel Hill. His research focuses on the intersection of metaphysics and political economy and, in particular, examines the form of information technologies and the techniques and social practices they enable. He has published over twenty-five articles and three books: *Digital Sensations: Space, Identity and Embodiment in Virtual Reality* (Minnesota University Press, 1999); *Everyday eBay: Culture, Collecting and Desire* (co-edited with Michael Petit) (Routledge, 2006); and *Online a Lot of the Time: Ritual, Fetish, Sign* (Duke University Press, 2009). He is currently researching a book on Google and metaphysics.

Albrecht Hofheinz

Albrecht Hofheinz is an Associate Professor of Arabic at the University of Oslo, Norway. Following years of research on Sudanese socio-religious history and humanitarian work in the Sudan, his interest turned to the use of the internet in the Arab world. He focuses on majority use and on long-term processes of social and intellectual change, tying recent developments to reformist tendencies in the Muslim world over the past two centuries. Publications include "Arab Internet Use: Popular Trends and Public Impact," in *Arab Media and Political Renewal* (ed. N. Sakr, London, 2007).

Jeremy Hunsinger

Jeremy Hunsinger is completing his Ph.D. in Science and Technology in Society at Virginia Tech. At Virginia Tech, he manages and was one of the founders of the Center for Digital Discourse and Culture. He is an Ethics Fellow at the Center for Information Policy Research at the University of Wisconsin, Milwaukee. He is co-editor of the *International Handbook of Internet Research, The International Handbook of Virtual Learning Environments* and has edited several special issues for journals.

Brent Jesiek

Brent K. Jesiek is an Assistant Professor in Engineering Education and Electrical and Computer Engineering at Purdue University. He holds a B.S. in Electrical Engineering from Michigan Tech and M.S. and Ph.D. degrees in Science and Technology Studies from Virginia Tech. His research interests are focused on the social, historical, global, and epistemological dimensions of engineering and computing, with particular emphasis on topics related to engineering education, computer engineering, and educational technology.

Iben Bredahl Jessen

Iben Bredahl Jessen is working on a Ph.D. dissertation on web advertising as part of the research project "Market Communication and Aesthetics" at the Department of Communication & Psychology, Aalborg University, Denmark. Her research concentrates on media aesthetics, multisemiotic analysis, and genre development. She has also (with N. Graakjaer) written on the communicative role of sound in web advertising.

Tomi Lindblom

Tomi Lindblom, Ph.D., is the editor-in-chief of SuomiTV, a commercial TV company in Finland. He has been working with the internet and other new media since the mid-1990s and has built up many large-scale online media portals and web pages. He has just finished his dissertation at the University of Helsinki on new media strategies. He authors many articles about the internet and new media.

Steven Schneider

Steve Schneider is a Professor and Interim Dean of the School of Arts & Sciences at SUNY Institute of Technology, Utica, New York, where he teaches in the Information Design & Technology program. As the co-director of the WebArchivist.org research group, he has organized studies and collections of web objects, including those associated with national elections and the September 11 terrorist attacks, and has designed web archive interfaces in collaboration with libraries and research laboratories. He is the co-author of *Web Campaigning* (MIT Press, 2006) and co-editor of *The Internet and National Elections* (Routledge, 2007). He has a Ph.D. in Political Science from MIT.

Dominika Szope

Dominika Szope studied Art-Science, Philosophy, and Architecture. Since 2006 she has been a scientific assistant at the University of Siegen, Department for Mediascience. She has published several articles in the arts and the book chapter "From Expanded Cinema to Future Cinema—Peter Weibel" in *Future Cinema* (Eds. J. Shaw & P. Weibel) (MIT Press, 2003) and has co-edited the book *Pragmatismus als Katalysator des kulturellen Wandels* (with Pius Freiburghaus, LIT Verlag, 2006). Her current research in the field of the social web and the role of the user and amateur in Web 2.0 will be published in 2009 in *Leitmedien* (Eds. Ligensa & Müller) (Transcript Press), and *Rezipient* (Eds. Ligensa & Müller) (Transcript Press).

Einar Thorsen

Einar Thorsen is a Lecturer in Journalism and Communication at Bournemouth University, UK, where he also completed a Ph.D. in Journalism Studies. His thesis, funded by the Arts and Humanities Research Council, focused on civic engagement and citizen voices on the BBC News website during the 2005 UK general election. He has co-edited (with Stuart Allan) *Citizen Journalism: Global Perspectives* (Peter Lang, 2009), in which he also contributed a chapter on the reporting of climate change. Other research includes articles on *Wikinews* and the neutral point of view and the development of public service policies in an online environment.

INDEX

A

B

C

D

E

F

G

H

I

J

K

L

M

N

Y

Z

General Editor: *Steve Jones*

Digital Formations is an essential source for critical, high-quality books on digital technologies and modern life. Volumes in the series break new ground by emphasizing multiple methodological and theoretical approaches to deeply probe the formation and reformation of lived experience as it is refracted through digital interaction. **Digital Formations** pushes forward our understanding of the intersections—and corresponding implications—between the digital technologies and everyday life. The series emphasizes critical studies in the context of emergent and existing digital technologies.

Other recent titles include:

Felicia Wu Song
Virtual Communities: Bowling Alone, Online Together

Edited by Sharon Kleinman
The Culture of Efficiency: Technology in Everyday Life

Edward Lee Lamoureux, Steven L. Baron, & Claire Stewart
Intellectual Property Law and Interactive Media: Free for a Fee

Edited by Adrienne Russell & Nabil Echchaibi
International Blogging: Identity, Politics and Networked Publics

Edited by Don Heider
Living Virtually: Researching New Worlds

Edited by Judith Burnett, Peter Senker & Kathy Walker
The Myths of Technology: Innovation and Inequality

Edited by Knut Lundby
Digital Storytelling, Mediatized Stories: Self-representations in New Media

Theresa M. Senft
Camgirls: Celebrity and Community in the Age of Social Networks

Edited by Chris Paterson & David Domingo
Making Online News: The Ethnography of New Media Production

To order other books in this series please contact our Customer Service Department:
(800) 770-LANG (within the US)
(212) 647-7706 (outside the US)
(212) 647-7707 FAX

To find out more about the series or browse a full list of titles, please visit our website:
WWW.PETERLANG.COM